Studies in Neuroscience, Consciousness and Spirituality

Volume 5

More information about this series at http://www.springer.com/series/10195

Farzad Goli
Editor

Biosemiotic Medicine

Healing in the World of Meaning

Springer

Editor
Farzad Goli
Head of Danesh-e Tandorosti Institute
Isfahan, Iran

Energy Medicine University
Mill Valley, CA, USA

ISSN 2211-8918 ISSN 2211-8926 (electronic)
Studies in Neuroscience, Consciousness and Spirituality
ISBN 978-3-319-35091-2 ISBN 978-3-319-35092-9 (eBook)
DOI 10.1007/978-3-319-35092-9

Library of Congress Control Number: 2016947722

Printed on acid-free paper

This Springer imprint is published by Springer Nature
The registered company is Springer International Publishing AG Switzerland

Preface

Life is nothing but information in practice
The more informed, the more alive one is (Rumi)

Numerous studies in the recent decades have revealed that we are experiencing a shift from the biological paradigm to systemic paradigm in medicine. However, this is not an omnipresent transformation; it is rather the average of diverse, and at times, opposite processes. On one hand, the accomplishments of genetic engineering in cloning, stem cells or screening, and genetic manipulations confirm the mechanical model of biomedicine that has provided the grounds for the selection and promotion of genetic programs or even mass production and change of the organs. On the other hand, multiple studies in other fields of science such as psychoneuroimmunology and epigenetics have deeply challenged the approach of biomedicine. Of course, reductionism seems still proves itself to be pragmatic for non-chronic conditions.

Although in emergency and acute conditions, the agency of the patient and coping strategies are less important and mechanical approaches are more efficient to a great extent, when moving towards the chronic conditions and planning macro health programs, the inefficiencies and insufficiencies of the mechanical approach reveal themselves more dramatically and the need for a systemic model becomes obvious. In order to establish such a systematic model, we need to develop interdisciplinary knowledge and the necessary methods.

Systematic clinical studies and the developing fields of medical anthropology, health psychology, and psychosomatic medicine make evident the interference of symbolic and physical worlds more and more. They uncover how our health and illnesses are formed in a multifaceted heterogeneous matrix of biological, emotional, social, cultural, and spiritual factors.

Tolerating this multilingual and interdisciplinary medium, after several centuries of attempts at establishing a single pure chemophysical language, is tremendously difficult for medical discourse. It seems that for explanation, clinical reasoning, and management in the systemic approach, we should prepare ourselves for a more complicated chaotic system with increasing uncertainty. This way we might be able to substitute human and societies' health for the diseases and their potential causes as the subject of medicine and move towards the development of sustainable happiness. It seems that we need a transdisciplinary groundwork to integrate such a vast anisotropic field of knowledge and practice.

Contemporary theorist scholars do not believe in a single metanarration that explains all levels of organization and all life worlds anymore. They do not believe people, similar to early Wittgenstein, should be silent about things that cannot be described with experimental and observable language anymore. These scholars concur more with the late Wittgenstein that accepted the interaction and coexistence of language games. From this vantage point, love is neither reduced to biochemical fluctuations, nor to a conditioned social pattern that people imitate in certain situations; not even a psychodynamic regression, and not necessarily a pure experience of selflessness and devotion. To understand these phenomena, we must first go beyond the objective and categorical level and explore the phenomenon itself: who actually experiences love. We should also be open to all the subjective and objective dimensions to be able to reframe these experiences in the bio-psycho-social framework. We should be aware that we are now part of that context and its result: an interpersonal interpretation that might lead us to the prescription of a remedial package including medicine therapy, psychotherapy, meditation, family therapy, and even environmental and social modifications. In order to integrate such a health service system that entails all of the intra/inter/transpersonal fields, we need something beyond a multidisciplinary approach that can trace the flow of signs in the body, mind, society, and culture and is also able to devise management plans.

Some psychosomatic medical theorists such as Thure von Uexküll have considered biosemiotics as an approach that can explain *the mental* and *the physical* in a single ground called semiosis, away from being limited in the Cartesian dualism boundaries. Speaking of mind from this perspective is actually talking from a self-organizing order, from a phenomenal world that perceives the world in a particular way and acts the same way: a differential system that differentiates stimulations in a systemic way. In other words, a mind is a specific way of *being* in the world. Now, if we return to the definition of life, we recognize that it has a similar domain with the mind as per the above definitions, and that all of the descriptions apply to the living body as well. In the systemic approach, mind is not only embodied in the form of the elemental body, but it is also embodied in the discourses and institutions.

Four different physical, emotional, cognitive, and social phenomena are indeed emergent recreations of mind in different levels of organization that has its own specific language and rules at each level. In Luhmann's opinion, each of these levels has its functional closure. At the same time, levels are structurally open to each other which is why the sign systems are interacting with each other, and the semiosis freely moving through and between the systems.

The inclusive phenomena of meaning response, that is traditionally called the *placebo effect*, is a distinguished example of relations between the levels of organization and one of the biomedicine anomalies that had made us think about the function of interpretation from symbolic components to physical components: to think about a language that can transform an idea or image to a chain of physiological changes. This phenomenon that accompanies all remedial interventions like a shadow – and is responsible for a large part of effectiveness of all psychological, chemical, and physical interventions – is not a fixed coefficient and not a non-specific effect, but rather a specific biosemiotic formula that acts in a special way and with a special amount in any psychosocial context.

Biosemiotic interpretation of the placebo response is our point of departure in this book. We have attempted to show how the process of meaning making and interpreting can play a role not only in symptom formation and psychoneuroimmunologic responses, but also in health/illness behavior, epigenetic patterns, and of course in psychosomatic treatments. In addition, through biosemiotic lenses, we observe that the direct mechanical or chemical agents do not result in healing symptoms, but actually, it is the organism interpretation of the chemical and physical signs that can lead to healing.

In the first chapter of this book, my colleagues, Dr. Rafieian, Dr. Atarodi, and I have initially aimed at addressing the complexities of the phenomena of placebo and stating that the explanation and conscious application of these phenomena with a pure biological behavioral approach would be an arduous task which would ultimately be inefficient. For this reason, we are addressing the methodological (noise vs. signal), the pragmatic (meaning-specific vs. non-specific), and the ethical (beneficence vs. autonomy) dilemmas. Later in the chapter, we explore some solutions in the systemic model for the dilemmas to convene these so-called heterogeneous dimensions. Finally, we will proceed with the semiotic approach to understand how it can explain and solve the psychosomatic phenomena.

In the second chapter of this book, Professor Brier, a science philosopher and a theorist of cybersemiotics, elaborates this transdisciplinary pattern rather extensively and explicates how this pattern can provide a common groundwork for social sciences, psychology, biology, chemistry, and physics: a context that seems to be essential for an interdisciplinary field like medicine.

With the combination of the two metalanguages – the cybernetic-informational approach, focusing on the bottom-up organization; and the semiotic-hermeneutic approach, explicating the top-down organization – Professor Brier has created this inspiring model that can illuminate the psychosomatic phenomena such as placebo responses convincingly.

In the third chapter of this book, psychoanalyst and psychosomatic specialist, Professor Scheidt focuses on how biology and biography intersect. Furthermore, he clarifies how the hermeneutic procedures can lead symptom formation, therapeutic relationship, and even bodily responses. He has gone further in illustrating the non-substance-bound healing effects in the narrative medicine framework in a quite elucidating and inspiring way. In this chapter, Professor Scheidt demonstrates how we narrate the self and the world with our body and language and how we construct our world in this way. When we experience a powerful, unpleasant event, and our previous narration loses its cohesion and consistency, we should reconstruct it more consistently with the other components. Each therapy, regardless of its verum effects, could be considered as a promising change in patients' narrations.

The truth is that we do not enter the patient's body with chemical and physical interventions solely. We intervene directly with inductions and interventions, and also indirectly, by entering the patient's narration and web of beliefs. It is evident that when therapeutic narrations are more compatible with patients' narrations, there is a more profound impact, therefore a greater motivation for the patient to change his/her narration. Entering the web of beliefs of an individual and a society, for creating a more congenial, salutogenic, and positive narration, is undoubtedly a delicate, complicated, and time-consuming task.

Clinical psychologist, Dr. Johari Fard, and I have attempted to present an outlook of intertwined webs of beliefs in the fourth chapter. We are displaying how the webs of belief of a person, a culture, and also a healing system interact with each other and their interventions could resonate or destroy a placebo effect. Globalized statistics alone will not suffice for optimizing the meaning effect, but also we have to consider the compliance, the individual's anticipations, and the culture or the subculture of the individual. This might be the solution for the resistances and chaotic phenomena in response to the various treatments. By using this model, more suitable, more effective, and more democratic clinical settings may be within reach.

Psychosomatic medicine practitioner, Dr. Farzanegan, and I decided to devote the fifth chapter of this book to the ritual effect and the structuralistic-anthropologic analysis of the treatment patterns and methods. We have presented how the form of each medical model and clinical setting, along with direct inductions (doctrines, prognoses, instructions) and indirect inductions (treatment metaphors, traditions, rituals, psalms), can systemically moderate the beliefs, behaviors, and psychoneuroimmune responses.

An important point that is frequently ignored in health training and medical advertisements, due to different reasons, is that information, similar to drugs, should be formulized and prescribed at certain measured doses; otherwise, it could lead to side effects and worse than that, without any positive effect, produce a nocebo effect. Increasing the risk of avoiding danger can cause increased health anxiety and, paradoxically, lead to the reduction of immune system functionality and the rise in being prone to illnesses which ultimately causes symptoms and even illnesses.

The discussion about the performing and metaphoric aspect of healing is continued in Chap. 6. Dr. Rafieian and social theorist, Professor Davis, address the role of performance and interpersonal interaction between health-care professionals and patients using examples of hypnosis and placebo research. They illustrate the health-care system from a sociocognitive view and show the importance of performance in medical practice.

The sociocognitive theorists of hypnosis believe that trance is not necessary for the experience of hypnotic phenomena. In their view, suggestions, belief, and expectancy are the key components for the development of hypnotic experiences. Consequently, they have described hypnosis as "believed-in imaginings" and defined it as a kind of role-taking. Placebos have also been used for a long time in medicine and are still used widely in medical practice. As the placebo itself is inert, it has been proposed that the mind–body mechanisms surrounding the prescription of placebo are instrumental in healing formation. As with hypnosis, suggestion, expectancy, and belief are the main components here as well. These findings cement the importance of performance practices and the verbal and non-verbal communication between the health-care professional and the care seeker in the clinical setting.

In the seventh chapter, practitioner and cognitive psychologist, Dr. Monajemi; psychiatrist having a fellowship in psychosomatic, Dr. Malekian; psychiatrist, Dr. Ahmad Zadeh; and I have addressed different dimensions of the medicalization and their context, personal impacts, and social effects. We illuminate the iatrogenic disorders of informational interventions. At the end of this chapter, we have tried to present practical solutions for optimizing the effects of informational drugs and minimizing their side effects.

In the eighth chapter, clinical psychologist, Professor Schmidt, and clinical psychologist and science philosopher, Professor Wallach, who have conducted several valuable studies in the placebo responses and parapsychology fields, address this topic from the mind–matter interactions perspective. Structural analysis of the previous chapter can be followed here to explore how a treatment process can correlate a group of symbols with specific psychosomatic changes systemically. Casual and mechanical patterns cannot explain such phenomena; hence, a correlational-semiotic pattern seems essential.

The concept of *pseudomachine* that authors have borrowed from von Lucadou is fully innovative and illuminative in the structural and semiotic explanation of the placebo response. From this point of view, any treatment process can be considered a pseudomachine that can make the expectation of psychophysical changes conditioned to behaviors (referring to the healer, drug consumption, therapeutic procedures, regimes), objects (doctor, drug, devices), locations (clinic, operation room, ashram), and of course, specific beliefs. Numerous studies reveal that even in the effective treatments, active agents commonly constitute a smaller portion of the treatment effects and the larger part of the treatment is due to semiotic factors.

In the final chapter, I have aimed at presenting a bigger picture of life and medicine from the biosemiotic perspective: a picture that can demonstrate a more profound and efficient meaning for life, health, illness, and medical practice. The human organism is a self-organizing and self-narrating stream of signs that lives in/with an infinite ocean of semiospheres. Throughout the history of evolution, unlimited semiosis has been inclined towards progressive coherence of signs and has gradually created more complicated and emergent characteristics. The omnipresent process, called *Agapism* by Peirce, entails the universal love that is present beyond Darwinian wars between organic systems and expands the meaning of the signs by creating more complex systems and higher levels of organizations. The evolutionary love is the tendency of life to form new and more complex forms and habits.

Medicine in such a world, where even the hard realities are nothing except natural habits, should be a more fluid, more creative, and more humane art. Semiotics not only interprets the psychological effects, but also the effects of the chemical and physical factors which depend on its interpretations. Ethics, psychology, and sociology in this view are as bodily and medical as drugs and surgeries, and all of them are semiotic agents. The healing responses are formed via interpretation of these meaningful agents by the whole organism. The formula of a patient–doctor relationship or a public health training program needs a great deal of precise semiotic accuracy, like the synthesis of a drug. The difference is that relationship formula is formed not only on the basis of predetermined instructions, but also it is constructed in live processes of relationship.

Thus, attuning to the developing cohesion of the signs towards the sustainable development of health – in addition to the meticulous psychological, sociological, anthropological, and semiotic studies – requires consideration of the qualitative, improvisational, and chaotic dimensions of therapeutic relations. As such, we need something more than medical science and technology; we need a hermeneutic participatory art of healing. "Doctor is medicine" as Michael Balint stated, and this medicine can heal properly when the healing system, healer, and client are attached to the evolutionary love.

Isfahan, Iran
2015

Farzad Goli

Acknowledgments

I would like to extend my thanks to the authors and our colleagues at Albert-Ludwigs University, Danesh-e Tandorosti Institute, Psychosomatic Research Center, IUMS, and German Academic Exchange Service (DAAD) who were actively and closely engaged with this project, especially Dr. Marsa Hashemi Jazi and Sepideh Motamedi. I am also grateful to Sara Ahangar Ahmadi for her kind cooperation in producing the English edition; Tom Levold and Sungchul Ji for their detailed reviews and very helpful comments. Ultimately, my special thanks go to my teacher, Carl Eduard Scheidt, and my kind friend, Stefan Schmidt, for their encouraging and inspiring presence in various stages in the production of this book.

Contents

1 **An Introduction to the Semiotic Approach to the Placebo Responses** 1
Farzad Goli, Shahram Rafieian, and Sima Atarodi

2 **Cybersemiotics as a Transdisciplinary Model for Interdisciplinary Biosemiotic Pharmacology and Medicine** 23
Søren Brier

3 **Some Reflections on Non-substance Bound Healing Effects and the Concept of Narrative Medicine** 85
Carl Eduard Scheidt

4 **How Can We Reconstruct the Health Anticipation?** 95
Farzad Goli and Reza Johari Fard

5 **The Ritual Effect: The Healing Response to Forms and Performs** 117
Farzad Goli and Mahboubeh Farzanegan

6 **Hypnosis, Placebo, and Performance: Recovering the Relational Aspects of Medicine** 133
Shahram Rafieian and Howard Davis

7 **How to Prescribe Information: Health Education Without Health Anxiety and Nocebo Effects** 151
Farzad Goli, Alireza Monajemi, Gholam Hossein Ahmadzadeh, and Azadeh Malekian

8 **Making Sense in the Medical System: Placebo, Biosemiotics, and the Pseudomachine** 195
Stefan Schmidt and Harald Walach

9 **Medical Practice in/with the Semiosphere** 217
Farzad Goli

Index 241

Chapter 1
An Introduction to the Semiotic Approach to the Placebo Responses

Farzad Goli, Shahram Rafieian, and Sima Atarodi

Today, the placebo effect is the Cinderella of a new medical world; a phenomenon which in one night turned from a platitudinous problem and paternalistic sham in practice and a disturber factor in clinical trials, to meaning response, spirit of practice and an extremely valuable subject for research. The word "placebo" is rooted in the Latin Psalm phrase "*placebo domino in regione vivorum*" – I will please the Lord in the land of the living (Kradin 2011). The word itself has been used in medical literature for centuries, but the first clinical trial was conducted in 1799, in which the author stated: "[A]n important lesson in physic is here to be learnt, the wonderful and powerful influence of the passions of the mind upon the state and disorder of the body" (Price et al. 2008). From the middle of the twentieth century, conventional medicine began using placebos as methodological tools to distinguish between specific and non-specific ingredients in treatment (Papakostas and Daras 2001). The placebo was first introduced as an inert agent solely prescribed for pleasing the patient. There was a paradoxical conceptualization in this way of thinking because doctors used placebos on one hand as an element with no therapeutic effect, but on the other hand, it did show some response in the patient. This paradox resulted

F. Goli (✉)
Head of Danesh-e Tandorosti Institute, Isfahan, Iran

Energy Medicine University, Mill Valley, CA, USA
e-mail: Dr.fgoli@yahoo.com

S. Rafieian
Danesh-e Tandorosti Institute, Isfahan, Iran

Bangor University, Bangor, UK
e-mail: rafieiansh@yahoo.com

S. Atarodi
Danesh-e Tandorosti Institute, Isfahan, Iran

University Hospital of South Manchester, Manchester, UK
e-mail: s.atarodi@yahoo.com

F. Goli (ed.), *Biosemiotic Medicine*, Studies in Neuroscience, Consciousness and Spirituality 5, DOI 10.1007/978-3-319-35092-9_1

in the shift from focusing on the inert content of placebos to the concept of an active therapeutic agent within a psychosomatic context.

Until recent years, placebos had a bad reputation amongst health care professionals. To some who are involved at the clinical level, it is a kind of "trick" to make the patient feel better by utilizing the power of positive expectations. It is the last chance of a doctor who has no other scientifically rational options for the treatment of the patient. In this situation, he or she administers an inert drug. The patient assumes that it is a potent pharmacological agent, and experience has shown that this belief in the potency of the drug is indeed beneficial and can lead to the patient experiencing an improvement in his/her condition.

As placebos are chemically inert and have no specific biological target in the body, it is generally believed that the effects induced are non-specific. But a closer look at the mechanisms involved shows that it is not as simple as it seems. In fact, placebo effects are specific to the therapist's latent and active inductions, and the patient's anticipations and interpretations, which can form the complex and hermeneutic response to the therapeutic communication which is called "Meaning Response" (Moerman 2006).

Now, more than ever, the paradoxical nature of the placebo response has manifested itself in the medical community. One can find a vast number of articles which introduce placebos as a chemo-physical (non-specific) effect or a semantic/cognitive (specific) effect; as noise of biomedical studies or a signal of doctor-patient communication; as a very beneficial, safe, and common therapeutic agent, or as immoral interventions which ignore the principle of autonomy. One can infer that there are very serious dilemmas in this field of practice and research: pragmatic (specific and non-specific), methodological (desired and undesired), and ethical (beneficence vs. autonomy). Introducing these dilemmas shows the paradoxical and complex nature of placebo responses and also addresses the clinical and paradigmatic opportunities and restrictions. We will discuss these topics and their biosemiotic explanations further in the following chapters.

1.1 The Pragmatic Dilemma: Non-specific or Meaning Response

As Moerman and Jonas (2002) explain, different elements of medicine are meaningful for the patients and are unrelated to the intentions of doctors and other health care professionals. A fitting example is the so-called "white-coat hypertension". Studies show that the environment of a hospital and the white coat typically worn by doctors is a trigger for elevated blood pressure in some patients (Pickering and Friedman 1991; Bügel 2004). Like doctors' white coats, many other elements present in the clinical context have meaning for the patient. Doctors' behavior, facial expressions, gestures and language, as well as the devices, colors and shapes in the hospital all are meaningful and play a part in the process of treatment. Paying attention to these factors and analyzing them make these seemingly non-specific effects specific. Moerman and Jonas also mention that the placebo response may even be stronger in surgical interventions because the rationale behind these interventions fit better to the mechanistic mind of modern man. In other words, it is easier to assign a meaning to a surgical intervention. A good example is a study that evaluated the effectiveness of binding internal mammary arteries to reinforce the blood supply to partially occluded coronary arteries. In the study, this technique was compared with a sham surgery in which only a skin incision was performed and the internal mammary arteries were not ligated. The results showed that there was no difference between the intervention group and the group of patients on which sham surgery was performed (Bügel 2004). Especially today, with new insights and developments in medical ethics, it is not possible to design such studies for evaluation of surgical procedures, because the patient would bear the burden of a surgical operation. There are, however, similar recent studies like one that evaluated the effectiveness of arthroscopic removal of osteophytes in patients with osteoarthritis. In this study, the results showed that there was no difference in the outcome between the patients with actual removal of osteophytes and the patients with a sham surgery, in which only an incision on the skin was done and no osteophyte removal was performed (Kradin 2011). Although the placebo effect is not specific like the effect of pharmaceutical agents designed to target specific chemical interactions or molecules, they have specific effects via the meaning that treatment produces for the patient and the direct and indire ct suggestions that they produce in the clinical setting.

To incorporate all of these facts into a comprehensive framework, we need a new perspective towards the fundamental concepts in medicine, such as health, diseases, and healing. A semiotic approach can provide such a framework, especially for explaining psychophysical events which could not be demonstrated by linear causal models. As Eco states, until a short time ago, medical semiotics was the only research project in the field of sign studies (Eco 1979). His interpretation of signs has been a central issue in medicine since its beginning, and there existed a close relationship between medicine and semiotics. The German thinker and physician, Thure von Uexküll, who is considered one of the founders of psychosomatic medicine, believed that the biosemiotic approach is a good tool for describing what happens in the process of clinical encounters in real life. Semiotics is the doctrine of signs developed by the American philosopher Charles Sanders Peirce. With this perspective, there are three important components: sign, object, and interpretant. A sign is something that stands for another thing which is the object that it signifies. The sign represents certain meaning or understanding in the recipient. The recipient then acts based on the meaning it perceives (Meyer 1984; Walach 2011).

In other words, an object which represents a sign itself can be a sign for recipients, and as recipients have minds, they interpret the signs. The meaning that can be produced by the object is called "interpretant" and the individual who analyses the sign and object is the "interpreter". When there is only one object that has one meaning, there is a causal relationship between the sign and meaning. A familiar example in the clinical setting is examining pulse. The pulse is a sign and the cardiac function is an object. For instance, in the case of tachycardia, the accelerated pulse shows the increased rate of heartbeats and only one meaning arises from this sign (cause and effect). Now, this accelerated heart rate could be considered as a sign and the object could be anxiety, the increase of certain hormones like thyroid hormones, etc. It is the context that reveals which object is connected to this clinical sign.

In the biomedical paradigm, man is considered a biophysiological machine. The medical interventions are based on the causal relationship between interventions and the change of the system toward recovery. The interventions could be pharmacological agents which block or activate certain receptors, or physical and surgical procedures that rectify the pathological deviations. In this perspective, the patients are considered to be a passive receiver of treatment, for which there is no place for their feelings, thoughts, believes, emotions, intentionality, and agency. In this model we cannot explain how a positive anticipation or a healing intention can initiate a complex chain of physiological procedures (Lewith et al. 2010).

In the biosemiotic perspective, patients are seen as active agents who have their individualized interpretations of different interventions. Each intervention is, as a result, considered as a sign, which is meaningful for the patient. This sign could be interpreted as belonging to different objects (Miller and Colloca 2010). For example, if the patient believes that more invasive and painful interventions are more effective, then an injection will have a more substantial effect than a pill. Also based on previous experiences of the patients, the meaning and effect of the injection will be different. It should be mentioned that in this approach, the causal element of pharmacological properties of the agent are not ignored, but that the meaning that is created in patients as an active agent is something above the causal element.

A good replacement for causal model of placebo is the biosemiotic approach. As Wallach (2011) explained: "Placebo effects are real physiological effects. But they are not caused by a physical intervention but arise from the intrinsic meaning-making of an active organism that interacts with the environment." (p. 1874)

It is clear that placebo responses are not matter-specific, and, from the mechanical and linear causality viewpoint, can only be deemed as non-specific effects. Yet from a biosemiotic point of view, they are meaning-specific responses, and the semiotic formulation of each can affect the psychoneuroimmunologic state in a distinct manner. There is no limitation for biosemiotic formulations; they could be even paradoxical mixtures of salutogenic agents (placebo-anticipated positive effects) and pathogenic agents (nocebo-anticipative negative effects) which arise from a communicative context.

1.2 The Methodological Dilemma: Placebo, Noise, or Signal

Apart from the clinical impacts, placebo effects are also a dilemma in research. They are problematic for the most important tools in evaluating new treatments, known as randomized clinical trials (RCTs). Historically, the American anesthesiologist, Henry Knowles Beecher, became familiar with the power of placebos during World War II. In a difficult situation in a hospital at Anzio Beachhead, Italy, there was shortage of morphine for the many wounded soldiers suffering from pain. Faced with a difficult situation, a nurse proposed injecting them with saline water, but telling them it was strong analgesic. Astonishingly, it was helpful. He was so fascinated by this that after the war, in 1955, he wrote a paper in the *Journal of American Medical Association* titled "The Powerful Placebo". He reviewed 15 published papers about different interventions and concluded that 35 % of the successes in treatment were due to the positive expectations of the patient, or the placebo effect. He later focused on the fact that in evaluations of effects of a pharmaceutical agent or a new procedure, it is difficult to differentiate between the healing effects of placebo responses and the genuine effects of the treatment. He went on to become one of the founders of double blind placebo-controlled RCTs, but did not explore the mechanisms of the placebo response itself. Today, RCT is the gold standard of evaluation of new treatments, and because researchers try to minimize the placebo response in their research, or in other words, get rid of it, it has also become a reason that there is a negative attitude towards placebo responses among health care professionals. In other words, Beecher experienced the placebo effect as a powerful healing tool in the clinical setting. He defined it as a noise in the research system of clinical trials; it is the unwanted and undesired part of research that should be eliminated in order to reveal the useful and therapeutic part of the study (Bensing and Verheul 2010). But in recent years, there are many studies that suggest this so-called "noise" or confounding factor that used to interfere with the physical and chemical interventions, which are designed based on precise pathophysiological knowledge, could be interpreted differently. Now, several studies show that placebo reaction is a healing message itself that can be explored and controlled for therapeutic purposes.

It is a well-known fact in medicine that the relationship between the therapist and the patient has a healing effect. Balint (1957) considers the doctor as the most potent drug. As Bensing and Verhul state (2010), although the two fields of placebo research and doctor-patient research were generally separate, they are converging and demand exploration of the mechanisms by which placebo effects are produced by doctor-patient communication. Adopting an analytic approach to medical investigations as a modern tradition obligates us to eliminate the placebo effect in order to distinguish the chemophysical effectors, but as a complementary approach, we can also have a synthetic approach to develop the psychosomatic dynamisms (an interactive intentional-physical network) for establishing more effective and humanistic health services which are not necessarily quantitatively and analytically well-defined. Knowing these healing pathways allows us to control and expand them, and make doctors aware of the powerful healing tools that they possess in clinical encounters.

1.3 The Ethical Dilemma: Beneficence vs. Autonomy

One of the barriers of using placebos in a clinical setting is the ethical problem associated with their administration. From the perspective of medical ethics, there is a conflict between two of the basic ethical principles in the treatment of the patients with placebos. These two principles are autonomy and beneficence (Miller and Colloca 2011).

On one hand, based on the autonomy principle, the patient has the right to accept or reject a treatment so he or she must clearly be informed of the safety, efficacy and nature of the therapy. And according to the beneficence principle, it is the duty of the therapist to act in the best interest of the patient in the safest and most efficient way possible. Upon first sight, placebo therapy seems fundamentally paternalistic and we also know that placebos are extremely safe, but here the question is: Are placebos really beneficial to the patients?

There is enough evidence in the literature that shows the power of placebos. A variety of health conditions, like heart failure, pain, Parkinson's disease, and schizophrenia, were responsive to placebos. There is a special interest in placebo responses to pain, and it has been shown that there is a real effect, the magnitude of which however differs vastly (Bensing and Verheul 2010).

Here, there is an ethical dilemma for doctors and health care professionals. On one side, the placebo effect is quite safe and helpful, and although it is chemically inert, the patient feels better after its administration. Alternatively, if a patient realizes that there was no rational and scientifically valid reason for the use of that drug (or procedure), a negative feeling will develop due to the patient feeling "deceived" or "tricked" which could possibly disturb the rapport. This issue in particular has become an increasingly serious concern, as it is generally accepted that the patient, from an ethical point of view, should be aware of the treatment process and informed about all drugs administered and procedures used by means of the autonomy principle; patients should give consent for all treatments applied. Another pertinent ethical issue in researching placebos is the potential danger for patients in specific circumstances. When effective treatment already exists, waiting for the assessment of the effectiveness of a placebo is unethical because it can cause irreversible changes in patients, placing them in danger. For instance, when treating myocardial infarction, some cancers and some infectious diseases, placebo-controlled trials are not moral.

In spite of these ethical concerns, placebos are used regularly in the clinical setting. In a study which has been done on American internists and rheumatologists, around half of the participants used placebos regularly in their practice. Most of them use placebos for chronic pain patients if they think it is beneficial, and a significant number believe that it is ethically justifiable (Tilburt et al. 2008). Many physicians prescribe antibiotics, sedatives, vitamins, and physiotherapy as placebos, and, especially in the case of antibiotics, it is the source of new problems like the development of multi-drug resistant types of bacteria (Miller and Colloca 2011).

Because of these issues surrounding the concept of placebos, it is a concept that doctors are not inclined to deal with. But is there a way out of this problem?

The point that helps us solve the problem is mentioned by Moerman and Jonas (2002) and is based on the fact that placebos are inert. We know that there are no chemically and physically-induced therapeutic changes by placebos, and that there are other related mechanisms that are involved in the formation of the healing response. That being said, the solution would then be to focus on psychoneuroimunological mechanisms of the placebo effect instead of focusing on the placebo itself. A great amount of empirical evidence supports three main mechanisms in the emergence of a placebo response: (1) conditioning, (2) The Expectancy Theory, and (3) The Affect Theory. It should be mentioned that there is an overlap among these mechanisms. Each of them can also be influenced by a doctor's behavior, which in turn shows the importance of doctor–patient relationship in this issue (Price et al. 2008; Bensing and Verheul 2010).

In conditioning, a natural stimulus is repeatedly associated with an unconditioned stimulus (e.g. active drug), where the natural stimulus with time can act as a conditioned stimulus. An empirical study shows the conditioning mechanism in the clinical setting very well. Goebel and his colleagues (2002) conditioned subjects in four sessions in a double-blind study. They paired an immunosuppressive drug (unconditioned stimulus) with a specific flavored drink (conditioned stimulus) and gave them to the subjects every 12 h. After 1 week, subjects again received the flavored drink that now contained merely placebo capsules that were free of active substances, and the patients interestingly showed further suppression of the immune system. The conditioning phenomenon happens in clinical settings consciously or unconsciously. Warm and empathic communication with a physician can be coupled with a patient's recovery from previous health problems, and this conditioning would also affect the patient's future experiences with medical problems and occurs often as an automatic unconscious process.

Expectancy is defined as a patient's expectation of response following administration and is, in contrary to conditioning, always a conscious process. Expectancies could be patients' beliefs about the effects of treatment or about the ability of themselves in fighting a disease and controlling or coping with it. It has been shown *in vivo* that this phenomenon induces endogenous opioid release. Additionally, expectation can be reinforced by previous experiences, verbal suggestions, and the stronger desire of a patient to reach positive goals. Several studies show that when patients are aware of the type and exact time of drug administration, the drug's effect is greater and faster (Meissner et al. 2011).

Bensing and Verheul (2010) consider affect manipulation as another mechanism in placebo response. It is defined as the influence of different methods of treatment administration on patients' affective state. They believe that positive affect and lower levels of stress and anxiety would facilitate the treatment. Affect can work through mediators like self-efficacy, adherence, and self-disclosure. The doctor's attitude towards the patient has also an important role in assembling a positive view in the patient about his/her disease.

Based on neurobiological studies on placebo effects, the reduction of neural activity can be seen in the parts of brain which are responsible for pain and anxiety. The increase in brain activity in emotion regulation areas has also been shown to exhibit a placebo response (Price et al. 2008; Flaten et al. 2011). Despite the fact that these mechanisms are well known in psychology, and many psychoneuroimmunological studies revealed the details of this phenomenain biological and physiological levels, the use of treatments based on these processes is not well established in clinical practice. In other words, although there is rich literature on the importance and efficacy of psychophysiological pathways, the application of these mechanisms is underscored in the clinical setting.

Now the question is: Why is this happening? The answer goes back to the ethical dilemma of placebos. As mentioned, the main problem in using placebos in the clinical setting is the contradiction between "beneficence" and "autonomy". This inconsistency emerges based on the assumption that if the doctor does not lie to the patient, there would be no placebo response. Based on this assumption, if the physician informs the patient that the prescribed drug has no pharmacological properties, the drug would then have no effect on the patient. Montgomery and Kirsch conducted a study on pain and analgesia with an artificially induced pain in their laboratory and showed that when the participants were merely told that they were not receiving a real drug and the cream applied was inert, the drug (placebo cream) had a little or no effect in pain reduction (Kirsch 1997). This study supports the idea that a placebo is only effective when the patients are not aware of the drug's contents.

In contrast, Kaptchuk and his colleagues (2010) questioned the explained assumption on the placebo effect. They did a randomized controlled trial on Irritable Bowel Syndrome (IBS). In their research, they presented the placebo pill to the patients as "placebo pills made of an inert substance, like sugar pills, that have been shown in clinical studies to produce significant improvement in IBS symptoms through mind-body self-healing processes" (p. 1) and the results showed significant improvement in the placebo group. Although in this study patients were aware that the drug has no pharmacological agent, identical to the former study, the results were completely different solely due to the different methods of information disclosure. In other words, the important point here is that the way that treatment intervention is described affects symptom relief and patient experiences (Miller and Colloca 2011). This point can facilitate solving the ethical dilemma of placebos regarding autonomy and beneficence.

The other way out of this ethical dilemma is by emphasizing the mechanisms involved in the formation of a placebo response instead of the placebo itself. We saw that the placebo itself is inert and that the psychological mechanisms in fact activate the process of healing. These mechanisms could be activated by doctor-patient communication and different psychological intervention (Miller and Colloca 2011). For example, expectancy is an important component of treatment in hypnosis. The suggestions which are given to the patient have a healing impact, and the physiological changes which occur through hypnosis are the result of expectancy mechanisms (Kirsch 1994, 1997).

Conditioning, which is the other proposed mechanism for placebo response, is well known in behavioral medicine (Mommaerts and Devroey 2012). Different treatment methods have been developed based on this mechanism in this approach.

Justman (2011) explored the relationship between psychotherapy and placebo effects. According to Justmsn, sychotherapy as one of the psychological treatments is widely used in the treatment of psychosomatic medicine. In psychotherapy, emotion is a core concept and affect manipulation plays an important role in the process of this treatment.

Generally, it can be said that in different mind–body interventions, like acupuncture, relaxation therapy, yoga, meditation etc., there is a component that is common with the mechanisms involved in the formation of the placebo effect (Brom 2012; Stefano et al. 2001). Considering these facts, it might be appropriate at this stage to think about the possible ways out of the mentioned dilemma.

1.4 The Way Out of the Dilemma

As discussed, the mechanisms by which a placebo effect is mediated are not unknown. But why, in today's medicine, is their use so limited? And why have they not become incorporated in the main treatment protocols of health problems?

The basic answer to these questions is hidden in the way that the current model of medicine; namely that biomedicine, defines the patient. As its name biomedicine implies, a person is a biological and at most physiological entity. The main focus of interventions in this approach lies in the subpersonal levels and the physiological mechanisms involved in the formation of different problems (Kihlstrom 2008; Gaines and Davis-Floyd 2004). In this framework, the mental phenomena like thoughts, feelings, beliefs, and imagination, although possibly considered as effective or functional in the process of treatment, are not considered tools for designing and developing new interventions. Any intervention developed based on these phenomena is in turn considered to be alternative, adjuvant, or secondary to biomedical interventions.

In biomedicine, the "signs and symptoms" have become separated and there is a split between hard and soft data in clinical encounters, but from a systemic viewpoint between hard and soft reality (Nessa 1996). As we see in the case of placebo responses, in the clinical context as well as in the very process of healing response formation, emotions, feelings, beliefs, and the patient's (and therapist's) personal experiences play a critical role. As a result, in order to solve these dilemmas, we require a broader perspective. A new framework is needed in which the phenomenal experiences of the individuals involved come to play their role and are considered as real but non-linear causal factors in the maintenance of health and development of disease. In this view, disease is not merely considered as a derangement of a physiological or biological organ, neither coincidentally nor because of an unknown reason. Instead, it is a malfunctioned pattern of behavior developed in the context of a person's life, world, and lifestyle with all of the biological, psychological, and interpersonal interactions and complexities.

There were clinicians who recognized this problem and attempted to develop an alternative framework to gain a broader view which could explain both the mechanical (hard) and semantic (soft) aspects of human systems; a systemic approach which would provide an integrated model for consciousness–information–energy–matter interchanges and interactions. Evidently, it would be an ambitious goal for us even now after decades of systemic speculations and trials, but nevertheless, the theoretical and practical impacts of systems theory have formed a vast variety of the current theories and methods in psychology and medicine. Developed by the American psychiatrist George Engel, the biopsychosocial model might be one of the most successful and influential views in this field.

1.5 Biopsychosocial (BPS) Model

In 1977, George Engel published a paper in Science Magazine titled: "The need for a new medical model: A challenge for biomedicine" and attempted to explain the deficiencies of the biomedical model and the advantages of the model he proposed (Engel 1977). His new model was developed based on Ludwig von Bertalanffy's General Systems Theory (GST) (1956). In GST, von Bertalanffy tried to develop a general model for the systems in different fields that is neither reductionist nor mechanistic. In this model, we deal with different systems from micro to macro with hierarchical organization. Engel applied this model to the human and considered a hierarchy of organization of the different systems in man. This hierarchy begins with the level of molecular interactions and extends to higher levels corresponding with cells, tissue, organs, nervous system, person, two persons, family, and community. He claims that in biomedicine, the emphasis is mainly on the subpersonal levels and the trend is to more highlight the cellular and molecular levels, with the assumption that all human phenomena can be reduced to underlying biological procedures.

He addresses that in practical clinical encounters, we deal with the personal level, of course by highlighting the behavioral aspect of this level and ignoring the experiential aspect; the mental activities and states such as intentions, thoughts, emotions, feelings and beliefs. These mental phenomena are the superimposed, emergent modes of this level which cannot be inferred or predicted from the underlying levels. As such, the analysis of a system, like that of a human being, could be helpful but insufficient. One must study the whole system's behavior and experience in order to synthesize a systemic approach and understand the irreducible properties of the system.

Engel explains that when a pathological change emerges in one of the levels of the hierarchy, the problem would not remain confined to that especial level; changes also take place in the upper and lower levels. For example, when a myocardial infarction occurs in the level of cardiac muscles, it raises concerns, fear, and anxiety in the personal level; new patterns in the relationship dynamic of the two person level; strain, anxiety, new tasks and roles in the family; and the use of medical-social and other recourses in the community level. Similarly, changes happen in the downward direction. The sympathetic system is activated and there is a neurotransmitter release from the nerve endings. There presents then a risk of damage to the other organs like the kidneys and liver. Ischemia and infarction are seen in tissues and the signs of cell damage could be seen in the cellular and molecular level. Engel suggests that in the treatment of a patient, all of these changes in the different levels should be considered and that the intervention should not be confined to the level in which the primary pathology has emerged. He also states that as we deal with the personal level of patients in the clinical setting, the communication skills and attention to patient concerns, emotions, feeling, and beliefs are of great importance.

On one hand, the Biopsychosocial model was very influential and many have tried to apply it to real life situations in medicine and other related fields like sociology and health psychology. This group believes that the assumptions of this model are still relevant and could be further developed with new ideas (Adler 2009). On the other hand, there are debates about its applicability and there are some who criticize it. For example, the psychiatrist Tavakoli (2009) believes that this model in practice confuses the students and residents who are learning the different psychiatric problems and creates an arbitrary separation between biology and psychology. Additionally, he claims that the use of this model in psychiatry and other fields of medicine such as surgery and medicine makes students frustrated and avoidant. We think that this confusion arises from a fundamental question which is not only present in medicine, but also in other fields which deal with mental phenomena like psychology and sociology. This is the basic philosophical question of the relationship between mind and body, or in a wider view, mind and matter. In medicine, it is a critical question, but in the field of psychiatry, it is more tangible because patients have mental problems. Developments in neuroscientific studies of psychiatric disorders have shown the patterns of change of neurotransmitters in the central nervous system in the course of these problems (Trimble and George 2010). On the other hand, plenty of studies show that the psychosocial context is important and plays a crucial role in the emergence of these disorders (Wallace and Gach 2008). At first sight, these findings seem controversial, but if we can rid ourselves of the linear causality framework, we will at least see a causality network in which the psychosocial and/or physical parameters can initiate and/or accelerate a mental/physical illness or healing response. The BPS model illustrates this circularity in the form of mutual interactions of different organizational levels. For example, a change in the personal level (experience and behavior) leads to changes in the subpersonal (molecular, cellular and vital systems) and also the suprapersonal levels (two-person, family, community, etc.). This model therefore shows interchangeability and merging of hard reality and soft reality, but what about translatability of these two worlds? How can we follow the energy-information flow through the levels of organization? And how should we manage the semiotic and mechanical consequences of each health event?

In any case, the BPS model is an appropriate theoretical framework for the enrichment of patient-doctor communication, elaboration of clinical reasoning, and also interdisciplinary research and development. Yet as an explanatory model, it has certain shortcomings and ambiguities, especially in exploring the mutual translation of the mental and the physical. The emergency principle of the BPS model is a good departure from the reductionism of biomedicine and its limitations but the embodiment principle of BPS model could not appropriately overcome the mind–body dichotomy (Schwartz 1982). According to the embodiment theory, mental procedures are embodied as neural processes (MacKay 1978). This means that mental activities are nothing but brain function, but it could not explain how a symbolic agent such as "this is a pill for pain relief" can control the neuroimmunologic responses (Sperry 1980). In fact this theory is not able to explain how signs flow through the human systems in the forms of molecules, cells, bioenergetic pulsations, sounds, writings, icons, and intentions, nor how these heterogeneous signs are transformed and translated to each other. For instance, pressure – point massage, NSAID pills, hypnotic suggestions, sugar pills, behavioral change, healing touch, mindfulness trainings, and corticosteroid injections can control or even treat inflammatory arthritis, but each one initiates the healing response from one level of organization with different sorts of signs and materials; energetic, informational and mindful.

We were unable to verifiably find a type–type identity between the mental and physical phenomena, and as one can see, a diversity of anisotropic, semiotic, and mechanical agents achieves an anti-inflammatory response in a joint. According to the American philosopher Donald Davidson (1970, 1994), there is no psychophysical law, and there are only token mental events which are identical to token physical events. From this viewpoint, it is not possible to produce a generalized map for all mind–body interactions in a case of a disorder because of anomalous monism in the mental and physical phenomena (Davidson 1970, 1994).

In addition to a bigger picture of human systems, which BPS has outlined, we need a common language to be able to translate signs of various levels of organization to each other; a systemic language which can illustrate microvessels of the psychophysical body, indeed in the unique, dynamic, and chaotic clinical contexts. Some of the BPS reformers such as Aviel Goodman have identified these pitfalls of the BPS model and tried to make light of them.

Goodman (1991), in the organic unity theory, tried to merge mental–physical identity and BPS theories to establish an integrative model which resolves the problem of translatability. This theory presents a satisfying solution for the problem, but relies on an abstract concept; "the pure psychophysical event," which is the unique reference of both the physical and mental phenomena. From this view, physical and mental events are originally psychophysical events which are described in mental and physical terms. The pure psychophysical event is a good assumption, but the parsimony principle necessitates lesser assumptions, of course if it were possible.

According to Thure von Uexküll and a great many pioneers of psychosomatic medicine, biosemiotics can play such a mediating role between the systems (Meyer 1984). They believe that the emerging discipline of biosemiotics is able to cast new light on the meaning and nature of biological survival and also properly translate the mind–matter interactions away from common-sense mind–body dichotomy. Could this be what we are looking for in a BPS model; a concrete psychophysical reference which can be expressed in the terms of the mental or the physical?

1.6 Biosemiotics and the Biopsychosocial Model

The German medical doctor and philosopher Thure von Uexküll was the son of eminent biologist and philosopher Jakob von Uexküll and, in fact, throughout his professional life, he tried to develop and apply his father's ideas in medicine (Kull and Hoffmeyer 2005). Jakob von Uexküll was a biologist who conducted numerous studies on the interaction of animals with their environments, the result of which was a theory that explains the development of the subjective internal world of an organism by the interaction with its environment. In this theory, his goal was to explain how an organism, based on its instinctual needs and biological structures, gives meaning to the perceptual cues from the world around it and acts according to this meaning. This action is the basis for the meaning that has been assigned to that entity, and this cycle occurs again and again, ultimately solidifying its meaning. This theory was called Umwelt, the German word for "environment" (Deely 2004; Rafieian 2010). Jakob von Uexküll did not use the terminology of semiotics in this model, but what he described as perceptual cues is practically the same as the signs in the semiotic framework. He was, because of this, called a cryptosemiotician by other semioticians like John Deely (1990, 2004). Among other efforts, like the development of the philosophical school of biosemiotics, Thure von Uexküll introduced the foundations of psychosomatic medicine in Germany. To provide a better theoretical framework, he tried to merge the principles of the biopsychosocial model and biosemiotics. As mentioned, the problem with the biopsychosocial model was that when one goes from the micro levels to the macro levels, it is not entirely evident how these different levels are connected. Uexküll believes that semiosis is the translator of the events from one level to another (Uexküll and Pauli 1986; Rafieian 2012). Semiosis is the milestone of life and life is actually defined by semiosis. To fully understand this idea about the process of translation, consider the following example: Imagine a situation in which a shouting person raises the heart rate of the other. In this situation, the shout is perceived as a sign and interpreted in his or her mind as a result of connections in the brain to memories and other signs coming from the context of the environment. Impulses are then sent to the heart and neurotransmitters are released from the nerve endings releasing hormones into the blood stream. The neurotransmitters and hormones attach to the receptors on the cells and convey a message. In today's biology, it has been shown that the metaphor of a key and lock is not an appropriate model for the way that hormones act at the level of receptors. Hormones are proteins with complex three-dimensional structures and the way they attach to the receptors and the affinity of the molecule for the receptor depends on the context in which this attachment occurs (Sivik and

Schoenfeld 2006). The message conveyed and the ultimate effect on the cell is then dependent on this context of attachment. Here again, the hormone acts as a sign which is interpreted by the cell resulting in a change inside the cell.

Uexküll with the help of Thomas Sebeok defined the Biosemiotic School in philosophy of biology. In this school, the ideas of the American philosopher Charles Sanders Peirce were used to provide a new framework for the biology. I n the biosemiotic perspective, all living organisms are living in a semiosphere and sign interpretation is present wherever life is present (Sebeok 2001). In addition, Peirce tried to categorize the signs and created detailed classification of signs, in which three main categories can be determined: indexes, icons and symbols (Colapietro and Olshewsky 1996).

Index is a sign that has an actual connection with the object. For example, when we see smoke we become aware of the presence of fire. Or in medicine, the pulse and its connection with the heart function is an indexical relationship. Icons are pictorial signs. The icon has a resemblance or likeness with the object. Any portrait of a person is an iconic sign of that person. In medicine, an X-ray radiography of an organ could be considered as the iconic sign of that organ. The third group of signs is symbols. Symbols are the signs that have an arbitrary connection with the object. For example, any word is a symbol of what it refers to in the real word. There is no real connection with the word "water" and the water that is present in the tap. There is only a convention that makes the connection with the word water and the real water in nature.

In biomedicine, every doctor is thoroughly familiar with the use of indexes and icons. Taking the pulse auscultation and percussion are all examples of the use of indexical signs of objects inside the body and cannot be directly visualized. With the advance of technology in medicine, many indexical items have been replaced by new iconic signs. Doctors today depend on echocardiography signs of a cardiac valve stricture or insufficiency for a diagnosis instead of merely confining themselves to findings in an auscultation. Similarly, new methods of imaging like the CT scan and the MRI give new information via the iconic signs they provide. In the case of symbols, doctors use symbolic words of language to provide information about the signs and symptoms in the process of history taking (Nessa 1996). But because in biomedicine, the focus is on the biologic and physiologic levels, the use of language and communication is limited to the process of gathering information related to the biological and physiological functions of the organ in which the pathology has been developed. Humans are the only animals that have the ability to use symbolic signs as means for assigning a new meaning to an entity in the physical world. This ability to create new meanings and the power of semiosis as the translator of events between the levels of hierarchy of existence provide a great therapeutic tool, freedom of action, and creativity in clinical context. As mentioned, Balint has noted that in clinical practice, the doctor himself or herself could be as effective as a drug and many physicians in their daily practice have indeed experienced this occurrence (Balint 1957). To be as effective as a drug, a good rapport with the patient is needed; communication here is a semiotic enterprise. The tone of the voice, facial expressions, and gestures could all be meaningful and could have a placebo or nocebo effect. In fact, here is the point of intersection between placebo research and research in the doctor–patient relationship as different types of meaning effect. It is the semiotic analysis of this relationship that makes non-specifics specific and provides the doctor with the power of healing in the very context of clinical encounters.

As one can infer from the above discussion, the framework thinking about key concepts in health such as health and disease, and, healing and person are very important. And in order to change the way in which medicine is practiced, these concepts need to be revised. As philosopher of science Ludwig Fleck stated, it is the thinking style of the scientists in a field that defines the rules and structures of that field (as cited in Cohen and Schnelle 1986; Zajicek 1995) and in a similar manner, Thomas Kuhn (1992) speaks about the concept of paradigms and the fact that a scientific discipline is developed based on a group of axioms. The paradigm of that discipline then emerges by the activities of the scientists in that field. The paradigm remains stable only until the time in which the number of unexplainable anomalies remains unsubstantial. After that, a paradigm shift occurs (Anderson and Funnell 2005). As discussed, the results of research on placebos and related fields show that there is a need for a more comprehensive perspective in medicine. Engel found the systemic view useful because it provides the possibility for the thinker to cover a wide range of aspects of the person but as mentioned, the need presents itself for the different levels to be connected. As previously stated, Uexküll incorporated semiotics into Engel's systemic model and developed it further but there were other new ideas emerging parallel to it in the twentieth century that could expand our understanding, making the model more comprehensive. Søren Brier tried to incorporate these concepts in the Cybersemiotics model and develop a non-reductionistic model of consciousness, cognition, communication, and meaning that has been applied in medicine (Brier 1999a, b, 2008, 2010; Rafieian 2010). Here we will briefly outline the model and its application to medicine.

1.7 Cybersemiotic Medicine

As discussed, the main deficiency of the biomedical approach in modern medicine which results in avoidance of researchers and clinicians from dealing with placebo responses and its mechanisms is that a placebo response and its mechanisms are mediated by patients' feelings, beliefs, and emotions. In other words, because patients' phenomenological and first person experiences are important in exploring the placebo response, biomedicine's dualistic and reductionist approach to the mind–body relationship and its framework cannot define a research project for exploration of this issue. In fact as Brier (2010) explains, this ignorance of the first-person experience and the consciousness of an embodied person can also be seen in other fields of the natural and social sciences and humanities in today's world. Cybersemiotics is an effort to incorporate this phenomenological experience in a theory of cognition, knowledge, and understanding.

The two pillars of the Cybersemiotic model are cybernetics and Peircian (bio)semiotics. Until now, we have discussed biosemiotics and its function for systemic thinking. Cybernetics is derived from a Greek word that means "the art of steering". Cybernetics was originally developed by the mathematician Norbert Wiener as the science of control of animals and machines (Masani 1990). His aim was to develop a science for prediction and control of complex systems. The main concept of cybernetics is a feedback mechanism that helps the system to self-regulate. Shortly after its development, researchers from different disciplines grew interested in it because it was able to explain the mechanisms of system control in different disciplines from micro to macro levels (François 1999).

An important thinker, Gregory Bateson (1972), made further developments. He was an interdisciplinary researcher with contributions in different fields from anthropology and linguistics to psychiatry. After the emergence of cybernetics, he became fascinated in its ideas, and in collaboration with others from other disciplines developed the second-order cybernetics. In cybernetics, there is an observer who studies the behavior of systems. In second-order cybernetics, it is the very observer that is considered as the system under study. In other words, here the observer is observed.

In his career, Bateson (1979) was searching for the "patterns that connect". Cybernetic rules were one kind of these connecting patterns (like semiosis as discussed above). He also has been considered as one of the pioneers of biosemiotics, as his research about communication and information has been influential in different field s. For example, his double bind theory for the development of schizophrenia was based on his understandings about different levels of communication. His achievements in this era led to the development of family therapy as a method of psychotherapy.

From a medical perspective, cybernetics concepts are familiar for doctors. Any medical student is familiar with the feedback mechanisms that control different physiological functions in the body, like the level of electrolytes and hormones and coordination of the muscles. But again, like the territory of signs, the territory of feedback loops is not confined to the body. The interactions of the people in the interpersonal space and family, groups and organizations are also regulated by circular recursive feedback loops.

Cybernetics, like biosemiotics, then provides the patterns observed by Bateson that connect different levels of the hierarchy of existence of human organization. The other aspect that has influenced thinking about the way we gain knowledge in the world was new achievements in modern physics. The philosophical consequences of quantum mechanics teach us that we cannot separate the observer from the observed. The knower is connected to any topic to be known in the world and any boundary between the subject and object is arbitrary. Bateson and other pioneers of second- order cybernetics like Heinz von Foerster explored this view (Pörksen 2003). To put it in a medical context, the relationship of the observer and the act of observing can be applied to the therapist-patient relationship in the clinical encounter. Placing any border between these two here is also arbitrary. There are ongoing feedback loops present that regulate the encounter; and the feelings, emotions, beliefs and intentions of both are influential in the outcome of the process. Biomedicine does not provide such a view. Instead the general picture is a patient, a broken machine to be repaired and the doctor, the person who knows what is wrong and tries to repair it by prescribing drugs or manipulating surgically.

Applying these views is the aim of Cybersemiotic medicine (Rafieian 2010); a framework in which different disciplines dealing with mankind, from physiology and anatomy to psychology and sociology, are equally important and relevant. The so-called "soft data" coming from the phenomenological experience of the patient exploring the psychosocial context of the emergence of the problem are as important as the "hard data" coming from physical examination, laboratory data, etc. The importance of the concept of information is well appreciated in the modern world, with different theories about the nature of the information. As Brier explains (2008), materialistic views of information consider it as a real entity in the world which is transferred from one place to the other. In contrast, the semiotic view of information takes an interpretative view and considers the semiosis as a translator of the message coming from one level to the other. He also considers the point that the materialistic view of information is more applicable in micro levels of physical and chemical interactions. Considering the example of a neurotransmitter or a hormone as a sign, the interpretation of the message is a semiotic process although the interaction between the ligand and receptor happens in the material world. In macro levels of interpersonal and social interactions, the semiotic view is more prominent although these interactions are ultimately based on the processes occurring at the molecular level.

Last but not least is the concept of culture-specific disorders, which are the health problems specific to a certain culture. Culture is developed in a network of semiotic interactions and sometimes health problems emerge out of that which is meaningful only in that context. These problems cannot then be generalized with those of other cultures. These health issues also could be explored in a cybersemiotic framework, as there are other issues such as medical ethics, health semiotics, and lifestyle modification that could be dealt with in this context (Rafieian 2010). These however remain open for more exploration and further research.

1.8 Applying the Biosemiotic Perspective: Towards an Integrative Medicine

The signs of dissatisfaction of the modern mainstream medicine have become evident in recent decades both in the public sphere and among health care professionals. Doctors who have been trained in modern conventional medicine are irritated by the rigid framework of biomedicine and are interested in alternatives. In a reactive manner to preserve its authority, the current paradigm has developed the new movement called Evidence Based Medicine (EBM). As Roberti di Sarsina and Iseppato (2011) explain, the pillars of this movement have been defined as: "1) medical knowledge and clinical skill, 2) (scientific) evidence through clinical investigations, and 3) patient preferences" (p. 5). But these aspects contradict themselves because the evidence that has come from biomedical research could not take the patients' preferences into consideration. Practically, patients' preferences are based on their beliefs, thoughts, and desire, and therefore cannot be categorized by the available methods of scientific research.

On the other hand, overemphasis on the anatomical and physiological aspects of human beings resulted in the focus of research and intervention production in these levels, and every day medicine is becoming increasingly drug and technology-based (Webster 2002; Conrad and Leiter 2004). The commercial benefits of the companies that produce these drugs and technologies have resulted in support of this trend and the development of a concept called medicalization. With medicalization, the medical system tries to define ordinary personal and social problems like shyness or baldness as medical problems and, instead of solving these problems in the context of life or simply accepting them as normal occurrences, tries to invent new drugs, technologies, or interventions to manipulate them (Conrad 2008; Rafieian 2010). As a result, the health care system is more and more becoming disease-centered with increased emphasis on new terms for new pathologies and developing new specialties instead of being saloutogenic, exploring prevention, and considering the person as a whole.

In fact, ordinary people noticed before professionals that there is something wrong with this approach. Re -emergence of pluralism in medicine and the interests of people in complementary alternative medicine (CAM) show that they do not trust the mainstream medicine as they did before. Studies have shown that roughly half of the population in industrialized countries and as high as eighty percent in developing countries use CAM (Bodeker and Kronenberg 2002). There are some alternative methods like acupuncture, the efficacy of which has been corroborated by the research methodology of biomedicine, but the rigidity of this paradigm does not allow them to be incorporated into the main body of health care systems and they have generally a marginal place and are applied as adjuvant methods. Accordingly, there is a need for innovation in designing research in this field (Pritzker and Hui 2012).

The reason behind these deficiencies is that medicine has applied a framework which has a much too narrow perspective and is unable to explain the seemingly anomalous phenomena like placebo responses and so-called alternative methods of treatment like energy medicine (Foss 1994). Because of the narrowness of this view, even when medical professionals try to search for the mechanisms underlying these phenomena or assess the validity of these methods, problems arise because of the paucity of available methodologies.

As discussed, biosemiotics provides a broader view that enables us to explain phenomena like placebo and to think creatively about healing and health in the semiosphere. The aim of this book is to explore different aspects of the placebo response from this perspective. The explosive rise in the research about placebo in different fields from philosophy and psychology to psychoneuroimmunology and neuroscience provides us with the raw material that could be incorporated in the framework of biosemiotics.

References

Adler, R. H. (2009). Engel's biopsychosocial model is still relevant today. *Journal of Psychosomatic Research, 67*, 607–611.

Anderson, R. M., & Funnell, M. M. (2005). Patient empowerment: Reflections on the challenge of fostering the adoption of a new paradigm. *Patient Education and Counseling, 57*, 153–157.

Balint, M. (1957). *The doctor, his patient and the illness*. Edinburgh: Churchill Livingstone.

Bateson, G. (1972). *Steps to an ecology of mind*. New York: Ballantine.

Bateson, G. (1979). *Mind and nature*. Glasgow: Fontana/Collins.

Beecher, H. K. (1955). The powerful placebo. *The Journal of the American Medical Association, 159*, 1602–1606.

Bensing, J. M., & VerheuL, W. (2010). The silent healer: The role of communication in placebo effects. *Patient Education and Counseling, 80*, 293–299.

Bodeker, G., & Kronenberg, F. (2002). A public health agenda for traditional, complementary, and alternative medicine. *American Journal of Public Health, 92*, 1582–1591.

Brier, S. (1999a). Biosemiotics and the foundation of cybersemiotics: Reconceptualizing the insights of ethology, second- order cybernetics, and Peirce's semiotics in biosemiotics to create a non-cartesian information science. *Semiotica, 127*, 169–198.

Brier, S. (1999b). What is a possible ontological and epistemological framework for a true universal information Science? The suggestion of a cybersemiotics. *World Futures General Evolution Studies, 13*, 79–100.

Brier, S. (2008). *Cybersemiotics: Why information is not enough!* Toronto: University of Toronto Press.

Brier, S. (2010). Cybersemiotics: An evolutionary world view going beyond entropy and information into the question of meaning. *Entropy, 12*, 1902–1920.

Brom, D. B. (2012). Body–mind medicine and the placebo response. *South African Family Practice, 13*(1).

Bügel, P. (2004). The many meanings of placebo. *Forschende Komplementärmedizin/Research in Complementary Medicine, 5*, 23–30.

Cohen, R. S., & Schnelle, T. (Eds.). (1986). *Cognition and fact: materials on Ludwik Fleck*. Dordrecht: Springer.

Colapietro, V. M., & Olshewsky, T. M. (1996). *Peirce's doctrine of signs: Theory, applications, and connections*. Berlin: Walter de Gruyter.

Conrad, P. (2008). *The medicalization of society: On the transformation of human conditions into treatable disorders*. Baltimore: JHU Press.

Conrad, P., & Leiter, V. (2004). Medicalization, markets and consumers. *Journal of Health and Social Behavior, 45*, 158–176.

Davidson, D. (1970). Mental events. In D. Davidson (Ed.), *Essays on actions and events* (pp. 207–223). New York: Oxford University Press.

Davidson, D. (1994). Psychology and philosophy. In M. Martin & L. McIntyre (Eds.), *Readings in the philosophy of social science*. Cambridge: MIT Press.

Deely, J. N. (1990). *Basics of semiotics*. Bloomington: Indiana University Press.

Deely, J. (2004). Semiotics and Jakob von Uexküll's concept of umwelt. *Sign Systems Studies, 32*(1/2), 11–34.

Eco, U. (1979). *A theory of semiotics*. Bloomington: Indiana University Press.

Engel, G. L. (1977). The need for a new medical model: A challenge for biomedicine. *Science, 196*, 129–136.

Flaten, M. A., Aslaksen, P. M., Lyby, P. S., & Bjørkedal, E. (2011). The relation of emotions to placebo responses. *Philosophical Transactions of the Royal Society, B: Biological Sciences, 366*, 1818–1827.

Foss, L. (1994). Putting the mind back into the body a successor scientific medical model. *Theoretical Medicine, 15*, 291–313.

Francois, C. (1999). Systemics and cybernetics in a historical perspective. *Systems Research and Behavioral Science, 16*, 203–219.
Gaines, A. D., & Davis-Floyd, R. (2004). Biomedicine. In M. Ember & C. Ember (Eds.), *Encyclopedia of medical anthropology*. New York: Springer.
Goebel, M. U., Trebst, A. E., Steiner, J., Xie, Y. F., Exton, M. S., Frede, S., Canbay, A. E., Michel, M. C., Heemann, U., & Schedlowski, M. (2002). Behavioral conditioning of immunosuppression is possible in humans. *The FASEB Journal, 16*, 1869–1873.
Goodman, A. (1991). Organic unity theory: The mind–body problem revisited. *The American Journal of Psychiatry, 148*, 553–563.
Justman, S. (2011). From medicine to psychotherapy: The placebo effect. *History of the Human Sciences, 24*, 95–107.
Kaptchuk, T. J., Friedlander, E., Kelley, J. M., Sanchez, M. N., Kokkotou, E., Singer, J. P., Kowalczykowski, M., Miller, F. G., Kirsch, I., & Lembo, A. J. (2010). Placebos without deception: A randomized controlled trial in irritable bowel syndrome. *PLoS One, 5*, e15591.
Kihlstrom, J. F. (2008). Placebo: Feeling better, getting better, and the problems of mind and body. *McGill Journal of Medicine MJM, 11*, 212.
Kirsch, I. (1994). Clinical hypnosis as a nondeceptive placebo: Empirically derived techniques. *American Journal of Clinical Hypnosis, 37*, 95–106.
Kirsch, I. (1997). Specifying nonspecifics: Psychological mechanisms of placebo effects. In A. Hrrington (Ed.), *The placebo effect: An interdisciplinary exploration* (pp. 166–186). Cambridge: Harvard University Press.
Kradin, R. L. (2011). Placebo response: A consideration of its role in therapeutics. *Current Psychiatry-Reports, 13*, 37–42.
Kuhn, T. S. (1992). *The trouble with the historical philosophy of science*. Cambridge, MA: Department of the History of Science/Harvard University Press.
Kull, K., & Hoffmeyer, J. (2005). Thure von Uexküll 1908–2004. *Sign Systems Studies, 33*, 487–494.
Lewith, G. T., Jonas, W. B., & Walach, H. (2010). *Clinical research in complementary therapies: Principles, problems and solutions*. Edinburgh/New York: Elsevier Health Sciences.
Mackay, R. (1978). Assembling a corpus: The nature and process of instruction. *Canadian Journal of Sociology/Cahiers Canadiens De Sociologie, 3*(1), 41–59.
Masani, P. R. (1990). *Vita Mathematica: Vol. 5. Norbert Wiener 1894–1964*. Basel: Birkhauser.
Meissner, K., Kohls, N., & Colloca, L. (2011). Introduction to placebo effects in medicine: Mechanisms and clinical implications. *Philosophical Transactions of the Royal Society, B: Biological Sciences, 366*, 1783–1789.
Meyer, A. E. (1984). Thure von Uexküll. *Psychotherapy and Psychosomatics, 42*, 97–99.
Miller, F. G., & Colloca, L. (2010). Semiotics and the placebo effect. *Perspectives in Biology and Medicine, 53*, 509–516.
Miller, F. G., & Colloca, L. (2011). The placebo phenomenon and medical ethics: Rethinking the relationship between informed consent and risk–benefit assessment. *Theoretical Medicine and Bioethics, 32*, 229–243.
Moerman, D. E. (2006). The meaning response: Thinking about placebos. *Pain Practice, 6*, 233–236.
Moerman, D. E., & Jonas, W. B. (2002). Deconstructing the placebo effect and finding the meaning response. *Annals of Internal Medicine, 136*, 471–476.
Mommaerts, J., & Devroey, D. (2012). The placebo effect: How the subconscious fits in. *Perspectives in Biology and Medicine, 55*, 43–58.
Nessa, J. (1996). About signs and symptoms: Can semiotics expand the view of clinical medicine? *Theoretical Medicine, 17*, 363–377.
Papakostas, Y. G., & Daras, M. D. (2001). Placebos, placebo effect, and the response to the healing situation: The evolution of a concept. *Epilepsia, 42*, 1614–1625.
Pickering, T. G., & Friedman, R. (1991). The white coat effect: A neglected role for behavioral factors in hypertension. In P. McCabe, N. Schneiderman, et al. (Eds.), *Stress, coping, and disease* (pp. 35–49). Hillsdale: Lawrence Erlbaum Associates Inc.

Pörksen, B. (2003). Heinz von Foerster (1911–2002): The Socrates of cybernetics. *International Journal of General Systems, 32*, 519–523.
Price, D. D., Finniss, D. G., & Benedetti, F. (2008). A comprehensive review of the placebo effect: Recent advances and current thought. *Annual Review of Psychology, 59*, 565–590.
Pritzker, S., & Hui, K. K. (2012). Building an evidence-base for TCM and integrative East-West medicine: A review of recent developments in innovative research design. *Journal of Traditional and Complementary Medicine, 2*(3), 158–163.
Rafieian, S. (2010). Cybersemiotic medicine: A framework for an interdisciplinary medicine. *Cybernetics & Human Knowing, 17*, 65–93.
Rafieian, S. (2012). A Biosemiotic approach to the problem of structure and agency. *Biosemiotics, 5*(1), 83–93.
Roberti di Sarsina, P., & Iseppato, I. (2011). Why we need integrative medicine. *The EPMA Journal, 2*, 5–7.
Schwartz, G. E. (1982). Testing the biopsychosocial model: The ultimate challenge facing behavioral medicine? *Journal of Consulting and Clinical Psychology, 50*, 1040.
Sebeok, T. A. (2001). Biosemiotics: Its roots, proliferation, and prospects. *Semiotica, 134*(1/4), 61–78.
Sivik, T., & Schoenfeld, R. (2006). Psychosomatology as a theoretical paradigm of modern psychosomatic medicine. *International Congress Series, Elsevier, 1287*, 23–28.
Sperry, R. W. (1980). Mind-brain interaction: Mentalism, yes; dualism, no. *Neuroscience, 5*, 195–206.
Stefano, G. B., Fricchione, G. L., Slingsby, B. T., & Benson, H. (2001). The placebo effect and relaxation response: Neural processes and their coupling to constitutive nitric oxide. *Brain Research Reviews, 35*, 1–19.
Tavakoli, H. R. (2009). A closer evaluation of current methods in psychiatric assessments: A challenge for the biopsychosocial model. *Psychiatry (Edgmont), 6*(2), 25–30.
Tilburt, J. C., Emanuel, E. J., Kaptchuk, T. J., Curlin, F. A., & Miller, F. G. (2008). Prescribing "placebo treatments": results of national survey of US internists and rheumatologists. *BMJ: British Medical Journal, 337*, a1938.
Trimble, M. R., & George, M. (2010). *Biological psychiatry*. Chichester/West Sussex/Hoboken: Wiley.
Uexküll, T. V., & Pauli, H. (1986). The mind-body problem in medicine. *Advances: Journal of the Institute for the Advancement of Health, 3*, 158–174.
von Bertalanffy, L. (1956). General system theory. *General systems, 1*, 1–10.
Walach, H. (2011). Placebo controls: Historical, methodological and general aspects. *Philosophical Transactions of the Royal Society, B: Biological Sciences, 366*, 1870–1878.
Wallace, E. R., & Gach, J. (Eds.). (2008). *History of psychiatry and medical psychology: With an epilogue on psychiatry and the mind-body relation*. New York: Springer.
Webster, A. (2002). Innovative health technologies and the social: Redefining health, medicine and the body. *Current Sociology, 50*, 443–457.
Zajicek, G. (1995). Normative medicine. *Medical Hypotheses, 45*, 331–334.

Chapter 2
Cybersemiotics as a Transdisciplinary Model for Interdisciplinary Biosemiotic Pharmacology and Medicine

Søren Brier

2.1 Introduction to the Biomedical Problem of Theoretically Integrating Awareness and Experience

This chapter is based on the hypothesis, that the major problem of the dominating biomedicine paradigm of western medicine lies in its inability to include the psychological and sociological realities in its theoretical foundation for describing the healthy embodied person and develop models of the causes of health and illness as a basis for finding ways to treat illness. I believe that this is most clearly shown in the lack of explanation of placebo effects. A way to understand placebo effects is that they are caused by psychological and sociological meaning producing effects and they are outside the scope of the biomedical paradigm's explanatory models.

A major reason to assume this is that the natural sciences – of which biology is one – in their foundation do not encompass theories of meaning and experiential mind in their basic assumption of causality in the world of bodies. It's because bodies ontologically are considered to be pure material and therefore fully explainable based on physical and chemical components through molecular biology. When looking into computers and robots, the structure and dynamics of language, or in the brain neuro-physiologically, no one has managed to find any qualia, experience, emotions, or awareness; only matter; or rather in technology, wires and bits; and in living bodies, electrochemical impulses and transmitter molecules, hormones, and functional structures of neurons, glial cells, muscles cells, and other functional structures.

But even in biology, we have to include in our models that living bodies have experiences and functions through instinctual meanings. In this aspect, they differ

S. Brier (✉)
Department of International Business Communication, Copenhagen Business School, Frederiksberg, Denmark
e-mail: sb.ibc@cbs.dk

F. Goli (ed.), *Biosemiotic Medicine*, Studies in Neuroscience, Consciousness and Spirituality 5, DOI 10.1007/978-3-319-35092-9_2

qualitatively from robots. When we come to the level of humans, this meaning processes are further enlarged by elaborate of culture of rules and artifact as well as oral language communications. When writing is invented, the cultural aspect in the form of religion, economics, law, literature and finally science, penetrates the human beings as permanent virus infections in cognition and communication much like our gut bacterial with which we cannot live without.

The mentioned missing areas are treated in biosemiotics for the basic characteristic of the embodied living systems, for the human living systems, cultural and hermeneutical studies, semiotics, discourse analysis and phenomenology paradigms, which on the other hand have great problems in establishing rational and meaningful bridges to the results of the natural sciences.

Many human diseases can be used to illustrate the interdisciplinarity of the workings of living, experiential self-conscious bodies, for instance, anorexia. In fact, anorexia has a pure physical-chemical aspect through metabolism and genetics, it bears a psychological experiential aspect of relief from stress, duties and ethics; and the condition of anorexia is culturally situated in society, which dictates how one's body ought to look according to gender.

The medical practitioner that wants to treat the whole person is in dire need of a transdisciplinary framework that can unite the data from these qualitative different fields or at least combine them in a consciously reflected way as a supplement to the practical way physicians have learned themselves to deal with the complex embodied human cultural reality.

Cybersemiotics offers one example of such a framework in its combination of cybernetic-informational theories of science with the biosemiotics based on C.S. Peirce's semiotics. This chapter will in a brief manner explain the main idea of the paradigm on which I have written the book: *Cybersemiotics: Why information is not Enough* and follow up papers (Brier 2008a, b, c, d). Here I propose an overview of different forms of cognition, communication and information in the shape of a visual model called the 'cybersemiotic star', which is an alternative to logical positivisms hierarchical unity science, with physics as the most basic one and the role model for all scientific endeavors. The star (see Fig. 2.1 below) is a diagram illustrating how human beings' rational knowledge systems can be mapped over four qualitatively different but evenly important and foundational ontological aspects: Energy/matter, life, consciousness, meaning. These specialized forms of knowledge are developed from a general area of embodied, enacted, embedded and extended social semiotic interaction. In this perspective, human communication is embodied and biologically situated (life); it also has a conscious experiential and intentional aspect (first person consciousness); the subjects consciousness is situated in cultural practice (meaning) through a shared language; lastly, it is environmentally naturally situated (energy/matter) in a universe whose shape and order is partly independent of our perception of it.

My philosophy of the scientific point of view is that "the brain" is a physiological concept or model of a complex material thing, which we believe is intricately connected with the production of our perception and subjective feelings. However, "the brain" is an objective thing; it is not a subjective experience. We do not experience

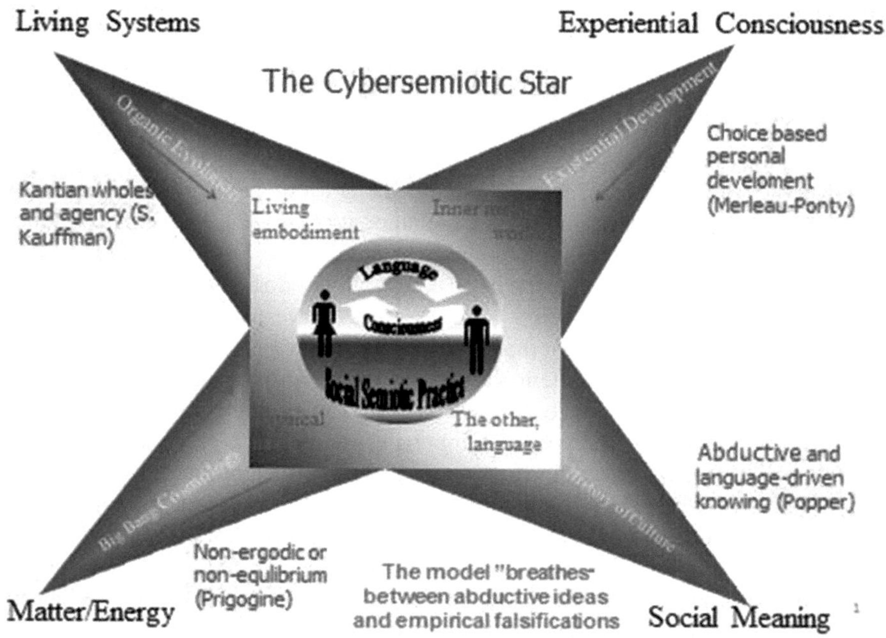

Fig. 2.1 Cybersemiotic star: a model of how the communicative social system of the embodied, enacted, extended and embedded mind produces four main areas of knowledge that can also be understood as the minimal prerequisites for interpersonal observation and knowing. Physical nature is usually explained as originating in energy and matter, living systems as emerging from the development of semiotic life processes (for the production of special proteins from DNA in the first cell). They differ from non-living system by being what Stuart Kauffman (2012) calls "Kantian wholes". Social culture is explained as founded on the development of new meaning and knowledge in language and practical habits; which is why the history of cultures and societies is not predictable. Finally, there is our experiential world, which in phenomenology is explained as deriving from the development of our individual life world and self-consciousness. All these types of knowledge, which are often considered incommensurable, are seen as having their origin in our primary semiotic intersubjective embodied world processing of observing and interpreting within social communication and action of which language is a part. The *arrows* in the arms signifies that interpretation of the worlds are produced intersubjectively and empirical put to falsification test and those which fails goes back into socio-communications semiotic net and get revised and thereafter tested again in an ever ongoing process of developing knowledge and skills. The model is developed from Brier 2008a and is still developing)

our brain. The brain is furthermore not I as a subject and the brain most likely does not "have experiences". If it did, then I do not know them, because I only know mine and I am not only my brain, but also my feeling body, senses, language and cultural meaning. I – whatever that phenomenon is – have the experiences; "the brain" does not. Therefore, "the brain" cannot have subjective thoughts and it cannot in itself be subjective; only I can. But there are on the other hand hardly any scientific definitions of the existence of an I, self or subject. Furthermore if you are modeling my subjective experiences through interview I do not believe you are modeling the – or my – brain.

This millennium's new focus on brain and consciousness shows the problem clearly when cognitive brain and psychology researchers wants to get writ of our free will and tell us that the bran and consciousness is the same. We are our brains. Brain scanning techniques make it possible to see which parts of the brain are used in what kinds of perceptions, actions, and moods by following increased blood flow to the active parts, as the brain uses a large amount of oxygen. We can also induce certain feelings, moods, sensation qualities or the memory of them, which people report orally, when we stimulate the brain electrically or do and say certain things to people. We can, through electrical stimulation of nerves, make limbs move and organs function. We can also externally register and describe the interaction between sense stimuli and behavior in meticulous experiments with humans and other living beings as it has been done since the hay days of Skinner's radical behaviorism and the European ethology of Lorenz and Tinbergen.

However, no matter how refined our empirical scientific approaches become, we cannot find any experiences in the brain, or the rest of the nervous system for that matter. The felt awareness seems to be found on another level of abstraction (Hinde 1970). Something central about the brain's function, as an organ, escapes us (McGinn 2000, pp. 66–68; Hofstadter 2007, p. 373; Searle 2007). The scientific tragedy is that our only access to the first person experience itself is indirectly through interpreting verbal or written testimonial as well as interpretation of behavior from the experiencing person. We seem to have no direct scientifically verifiable access to the quality of experience and meaning of other people. Thus far, our most direct access to first-person experiences is through meaningful verbal or written communication from the experiencing person (Heil 2004, p. 3).

But the body seems to have its own non-verbal experiences. How do we measure how much of it we are able to be conscious of, and able to give verbal report of? This is our main problem in philosophy as well as in medicine, especially in psychosomatics. We have no idea how the concept of standing up from the chair we sit in to fetch a cup of coffee is able to translate into the physiological processes that create the movements of the body. But the problem is also obvious when diagnosing children. Here we see limited ability to verbally precisely describe location and flavor of pain in the body. Through language acquisition, cultural knowledge and experience we learn to be more and more precise in the description of our experiences and to relate them to one of the accepted classification systems of our culture. When you are in love for the first time for instance it can be difficult to distinguish it from fever or stomach ache.

It is a very fundamental point and one of the reasons that I believe there are limits to what science can come to know about conscious experience and why phenomenology is unavoidable transdisciplinary. The subjective experiences are not a part of the scientific universe, because it only deals with objective phenomena. We can say that the existence of subjects is somewhat of an inductive objective theory, though we cannot measure subjectivity in itself. It does not have a mass, energy contend or a momentum, or any kind of movement we can measure. It is outside objective measurement, like most doctors and dentist have had to realize, when dealing with patients reports of pain (Firestoner et. al. 1999). Not only is it individual but it also seems to be situational and related to the meaning, we attach to the causes of the pain (Jong et al 2006). I find this point crucial in understanding the limits of the scientific way of producing knowledge.

It means that language and culture are "in the way" for the natural science approach. We cannot experience other people's experiences directly. What people experience when doing certain behaviors, we only know from their own reports, though we can see physiologically what part of the brain they use or how they behave externally as well as internally. The paradox of modern attempts to work towards a "science of consciousness" is that we have no direct scientific empirical access to the experiential qualities of will, intentions and meaning on which to build such a science (Edelman and Tononi 2000). As a philosopher of science, it seems to me that this is why we have the qualitative phenomenological, hermeneutical and discourse theoretical methods of humanities and the social sciences. However, they are not really considered to be truly scientific by the natural sciences; only the brain sciences are (Bennett et al. 2007).

Nonetheless, as responsible and experientially aware social citizens, we are not identical to our brains (Edelman and Tononi 2000, p. 1) although we do need them in order to stay conscious. Free will is a prerequisite for Democracy and science! So we seem to be a more complex integrative product of physical, chemical, biological, social, mental, semiotic and communicative systems producing and being produced by culture and language. Language and culture seems to be a kind of rather permanent infection of the nervous system and our whole cognitive apparatus. The brain and the body surely are important components of this product we call the self-conscious feeling subject, but so are the ability of living systems to produce experiences, and think about them and communicate them in language and paralinguistic signs. This is the problem, which some formulate as an explanatory gap (Thompson 2003; Levine 1983). The way we speak about our experiences and feeling seem to be able to change them, but we really do not know how. What is the physiology of being brave or having faith in the justice and healing power of God?

There is no agreement on how to formulate this explanatory gap problem (Rorty 1980). Therefore I will suggest a working hypothesis here: The attempt to explain consciousness through scientific physicochemical, informational and computational objective realistic paradigm leads to counter claims of phenomenological paradigms that our knowledge or process of knowing is based on an experiential world (what Husserl called a 'life world') prior to any culturally-developed scientific explanations. Husserl's method was an attempt to put these influences in parenthesis, or bracketing (Epochè), in his attempt to get to the pure phenomena or the "thing in itself" through a systematic peeling away of their symbolic layers of meanings until only the thing itself as "originally" meant and experienced remains (Husserl and Bundgard 1997; Husserl 1999). That is radically different method from the scientific attempts to explain our experiences from how we use our brains.

Husserl's (1970, 1999) problem was that our consciousness and intentionality are always contaminated with intersubjective linguistic and cultural mentality conceptions, and ontological assumptions of the situation at hand. Consequentially, in order to get to the pure phenomenon he must seek beyond those disturbing intersubjective influences. Thus even phenomenology has trouble getting to the experience itself. But anyhow this basic phenomenological position is shared by Edmund Husserl, Maurice Merleau-Ponty and Charles Sanders Peirce. But Peirce deviates from the European phenomenologist in the development of a triadic phaneroscophy – as he calls his brand of phenomenology – as the point of departure for his semiotics. I find these three authors most relevant for the problem I want to discuss here. I have selected them as the most interesting defenders of the phenomenological transdisciplinary view. It is from his phenomenologically-based work, that C. S. Peirce develops his three basic categories Firstness, Secondness and Thirdness, which are foundational to his entire semiotic and pragmatic paradigm. Joseph J. Esposito's *Evolutionary Metaphysics: The development of Peirce's Theory of Categories* (1980) describes this quest in a most profound way. From this new form of phenomenology, Peirce attempted to prove mathematically that triadic relations cannot be broken down to duals.

I do find the phenomenological argumentation very convincing and they have recently been supported by many other developments in science. Nonetheless, the fundamentality of triadic thinking has been the stumbling block for many scholars failing to accept Peirce's paradigm's basis for developing the empirical and fallibilist pragmaticism that I have used as part of the foundation for my transdisciplinary framework. The combination of phenomenology, empiricism and fallibilism is unique for Peirce. It is therefore important not to under-estimate how deep reflections of logic – including the logic of relations, time, reality, continuity, the moment, perception and meaning – are connected to this path-breaking invention of Piece. Here is one of Peirce's paradigmatic foundational formulations on his type of phenomenology:

> Phaneroscopy is the description of the *phaneron*; and by the *phaneron* I mean the collective total of all that is in any way or in any sense present to the mind, quite regardless of whether it corresponds to any real thing or not. If you ask present *when*, and to *whose* mind, I reply that I leave these questions unanswered, never having entertained a doubt that those features of the phaneron that I have found in my mind are present at all times and to all minds. So far as I have developed this science of phaneroscopy, it is occupied with the formal elements of the phaneron. (Peirce, CP 1.289[1])

The formal phaneroscopic elements inspired from pure (abstract) mathematics can then be derived from this combination of a phenomenological and mathematical analysis:

> It seems, then, that the true categories of consciousness are: first, feeling, the consciousness which can be included with an instant of time, passive consciousness of quality, without recognition or analysis; second, consciousness of an interruption into the field of

[1] By convention CP refers to Peirce (1931–35 + 1958) Collected Papers of Charles Sanders Peirce, Volumes I–VI and VII + VIII collected in a CDROM or books, citations give volume and paragraph number, separated by a period.

consciousness, sense of resistance, of an external fact, of another something; third, synthetic consciousness, binding time together, sense of learning, thought. (Peirce, CP 1.377)

Our gap-problem is that these, the natural scientific and the phenomenological paradigms, are in Kuhn's (1996) terms incommensurable. They do not have the same epistemological and ontological conceptions. They have two different maps of reality, and either one's map is not on the others. The life sciences and social sciences again cover different areas of reality. But it is important to acknowledge that the structures of language or market economies are as real as the structures of a rock, but they are different aspects of reality. How can we grasp the material, the living, the social and the experiential in a framework that does not reduce one to another, as the logical positivists tried and thus failed to integrate the phenomenological and hermeneutical aspects of reality. I think that semiotics and systems thinking can provide a more comprehensive foundation for transdisciplinarity than a positivist inspired unity science like for instance E. O. Wilson's *Consilience: The Unity of Knowledge* from 1999 as the most advanced example.

Thus, this is my philosophy of a science-working hypothesis of what is the root of the explanatory gap. My suggestion of a solution is to contribute to the crafting of a transdisciplinary framework – inspired by Popper, Luhmann, and Peirce – wide and deep enough to contain paradigms of all four areas and thus enlarge our ontological conception of reality.

Peirce's phenomenological and pragmaticist semiotics has to be an important part of our transdisciplinary framework. After all medicine is the oldest semiotic and informational science as it base its diagnostic system on signs in what Hippocrates called symptoms. Thus the real illness is a sort of form creating the informational-semiotic physically visible symptoms. They are only the signs of the illness. The causes have to be found on deeper ontological levels through specialized sciences such as biochemistry, physiology, genetics, sociology and psychology.

But not even psychology can be said to be one science as it also has as aspects that can only be described by natural, as well as human and social aspects. The problem is – if we want to develop further from our present position – that we have no clue of how to integrate the different types of knowledge. Why can a man in African culture fall physiologically dead when he believes he has been cursed by a strong and frightening wizard, witch or medicine man? Why can some girls not eat or hold their food in them because they experience themselves to be fat in spite of the perception of their friends and family? The natural social and human sciences have been established independently and have different mutual incompatible foundations. Especially hermeneutics and phenomenology seems to be incommensurable to the sciences, as the two firsts are qualitative "sciences" almost completely based outside statistics and the natural sciences are quantitatively and mathematically based. In order to make better integration of the interdisciplinary knowledge we have collected and systematized we need to develop new transdisciplinary frameworks.

I have named the framework I have developed for Cybersemiotics, as it attempts to combine the two major attempts to unify theories for the area of life, cognition and communication in the form of intersubjective, systematic and consistent systems of knowledge: (1) the informational-cybernetic-systemic and (2) the semiotics-phenomenological-hermeneutical meta-paradigms.

If we – for instance, for the sake of medicine – want to create a transdisciplinary scientific theory of information, cognition, consciousness and meaningful communication (see Cowley et al. 2010 for starting such an attempt), then it seems that the first problem that we need to attend to, is to adjust or rather deepen the ontology in the theoretical framework, when our purpose is to make the integration of the different subject areas possible. Furthermore, the various subject areas and paradigms, self-description and concepts of knowledge and truth have to be made compatible in a larger context. This view can also be found in Roy Bhaskar's *Critical Realism* (Bhaskar 1997[1975]). For lack of a better word a "transdisciplinary paradigm" is what I will call that framework which in the long run should make it possible for us to deal with the psycho-somatic-cultural interactions in illness.

The concept transdisciplinary science is supposed to cover the sciences, humanities and social sciences, much like the German word 'Wissenschaft' or the Danish word 'videnskab'. Basarab Nicolescu (2002) wrote a profound work on the meaning and consequences of transdisciplinarity, titled *Manifesto of Transdisciplinary* where he attempt to lay the ground rules for transdisciplinary work.

2.2 Is Consciousness a Part of Reality?

A basic problem in our culture's systematic knowledge production is that the natural and social sciences, and the humanities do not agree on a common definition of reality. We talk about the physical, mental, and social reality, but do not really know how to bring them together into a larger conception of reality. Instead, each of them is often individually attempting to take over the power of defining reality. In medicine it has been most obvious in the antipsychiatry movements fight against biomedicines psychiatry. Is mental illness physiologically, sociologically or psychologically caused? It seems more and more that the answer is yes, but we have not worked very much on what the theoretical consequences of that is.

Finding the right basis for transdisciplinarity has been a problem ever since Otto Neurath (1983) introduced the logical positivistic idea of a unity of science based on physicalism. The physical world is here considered to be the given, the cause of all phenomena and from which all causes have to be extracted. The critique from the social sciences and the humanities has never stopped since. Its most alternative reaction has been to produce radical forms of social constructivism disclaiming any kind of positivistic truth claims for instance in the explanation of the processes creating psychosomatic diseases.

Most radical social constructivists consider political-ideological as well as cultural conceptions of reality to be the primary reality, of which science and the phenomenological life world is only one product of many. There are many theories in social medicine why different cultures have different patterns and frequencies of diseases, especially mental diseases. Thus we have many explanatory systems that take departure either in the material or in the social world. But phenomenology from Husserlian and Peircean traditions insist on a third view; namely that the experiential phenomenal world is the given reality and the truth is to be found in analyzing its structure, be it as intentionality schemata (Husserlian tradition) or basic categories of cognition in the form of sign types, which is then developed into a semiotics in Peircean tradition. The eternal foundation that Husserl (1970, 1999) was seeking in the pure intentional structures or forms of conscious awareness became independently for C. S. Peirce semiotic dynamical ways of knowing that emerged from Firstness as 'may-bes' and developed into 'would-be's' in Thirdness through the evolution of reasonableness in a continuity ontology, which he called Synechism. That was a great break with classical physical explanations based on universal natural unchangeable laws. He wrote:

> Once you have embraced the principle of continuity no kind of explanation of things will satisfy you except that they grew. The infallibilist naturally thinks that everything always was substantially as it is now. Laws at any rate being absolute could not grow. They either always were, or they sprang instantaneously into being by a sudden fiat like the drill of a company of soldiers. This makes the laws of nature absolutely blind and inexplicable. Their why and wherefore can't be asked. This absolutely blocks the road of inquiry. The fallibilist will not do this. He asks: may these forces of nature not be somehow amenable to reason? May they not have naturally grown up? After all, there is no reason to think they are absolute. If all things are continuous, the universe must be undergoing a continuous growth from non-existence to existence. There is no difficulty in conceiving existence as a matter of degree. The reality of things consists in their persistent forcing themselves upon our recognition. If a thing has no such persistence, it is a mere dream. Reality, then, is persistence, is regularity. In the original chaos, where there was no regularity, there was no existence. It was all a confused dream. This we may suppose was in the infinitely distant past. But as things are getting more regular, more persistent, they are getting less dreamy and more real. (Peirce, CP, 1.175).

To Peirce, Firstness is an unbroken continuity of pure mind or feeling, quality and tendencies to become existent in what Peirce called Secondness.

The social sciences and humanities have felt dominated by biologistic scientistic reductionist explanations of experience and behavior of human beings like Dawkins' selfish genes, memetics, E. O. Wilson's socio-biology and his later attempt to make a unified view from it (Blackmore 2000; Dawkins 2006; Wilson 1999). What this reductionist meta-scientific paradigm is supposed to mean is most clearly spelled out in Wilson (1999). Taking up the torch from logical positivism, Wilson predicts that most of the humanities will be replaced by hard scientific knowledge, and just like neuroscience, will eventually tell us what conscious experience is. Consilience, literally a "jumping together" of knowledge, has its roots in the ancient Greek concept of logos, which is the vision of an intrinsic orderliness governing the Cosmos. This seems also to be a prevailing view in much medical research and thinking. The problematic view inherent in much of science and analytic philosophy is that Logos is comprehensible by formal logical process guiding material processes and organization only, which often lead to a strong genetic dominance when constructing causal model for illnesses. A reason to believe that Peirce's semiotics can move us out of this predicament is that he combines his view of semiotics and logic in an evolutionary pragmatic framework. He writes:

> Logic will here be defined as formal semiotic. A definition of a sign will be given which no more refers to human thought than does the definition of a line as the place which a particle occupies, part by part, during a lapse of time. Namely, a sign is something, A, which brings something, B, its interpretant sign determined or created by it, into the same sort of correspondence with something, C, its object, as that in which itself stands to C. It is from this definition, together with a definition of 'formal', that I deduce mathematically the principles of logic. (Peirce 1980, pp. 20–21, p. 54)

For Peirce, pure mathematics is more fundamental than logic and in combination with phenomenology is the foundation of his metaphysics, as we have already shown. According to Peirce, logic is developed from mathematics and not the other way around, as some researchers and philosophers believe. This view clashes with the received view of science, which does not include phenomenology. As a function of the logos and unity of science view, the received version of science denies the validity of all claims and practices other than its own. In this way it turns science into a kind of war machine, destroying all other discourses and points of view, a process which the physicist and philosopher Paul Feyerabend (1975) was already aware of.

The same critique applies to the information and computer science-based cognitivist explanations of human social coordination and communication (Brier 2008a). However, natural science was confronted by the social sciences in what is called the "linguistic turn" in the philosophy of science and various forms of constructivism, from solipsistic radical ones to social constructivism; all undermining the objective authority of science's explanations of how the world works (Brier 2009a). This ignited what has so often been called the 'science wars', from which not much good emerged aside from a realization among some researchers of the necessity to construct a new integrative transdisciplinary framework, in which all can work together in a fruitful way.

Nicolescu (2002) is one of the rare examples of a quantum physicist practicing a non-reductionist transdisciplinary philosophy of Wissenschaft. One fact that has been emerging from the science wars with the social science and humanities is the realization that the natural sciences were dependent on the language they were formulated in, and that language, worldview, and mentality were deeply interconnected. Thus, we are back to Otto Neurath's basic ideas, since we have given up on the idea of a special objective scientific language combining logic and mathematics to unite all Wissenschaft. Thus, theories of language, cognition, and the conditions for signification had to be integrated into the interpretation of scientific data. This is another reason for introducing Peirce's semiotics (Peirce 1958).

The research of semiotics was a project mainly conducted from 1865 to 1910 by Charles Peirce to provide an understanding of the logic of the scientific method, which he informed, by a semiotic, phenomenological and pragmaticist view of knowledge, aimed at providing insight into the methodological commonalities found in all attempts to produce scientific knowledge, or one could say the semiotic processes of science. The project ended as a semiotic paradigm with a new transdisciplinary ontology and epistemology. As Emmeche writes:

> A logical implication of the ontological-phenomenological basis of Peirce's semiotics ... points to an interesting continuity between matter, life and mind, or, to phrase it more precise, between sign vehicles as material possibilities for life, sign action as actual information processing, and the experiential nature of any interpretant of a sign, i.e., the effects of the sign upon a wider mind-like system. (Emmeche 2013, p. 118)

The issue of what awareness of sensory information and its qualia is and how we come to interpret sense experience, and furthermore how it is connected to subjectivity is also a problem at the basis of the philosophy of science, as well as questions of truth and meaning, and how science is placed between them or may contribute to integrating them. Thus, the difficult problem of why we have qualitative phenomenal experiences, is not a superficial question; rather, it is one that demands that we dig deep down to the prerequisite for our way of producing knowledge, worldviews and explanations (Bennett et al. 2007).

Thus, in this chapter I will suggest a way to address these problems through a philosophy of science reflection on the limitation of coherence and consistency in our generally accepted but specialized epistemological and ontological frameworks in the natural, life, information, social sciences and humanities.

The first move towards constructing a transdisciplinary framework (or meta-paradigm) including the natural sciences, phenomenology and a paradigm of semiotic-linguistic constructionism is to accept that natural, life and social scientific knowledge, and knowledge in the humanities is created in intersubjective meaningful communicative action by embodied living systems, and that we are unable to give any final proof of its truth.

This is in accordance with Popper's (1972) critical rationalism and with Peirce's (1958) original idea of fallible objective knowledge. This view is also based on the fact that meaningful intersubjective communication is still – like first person consciousness – not (yet?) scientifically explainable. Furthermore, we need to be aware that the life sciences have their own perspective, which we also need to integrate, since all the conscious beings we know today are embodied in living, autopoietic systems. No computers, artificial intelligence (AI), or robots can produce conscious awareness presently. AI is still not artificial consciousness (AC). All scientific knowledge demands embodied minds meaningfully sharing interpretation of sense experiences through signs. Robots do not make science. Bits and signals between machines (still) do not produce communication that has vital meaning for living embodied systems of awareness.

Meaning is, thus, in a way created before and outside the realm of natural science as we know it today in ordinary social language, since subjective and intersubjective cultural meaning is explicitly removed from the foundational framework of the classical positivistic- influenced concept of science in order for its strive towards knowledge of universal character mostly in the form of deterministic or statistical laws. In order to obtain objectivity in the empirical sciences, it is usually taken for granted that one must remove any influence of the subjective and cultural ideas of reality. This fact presents one aspect of the problem of a scientific explanation of consciousness, as subjective awareness and meaningful communication are not really deeply reflected into the concept of scientific objective knowledge. Heelan (1987, 1988) has spent a lifetime investigating and arguing for the relevance of hermeneutics and phenomenology for the understanding of scientific observation and interpretation of data, which is also the main point of Gadamer's (1989) main work.

2.3 Integrating the Four Views on Consciousness in the Cybersemiotic Star

Cybersemiotics suggests then, that we have four different approaches to the understanding of cognition, communication, meaning and consciousness: first, are the exact natural sciences; second, the life sciences; third, are the phenomenological-hermeneutic interpretational qualitative "sciences"; and fourth, is the sociological discursive-linguistic cultural view. We are here inspired by Wittgenstein's (1958) pragmatic linguistic view, but not only that; the Cybersemiotic paradigm views the production of knowledge from the middle, where we, as embodied aware semiotic and communicating living systems, create knowledge in a cultural and an ecological surrounding. This means that we cannot attribute more importance to one of the four systems of knowledge than any of the others without committing a reductionism or an unfounded one-sided simplification of reality. Thus, the four approaches are all equally important. This philosophy is parallel to Bruno Latour's break with modernity in his book *We Have Never Been Modern* (Latour 1993). Somewhat earlier he wrote: "Nothing can be reduced to anything else, nothing can be deduced from anything else, everything may be allied to everything else" (Latour 1988, p. 163), but also inspired by Merleau-Ponty (1962). I work with four main paradigms, where Latour works primarily with the dichotomy between nature and culture.

In Latour's *Actor-network theory* and philosophy of science (Latour 1993, 2004), he points out that explaining consciousness only through the brain as a natural entity is nearly an impossible idea. Because, what are considered "natural entities" by science are "hybrids" for Latour and they achieve their existence for us through a semiotic network of actants. Nevertheless, Latour does not deny that they have a "Ding an sich" existence. We should not forget that Bruno Latour's (1993, 2004) theory of hybrids and actor-network theory is based on a semiotics inspired by Greimas's actantial model that is a semiotic combination of material existence and social role created by a narrative. Latour views science as one narrative of the working of nature among many possible narratives based on the data we have so far. However, not all stories about nature have been shown to be viable. Latour's view is thus of a semiotic processual kind. Its semiotics is not really a Peircean version (Brier 2008d), but a special brand of Saussurian semiology developed by Greimas and further formed by its inclusion in Latour's realistic vision of a communicative/

semiotic network of humans and things (including technology and cultural artefacts) viewed as 'hybrids'. It is a realistic vision of living and dead natural entities to which we relate and which act back on society, changing it (the HIV-virus is an example) (Latour 2007, pp. 10–11). Despite the fact that many call Latour a social constructivist and a postmodernist, he insists on being a realist. Moreover, the normative view of ANT (actor-network theory) is that it should contribute to a better social order, not break things down (Latour 2007). This places him closer to Peircean semiotics than Saussurian semiology.

Science is a cultural product, a method to find truth in the empirical world. It is a technology which we use in order to see, understand and manipulate the nature on which our existence is dependent. The tool of scientific discourse based on empirical investigations makes us able to describe the part of reality we need to understand and in that process ascribe meaning to it and its processes. That certainly does not mean we are able to describe all of nature or give consistent meaning to all we have described so far, such as the relation between brain, culture and consciousness.

The idea of Fig. 2.1, called the cybersemiotic star, and the epistemological turn it is illustrating, is to escape the great explanatory burden of reductionist mainstream science, wanting to explain both life and consciousness from its basic assumption of energy and mathematical mechanistic laws. The Cybersemiotic philosophy of natural life, social sciences, and humanities sees their different types of explanations moving from our present state of socio -linguistically common- sense-based conscious semiosis towards self-organized and highly specialized knowledge systems. Each of them develops towards a better understanding of its subject area but at the same time also about the prerequisites of language, culture and our self-conscious subject, and their production of systematic knowledge in a time perspective.

There are four forms of historical explanations occurring: (1) the cosmological (physic-chemical), (2) the biological (biosemiotics and biosciences), (3) the historical (socio-cultural), and (4) the subjective perception of a lifetime; the experienced time.

The Cybersemiotic star illustrates the equal importance of the four basic approaches, and from the model a few other points can be made. To be a realist about the possibility of science providing us with usable knowledge about reality is to accept the reality of language, embodied minds, culture and non-cultural environments However, that is something quite different from believing in reductionist explanations from one of the arms of the star. I agree with Steffensen and Cowley (2010, p. 348) that we must move towards a much more non-local understanding of mind. What they call "a transdisciplinary non-local approach to bodily, cognitive and interactional processes". But my model adds the subjective experiential aspect of reality to in the form of an embodied phenomenological basis developed into a triadic semiotics by Peirce.

The natural sciences work towards making one grand cosmogonist explanation, but lack theories of individual experiential awareness, the semiotic aspects of life and the workings of social-linguistic reality. Interestingly, Ellis (2004) also accepts in his framework that there are four different worlds, though his fourth is mathematical abstract reality and not linguistic intersubjectivity, as it is stipulated in Cybersemiotics. However, so far the sciences have not resolved the problem of the emergence of life and consciousness in evolution. Until they do, we might have to accept that an all-encompassing explanation of the meaningful conscious communication process cannot be provided from any one of the corners of the model alone. I argue further for this in the rest of the chapter. As we cannot reduce our scientific explanations to one grand story and claim it to be the one and only reality, my theory is that we have to juggle and combine all four types of knowing at the same time. This puts us in a new situation and changes the research questions about consciousness, as we will argue further in the rest of the article.

Science works on the assumption that the material world has no sense experience or meaning at all, but only natural laws. The reason for this is that scientists are brought up to think that to indulge in other kinds of ontological assumptions would make our search for knowledge religious or political, as these are the two major meaning-producing systems we know. Science fought its way out of the powerful grip of religion in the Enlightenment, and later out of totalitarian political ideologies, like Nazism and Communism.

If we stay clear of religion and political worldviews, what are we then to call the meaning interpreting disciplines in the social sciences and humanities? This problem is well known, and answers have been developed within phenomenology, phaneroscopy (Peircean triadic semiotic phenomenology) and hermeneutics. The ultimate philosophical version of hermeneutics was developed by Gadamer (1989). Gadamer's book is clearly developing a philosophy for the humanities and the qualitative social sciences, but also a foundation for the methodological argumentation for scientific methods. Are we going to accept meaningful interpretation as part of our view of consciousness and legitimate objective knowledge? I cannot see how we can ignore this fundamental human process of cognition, since meaningful human communication is a prerequisite for the possibility of science. If we want scientific answers about the nature of consciousness, we must integrate some version of hermeneutics into a transdisciplinary theory of knowing.

In this case we need to move from talking about a science of consciousness to call what we deal with for Wissenschaft, as this German concept includes natural and social sciences, and humanities in a single concept. How can we view qualia and meaning as coming from the culturally-embodied languaging mind and understand it in a grander scientific, evolutionary and ecological view?

This is where I think only a Peircean biosemiotics (Favareau 2010) can answer "yes". A realistic and pragmatic conceptualization of sign processes, in all their variations, could be seen as the unitary phenomenon which connects all living natural systems with human cultures and distinguishes them both from inanimate nature. It could serve as the framework that provides the human, social, engineering, business, life, and natural sciences with a common theoretical basis for empirical research. Peirce's realism is, among other things, based on his belief in the aspect of reality he characterizes as Secondness, or the unexplainable or random facts. There are immediate differences and resistances between phenomena (Haecceities). Peirce adopts Duns Scotus's term haecceity to designate the arbitrary here-and-now-ness of existence, a person or object's "this-ness", that is, the brutal facts based on relations. Peirce identified this haecceity as 'pure Secondness'. He writes about this fundamental concept in his phaneroscopic semiotics:

> Most systems of philosophy maintain certain facts or principles as ultimate. In truth, any fact is in one sense ultimate – that is to say, in its isolated aggressive stubbornness and individual reality. What Scotus calls the haecceities of things, the hereness and nowness of their existence, is indeed ultimate. Why this phenomenon, which is here as we pick it up – like for instance one grain of sand out of billions on a beach – is such as it is and is where it is in space and time we can ask; but the explanation in this case will merely carry us back to the fact that it was once in some other place, where similar things might naturally be expected to be. It is simply an ultimate fact. There is also another class of facts of which it is not reasonable to expect an explanation, namely, facts of indeterminacy or variety. Why one definite kind of event is frequent and another rare, is a question to be asked, but a reason for the general fact that of events some kinds are common and some rare, it would be unfair to demand. If all births took place on a given day of the week, or if there were always more on Sundays than on Mondays that would be a fact to be accounted for, but that they happen in about equal proportions on all the days requires no particular explanation. If we were to find that all the grains of sand on a certain beach separated themselves into two or more sharply discrete classes, as spherical and cubical ones, there would be something to be explained, but that they are of various sizes and shapes, of no definable character, can only be referred to the general manifoldness of nature. Indeterminacy, then, or pure Firstness, and individual existence as haecceity, or pure Secondness, are facts not calling for and not capable of explanation. Indeterminacy affords us nothing to ask a question about; haecceity is the ultima ratio, the brutal fact that will not be questioned. (Peirce, CP 3.405)

Peirce's view of haecceities as being unexplainable as singular events is close to the modern understanding of quantum events. Quantum physics cannot deduce the singular event; it can only make a probability model from thousands of them. There is an undetermined spontaneity – what Peirce calls Firstness – of the single event that is not explainable in itself from a scientific point of view. Quantum mechanics, like Peirce's pragmaticist semiotics, thereby, breaks with classical deterministic mechanicism.

How does the mind collect all these haecceities to one quale experience? One way of formulating this question is in the form of the binding problem widely discussed in brain and consciousness studies (Chalmers 1996). It asks how the unity of conscious perception is created in the neurological processes that make up the central nervous system. Some researchers see this as only a neuro-physiological question. In fact, it is a question that demands types of answers that extend beyond the realm of physical science alone, since it concerns meaningful subjective and intersubjective experiences that point beyond physical explanations. Searle defends the view "that consciousness consists of unified, qualitative subjectivity caused by brain processes and realized in the brain" (Searle 2007, p. 102). In that case, how do we integrate all those different perceptual inputs from inside and outside the body into a life world or a conscious horizon, with ourselves in the center? The question from science should be: How can we systematically work with any reality beyond the physical? It is a foundational philosophical problem before any empirical science.

I will here argue that giving answers beyond physics and physicalism does not need to lead to the introduction of elements or worlds outside nature in the way in which Cartesian dualism, for instance, can be interpreted to do in its postulation of a res cogitans. The ontological idea is to not place consciousness and the world of thought outside nature in a special mental world. It is rather to expand our ontological views of living nature to a biosemiotic-based interdependency thinking of lived sense making (Cowley et al. 2010).

Husserl's works as well as Gadamer's hermeneutical philosophy (Gadamer 1989), are attempts to provide another more comprehensive model for reality, including the sciences and a theory of understanding, communication and history of culture. Gadamer's theory of interpretation and understanding goes through pre-understanding and the process of the hermeneutical circle in order to integrate parts, including the subjects' and objects' horizons. The object can be another subject's mind, an artifact, a piece of art, or a text or a person's story about his psychological problems.

Gadamer's view is that truth does not spring automatically from using one type of method and deeming it "scientific", "mathematical-logical", "empirical", or a combination of them; one has to reflect on the knowledge horizon or metaphysics from which one produces knowledge. This is done in order to create understanding in the form of fusing knowledge and experiential horizons (Heelan 1987, 1988) for all living beings with conscious awareness. Thus, consciousness in the form of awareness and the ability to have sense experiences needs to be conceptualized within an understanding of a natural reality larger than physics, unless one wants to deny that animals have sense experience and that our own animal body is a prerequisite for self-consciousness. We will, therefore, assume that consciousness, matter and information coexist in, or make up, nature and culture.

To go one step further, we might add the work of David Chalmers. Chalmers (1995, pp. 201–202, 1996) is well-known for defining what he calls "the easy and the hard problems of consciousness". The easy problem has to do with the inner workings of consciousness, such as the ability to discriminate, categorize, and react to environmental stimuli, to be able to report mental states by accessing internal states, to focus attention, deliberately control behavior, and distinguish between mental states. The hard problem, which is the one we are speaking about here, has to do with solving the problem of what the nature of sense experiences and their different qualia –such as pleasure and pain, sweet and sour, colors, and mental images in themselves – might ontologically be determined to be? That is the problem we are dealing with here in a naturalistic, and therefore, also evolutionary frame. Thus, our question has now developed into: How can the ability to experience arise from what in natural and technical sciences is presumed to be a material world?

Collin McGinn (2000) poses this very question in his famous book on consciousness: *The Mysterious Flame: Conscious Minds in a Material World*. McGinn is skeptical towards our ability to explain the phenomenon of consciousness, at least with the present vocabulary in our possession. How it is possible in a natural world, which we so far have defined as "material", to "feel like someone" in the way it is framed in Nagel's famous article *What Does it Feel Like to be a Bat?*, or what it means to experience the qualities of, say red or blue (Nagel 1974). The problem of explaining and modeling the ability to experience qualitative differences in sense experiences in a scientific way is formulated as the question of qualia (Jackendoff 1987). How do nervous systems produce sense experiences? Those opposing the importance of qualia, however, are functionalists. They argue that in understanding the function of a system, it is not its materiality or its experiential quality that matters; there is no need to ascribe causal powers to experience. An example of this denial is Bennett et al. (2007) in their Wittgensteinian-inspired pragmatic linguistic theory of mind. However, here I will side with Searle (2007) and argue that this functionalistic view does not make the ontological dimension of this problem go away. This functionalistic view often leads to the assumption that computers have minds (Harman 1990). But this is then not an experiential mind in the way I speak about it here.

Another view towards the problem of the limitations of computers for our theories of experiential consciousness is that of Roger Penrose's work. In *The Emperor's New Mind* (1989) and *Shadows of the Mind* (1994), he shows that even in mathematics, human minds are capable of non-computable or non-algorithmic processes that go beyond the present capabilities of computers. Based on this, my position in this chapter will be that computers or algorithms only simulate certain limited aspects of mind processes since most researchers presently agree that computers – as we presently know them – cannot compute awareness, qualia and meaning. Based on Peircean biosemiotics I side with Searle (1980) against the view of hard AI, when he is denying that symbol manipulation in itself is the core of intentionality. I fail to see how automatic symbol manipulation in computers has anything to do with the production of qualia, intentionality and meaning. Jackendoff (1987) has very precisely framed the problem in the form of the concept of "the mind–mind problem". I agree with him when he formulates the gap problem as the relationship between "the computational" and the "phenomenological mind".

Thus, if we do not believe that the brain is only a computer and that informational computation is what creates consciousness in the human body, then it must be something else. Searle (1980, 1989, 2007) argues that it has something to do with our biology. Consciousness and intentionality must be biological products (Searle 1980, 1989, 2007; Searle et al. 1997). The secret of consciousness is also the secret of life, one could say. The tragedy is that biology so far has only been able to give functional definitions of life. Searle (1980) believes that the brain's production of intentionality is like chlorophyll's production of carbohydrates through photosynthesis. Boden (1990) points out correctly in a critique that experience is a qualitatively different product than carbohydrates. We can describe and measure carbohydrates scientifically, but that is not the case with the quality of experience. As far as we know today, only living bodies can produce the awareness necessary for having experience. To live is to experience, but the living experiencing flesh is still a mystery to the physic-chemical sciences as well as the life sciences in their present non-semiotic form, as Merleau-Ponty (1962, 1963, 2003) has thoroughly argued from the philosophy of embodied phenomenology. As experience is a prerequisite for science, science may not be able to explain it. As Favareau (2010) points out, the problem is rather a triplet; "What is the relation between mental experience, biological organization, and the law-like processes of inanimate matter?" Scientific biology in the form of physics, chemistry and physiology is unable to describe important aspects of the processes of living systems. The suggestion here is that we supplement our physico-chemical knowledge with a semiotic view.

Thus we need to find other framework to be able to continue fruitful searching after formulation of the hard problem that can bring us forward. An alternative frame of investigation would be biosemiotics founded as the transdisciplinary semiotic study of the biological as well as the human significance of codes and sign processes, such as genetic code sequences and intercellular semiotic processes in the nervous, hormone, and immune systems, animal display behavior, human language, and abstract symbolic thought. Biosemiotics, including human and cultural semiotics, can be defined as the study of how meanings are created in living systems between signs and the information they encode in the perceptual and cognitive apparatus (Hoffmeyer 2010). As one of the contributors to biosemiotics, I find that, especially in its stringent Peircean formulation with its triadic phaneroscopic categories, it represents a promising way out of dualism, monistic eliminative materialism and other types of physicalism and radical forms of constructivism (Brier 2008d). Here is one of Peirce's formulations:

> But by "semiosis" I mean, on the contrary, an action, or influence, which is, or involves, a cooperation of three subjects, such as a sign, its object, and its interpretant, this tri-relative influence not being in any way resolvable into actions between pairs. {Sêmeiösis} in Greek of the Roman period, as early as Cicero's time, if I remember rightly, meant the action of almost any kind of sign; and my definition confers on anything that so acts the title of a "sign" (Peirce, CP 5.484).

Such a paradigm was originally formulated as Umweltslehre by Jacob von Uexküll (1957, 1982), and later, inspired by him, as ethology by Konrad Lorenz (1971) and Niko Tinbergen (1973) (see Brier 1980, 1999, 2000, 2001). Connected to these questions is also the problem of how living systems perceive sense experiences and communicate in the frame of meaning, and why and how they seem to have intentionality. Furthermore, it is a scientific enigma how signs and the grammatically ordered symbols of language can evoke feelings, qualia and images from the body. How can individual emotional purpose enter the nervous system and create semiotic interpretations? How can free will have causal influence when physics believes that causality is primarily based in initial conditions and universal laws? How can my subjective experience of anxiety of doctors and hospitals make the pain of a medical procedure bigger, than if I was in a completely calm trust in the doctor, if it is all going on in the nervous system only?

As already mentioned, meaning is not part of the paradigmatic foundation for the exact natural sciences. How can, then, the life sciences, of which biology is the most prominent pure (as opposed to applied) one, avoid working with the reality of emotions, intentionality and meaning? This is a problem Konrad Lorenz struggled with for over 30 years and could not solve within the natural scientific paradigm (Brier 2008a; Lorenz 1971). Most recently, Ellis and Newton (1998), inspired by phenomenology, attempted to integrate it with biology:

> It is the organism's emotions that motivate it to act on its environment rather than merely react; the phenomenal aspect of conscious experience requires the organism's emotionally motivated action in relation to the perceived world, particularly in its interest in selecting for attentional focus. If the organism's knowledge of its environment is to involve a "felt" dimension, in the sense that there is "something it feels like" to have a state of consciousness, the conscious processing must first flow from an emotional process within the organism, which pre-exists to any particular input, and puts its qualitative stamp on each selected input.
>
> We are suggesting that the "felt" aspect of experiencing is tied in with the fact that organisms are emotionally motivated to "look for" elements of the environment that are significant with respect to the organism's motivational purposes; that the organism "anticipates" experience in terms of motivational categories which preselect for attention; and that the emotions that guide this anticipation and selection process are a major contributor to the conscious feeling of 'what the consciousness of such-and-such an object is like'. (Ellis & Newton 1998, p. 431)

The point is, again, that if biology is to encompass the felt experience of animals, its foundation has to differ from that of physics and chemistry. Present biology is, therefore, not enough. Hoffmeyer (2008) writes that scientific description in gene-fixed reductionist biology,

> exclusively deals with phenomena that may be described in the language of third-person phenomena, and thus ... excludes this science from arriving at a theoretical understanding of the human biosystem as a first-person being. (Hoffmeyer 2008, pp. 333–334)

Thus, we need a Wissenschaft which includes a theory of signification and meaning; this is exactly what biosemiotics attempts to create. Emmeche (2013) writes:

> The semiotic approach means that cells and organisms are not primarily seen as complex assembles of molecules, as far as these molecules rightly described by chemistry and molecular biology – are sign vehicles for informational and interpretation processes, briefly, sign processes or *semiosis*. (Emmeche 2013, p. 118)

However, this view is not a possibility for energetic, molecular, or even informationally founded biology. Kull et al. (2009) discuss what this kind of Wissenschaft biosemiotics could and should be.

From a biological evolutionary viewpoint, meaning has a history of millions of years in the development of living systems. This is a story biosemiotics attempts to tell, since our conclusion here is that ordinary science is not conceptually equipped for it (Emmeche 2013). Thus, we should encompass the social and the individual experiential reality and their history in nature. Nonetheless, how are we going to connect them? Where do we put the brain in experience?

Chalmers' *The Conscious mind: In Search of a Fundamental Theory* (1996) collects nearly all the material in science and philosophy available on the subject at that time, except Peirce's semiotic philosophy. His suggestion of a solution is a type of double aspect theory, where the experiential is seen inside the information we can measure in the processes of the brain. However, viewing objectively-defined information and experiential meaning as two aspects of "the same" does not solve the deep troublesome problem lying in the obvious observation that I am not my brain. One should not commit the merological fallacy to contribute to the part that only makes sense when attributed to the whole. It is not the brain that experiences; it is embodied human persons in a culture with a language (Bennett et al. 2007; Cowley et al. 2010).

Is it then possible that conscious awareness and experience is something we are missing in our scientific explanations of living systems perception, cognition, and communication as we know them? For instance, black matter and energy were missing in cosmogonic descriptions of the universe. They were concepts introduced because we were lacking something to harmonize what we observed astronomically with the physical laws we have developed until today. What we saw and measured did not fit with the laws we believed were universal. After introducing the new aspects of physical reality christened "black energy" and "black matter", what we before had considered to be the whole of material reality, now showed to be 3–4 % of the whole. Thus, introducing new ontological elements created a revolutionary new cosmology.

Maybe this is a parallel to our present state of the lack of connection between the natural, technical, social and humanistic sciences and their use as the foundation for medical research and modelling. The parallel I am arguing for is, what we now consider the material reality of biological systems could turn out to be just a small percentage of the whole of the living system because we missed something vital for the function of living systems; namely signs and sign functions.

Materialistically-based evolutionary and ecological theories beg the question that if culture comes out of nature, how does experiential subjects emerge from an objective world? Or do they? Is that the right way to ask the question? Is reality not a much more entangled non-local affair? Here, I am not thinking about research, which accepts the experiential aspect of life in the living, and therefore describes how it has developed through evolution. It is works like Donald (1991) that describe the evolution of consciousness and its forms from a bio-psychological platform, and Sonesson (2009) who bases his work on phenomenology, Piaget and aspects of Peircean semiotics, or the work of Zlatev (2009a, b), that in an evolutionary framework uses aspects of Peircean semiotic terminology, but not his ontological foundation. Nor am I thinking of Deacon (1998), or his later articles (2007, 2008), which increasingly stray from a Peircean foundation (Deacon 2011). None of these use the Peircean philosophical framework in full.

Thus – in my view – a pure materialistic and scientific theory cannot answer the question I am asking, because it cannot describe the feeling of being aware, experiencing qualia, will, and intentionality. They can only describe physiological and behavioral consequences. Therefore, a new ontological reflection creating a new transdisciplinary framework behind physics and scientific knowledge in general seems to be required, because the unity of conscious experience, in spite of the numerous neuro-physiological systems that underpin it, does not really seem to have a scientific meaning. But science is not all. We cannot expect science to be able to describe all of reality. It can only describe those phenomena that its public intersubjective empirical methods can be used on.

My claim from the Cybersemiotic star is that the physical is only one aspect of our world, as are the formal and computational. These aspect are these days attempted integrated from a new metaphysical frame we call the info- computational paradigm (Dodig-Crnkovic 2010). But the problem of describing the living embodied beings intentionality, emotions and qualia in social interaction is that we have no artificial instantiation of these phenomena.

2.4 The Idea of Cybersemiotics

The transdisciplinary frame for information, cognition and communication science called Cybersemiotics (Brier 2008a, 2010a) is an attempt to show, using Peircean biosemiotics, how to combine knowledge produced in the natural, the life sciences, the social sciences and the humanities, as each describes an aspect of consciousness.

But first, we have to consider the incompatibility between the two transdisciplinary paradigms attempting to create a theory of consciousness. With an expression from Kuhn's (1996) paradigm theory, their theories on thinking and communication suffer from incommensurability. The first paradigms are cybernetic information theory and cognitive science, which are actually technology-oriented paradigms. Many members of this worldview have the deep problem that they usually do not consider their views to be founded on metaphysical postulates at all, but only on common sense reality. Therefore, they do not want to be drawn into metaphysical speculation or philosophy. Many people have the misconception that modern physics is concerned with the world as we know it in our daily life. Nothing can be further from the truth. Quantum field theory and the special and general relativity theory, super string theory, black holes, dark matter, and the likes are entirely outside of our common sense. If you ask people to interpret everyday physical processes, most give explanations close to Aristotelian physics. Thus, the majority of human beings have not even moved into a Newtonian paradigm, not to mention Einstein's, Bohr's, Feynman's, or Hawking's. Modern physics have no direct bearing on awareness, meaning, and common sense. Still, many researchers after World War II inspired by cybernetics attempted to add information and computation to this physicalistic worldview to explain the emergence of conscious awareness.

Building an enlarged new worldview by adding the concept of information to energy, space, time, and force, and imagining that all natural processes including consciousness and emotion can be fruitfully described and understood in a grand theory of natural computation (Dodig-Crnkovic 2010; Dodig-Crnkovic and Muller 2011). This pancomputational/paninformational view is an interesting scientific endeavor as such; however, I fail to see how it will ever be able to solve the experiential and qualia aspects of conscious feeling experience as it lacks the experiential aspect of reality. Thus, like Peirce, I want to enlarge our wissenschaftliches concept of reality. I am not only talking about that aspect of it that can be described by physics (often reified as the physical world, turning an epistemological concept into an ontological one and reifying it), but also what can be described by life sciences, communication sciences and psychology. Thus, reality includes at least a material environment, a living body, a life world of experience and a social communicative world, all being necessary in producing experiential knowing. Science is based on intersubjectively well- functioning communication in a field of meaning coordinating knowledge and practice in the real world. I am, therefore, asking what kind of transdisciplinary ontology and epistemology do we need to construct a theory of the evolution of meaning and conscious lived experience coherent with the natural, life and social sciences?

2.5 Phenomenology and the Life World

What is the rational basis of my insistence that the physical aspect of the world is not the paramount foundation of reality? It is basically the acceptance of the main point of the whole phenomenological movement, the history of which Spiegelberg (1965) has made a highly recognized history of, including Peirce. We will not go into that grand history here. Let us just accept that many researchers are taking departure in the work of the father of modern European phenomenology Husserl (1970, 1999). Many other researchers take departure in the work of the father of C.S. Peirce (1958), whose American variant of phenomenology is called phaneroscophy. He is also the father of the pragmatic, triadic transdisciplinary semiotics, which much of biosemiotic is built on.

Husserlian phenomenology claims that the so-called life world is a unit of reality before science splits the world into subjects and objects, or interior and exterior. The dualism of subject and object is really not essentially relevant for the phenomenological paradigms, which like hermeneutics claim to be concerned with the cognitive processes that are prerequisites for the invention of science in our cultures. This is the area where the philosophical grounding for the natural, life and social sciences becomes relevant for the analysis. Thus, in phenomenology the percept is a primary reality, before scientists try to explain the origin of sense perception and its information and meaning from a combination of interior physiological processes and exterior physical information disturbing the sense organs, or as biology tries to explain it: the function of the sense organs and the nervous system from evolutionary and eco-physiological theories.

Even modern biology cannot explain why and how we see, hear and smell the world (Edelman and Tononi 2000, p. 222). It can only model the physiological way the organ works, but it has nothing to say about how it produces experience. This is a stumbling block for a neuro-and behavioral scientist studying philosophy of science. However, it is only a problem for those scientists that have taken philosophy of science seriously, and their numbers are fairly low. Many empirical researchers do not see the problem and believe that more empirical research will solve any problem. I am arguing for another more philosophical reflective view here.

In phenomenology the knower, the known and knowing is viewed as one living whole in the *life world*. The knowing consciousness contains the known objects (Drummon 2003, p. 65). Thus, phenomenology considers life world experiential first-person awareness to be producing knowledge more foundational than the natural and social sciences can produce. For a traditional scientist the view is shocking. But Merleau-Ponty (1962) for instance writes:

> Phenomenology is the study of essences; and according to it, all problems amount to finding definitions of essences: the essence of perception, or the essence of consciousness, for example. But phenomenology is also a philosophy which puts essences back into existence, and does not expect to arrive at an understanding of man and the world from any starting point other than that of their 'facticity'. It is a transcendental philosophy which places in abeyance the assertions arising out of the natural attitude, the better to understand them; but it is also a philosophy for which the world is always 'already there' before reflection begins—as 'an inalienable presence; and all its efforts are concentrated upon re-achieving a direct and primitive contact with the world, and endowing that contact with a philosophical status. (Merleau-Ponty 1962, p. vii)

Phenomenology holds that conscious experience, in both its subjective and intersubjective versions, comes before science, and is, therefore, not something that is in need of or can possibly be scientifically (materialistic or informationalist) explained. This is in direct confrontation with scientism and the physicalist philosophy that scientific knowledge is the sole foundation of a rational worldview. I do not believe that anybody has in a short form expressed it clearer than Merleau-Ponty, who in the following quote, views the natural and the social sciences as secondary to the phenomenological stance:

> Science has not and never will have, by its nature, the same significance qua form of being as the world which we perceive, for the simple reason that it is a rationale or explanation of that world. I am not a 'living creature' nor even a 'man', nor again even 'a consciousness' endowed with all the characteristics which zoology, social anatomy or inductive psychology recognize in these various products of the natural or historical process. I am the absolute source, my existence does not stem from my antecedents, from my physical and social environment; instead it moves out towards them and sustains them, for I alone bring into being for myself ...the tradition which I elect to carry on. (Merleau-Ponty 1962, p. ix)

It is one of the clearest arguments for the necessity of philosophy when determining how to evaluate and use the knowledge from the natural and the social sciences. It is especially Husserlian phenomenology upon which Merleau- Ponty draws, which considers the life world as more fundamental than natural and social scientific knowledge, and therefore claiming that there is no scientific explanation for consciousness as it is the primary given. Consciousness is not viewed as a product of the brain or of culture and language in Husserl (1970, 1999), only its content and way of expressing itself are. On the other hand, Merleau-Ponty does not privilege the body over the mind; the body is the mind and vice versa, in that they are one whole synthesis. The phenomenological 'I' is a universal, natural, human sense-perceiving 'I' that brings things into existence for oneself through one's intentionality; this includes "the other". Merleau- Ponty writes: "Perception is not a science of the world, it is not even an act, a deliberate taking up of a position; it is the background from which all acts stand out, and is presupposed by them." (Merleau-Ponty 1962, p. xi).

It is through being in the world and experiencing the world that we have consciousness, but that world is ontologically not the same as the "physical world" as it also includes the subjective and intersubjective world of living and communicating with other living, embodied and conscious linguistic beings. Thus, the physicalistic and/or computationalist brain science, on the one hand, and phenomenology, on the other, operate in two different worlds that see the other as only describing a small part of reality that is only marginally important for the big picture. Both claim to be the most fundamental description of reality. They each have their own map of the world on which the other almost does not exist or is not represented in a way they will themselves accept.

One of the deepest conundrums for the sciences is the undeniable fact of our own ability to undergo qualitatively varied sense experiences and experience internal drives urges states of feelings and will to alter body processes according to that. These lead to the ability to make our body carry out goal-directed movements, which in turn fulfill goals, of which some can be bodily and psychological desires. Furthermore, this poses a very general problem for the sciences, because this experiential aspect of reality is not just a matter of the special category of human consciousness; all living beings have these abilities to varying degrees. This is one of the reasons why biosemiotics is a necessary supplement to ordinary scientific biology as well as cultural semiotics.

One can try to avoid the problem; of course, by claiming that our experience of making decisions on the basis of analysis of our qualitative experiences is an illusion or folk psychology (Churchland 2004; Dennett 1991, 2007) and that consciousness has no causal effect in the world as we know it. However I refuse to take this eliminative materialism seriously, as I consider it to be a self-defeating paradigm, since it, by its elimination, denies the prerequisites for that scientific knowledge it claims to produce. Therefore, it must follow the assumption that the same science that eliminative materialism wants to credit for its arguments is also a pure hallucination without any effects on the world.

The position is therefore inconsistent. It ignores the fact that science has sense experience, and the ability to think, create and communicate meaningful theories, and the ability to make purposeful experiments as a prerequisite. As Gadamer (1989) shows in his hermeneutics, science also has meaning and interpretation, based on a cultural historical horizon as a prerequisite, because it is dependent on the ability to make linguistic concepts and interpret them through one of the many natural languages produced by cultures and their worldviews. That is very much the insight on which Kuhn's paradigm theory builds (Kuhn 1996). Simply put, science is a cultural product, a method for collective development of knowledge to improve society and human life through the pursuit of truth. Medicine is a very intricate part of this process.

To sum up, then eliminativism – also in the form of a pure materialist biomedicine – is self-refuting, because the same consciousness that makes our knowledge and science possible is denied any real existence and causality in the world; as such, the theory is a poor philosophy of science much in the way that it's opposing paradigm, radical constructivism, is, as it can make no truth claims in that its philosophy of Wissenschaft denies the possibility of truth (Churchland 2004). Another point of departure, therefore, for my argumentation is Karl Popper's critical rationalistic and fallibilist philosophy of science and knowing, which aligns so beautifully with

Peirce's that one should think Popper got his views from there (Niiniluoto 1984, Chap. 2). He certainly thought that Peirce was "one of the greatest philosophers of all times" (Popper 1972, p. 212).

2.6 Popper's Three Worlds and Evolutionary Theory of Knowing

The problem of how knowledge is possible goes back to the beginning of the observer and of the world. Karl Popper (1972) built his general theory of evolution on Darwin's theory and tried to integrate it with an evolutionary epistemology. Popper notes in passing that knowledge is a property of living things, and he asserts that natural selection can build primitive forms of knowledge even into single-celled organisms like amoebas. His epistemology is here very compatible with biosemiotics. Knowledge is something that has to be considered in a time perspective. I believe, like Peirce and Prigogine, that the arrow of time and irreversibility is foundational to all human knowing, but not to the information in computational systems including IA and robots. Therefore, we are forced to view the production of knowledge and consciousness not only in the subjects lifetime experience, but also its culture's natural language, knowledge traditions, historical development, and finally – because of the unavoidability of embodiment – to include theories of our living systems evolutionary origin and ecological connectedness (Edelman and Tononi 2000).

Like Peirce, Popper saw knowledge as a subjective and intersubjective construct and underlined in his falsification theory that knowing entities can never prove – logically or through induction based on empirically collected experiences – that their general theories truly represent a universal aspect of reality. Although claims about external reality are more and less viable and work to our satisfaction for certain purposes, there is no way to prove that a claim will not be falsified the next time it is tested (Popper 1972). Still, we have pretty workable theories for various micro-organisms as – almost material – causes of diseases. But we also know that individual bodies vary in responses and some are inherently immune and do not get ill and produce the symptoms we have classified as signifying the illness. However, our knowledge of the world becomes more encompassing, accurate, and useful through an iterated process of making tentative claims and empirical testing to allow erroneous claims to be selectively eliminated (hypothetical deductive method). Popper (1972) defines the origins of knowing as living individual's and species solutions, or at least claims to solutions, for survival problems. This view is essential for the Cybersemiotic Star.

This is especially the case since Karl Popper's critical analysis (Popper 1959, 76) and argumentation for a falsificationist view of scientific knowledge has been accepted as a turning point in the break with the positivist unity of science on one hand, but on the other not as providing any final solution to the problem. To allow for the subjective aspect of knowledge in an evolutionary context, Popper found it necessary to embed his theory of knowledge in the transdisciplinary ontology of three worlds. World 1 (W1) is physical reality, World 2 (W2) includes the subjective aspects of mind and living knowledge, and World 3 (W3) includes intersubjective knowledge (his understanding of objective fallibilist knowledge) that can exist over time independent of the knowing individuals who made it. However, it can be interpreted, modified, and used by other living individuals at other times and places (Popper 1979). But Popper – whose conception of wissenschaft was so close to Peirce's – did not use Peirce's full triadic pragmaticist semiotic theory, which I will argue put severe limitations on his theory in the area of knowing.

Popper called knowledge in W2 – in the subjective consciousness – subjective or "dispositional" (Popper and Eccles 1977). W3 knowledge on the other hand, could be encoded in books and documents, DNA molecules, in computer memories, and as manufactured objects. The three worlds are aspects of reality that have to interact in order to produce objective knowledge. Knowledge about W1 held by living entities in W2 can be applied to W1 via action, or it may be persistently stored in W3 in various forms, for instance computer programs and models, and literary books. The reality of W3 knowledge is demonstrable when other entities decode the knowledge and can then apply it to W1 via their actions, for instance by building a piece of technology. However, Popper – in my opinion – lacked the deep understanding of living systems and the evolutionary semiotic development, which biosemiotics deals with (Emmeche 1992, 2003). This would have provided his theory with a deeper clarity and a better justification for his ontology, and of course a theory of how communication and language are foundational to understand the intersubjective production of knowledge. In my view, C.S. Piece did much of this work, as we will see (Peirce 1958). This is, in my view, where Peirce's pragmaticistic semiotic philosophy of Wissenschaft can add value to the modern quest of the understanding of mind.

As we all know, the positivistic unity of science ideas and its modern version in Wilson's (1999) *Consilience* theory is a very idealistic, but at the same time very reductionist attempt to establish a subject-free objective knowledge. They have in my view, failed in their attempts of universality, at the least in its inability to incorporate the original grounding of knowing in an experiential world of perception and emotion; not to mention the problem explaining how we decide to move our bodies from mere intensions or experiences (such as pain or pleasure) and as such create movement in our body, which indicates some kind of deep connection between mind and matter. The positivist also lacked to establish an evolutionary foundation for their epistemology. When evolutionary theory appeared both for the living and dead world of nature in the form of a Cosmogony in the middle of the nineteenth century the paradox became worse. Because, if inert matter was first, and sense experiences appeared next in the evolution of life, then meaningful cognition and consciousness must have its origins in matter; the mental must spring from the development of the physical. Thus theories of self-organizing systems, which through emergence create new, unforeseeable stable phenomena like water out of hydrogen and oxygen. But we have really made a commonly accepted theory of emergence producing a model with precise predictability skill?

The second problem is finding a new ontological and epistemological framework that provides the possibility to integrate the knowledge we have into a bigger picture. That I am not alone in this diagnosis and attempt, as can be seen in works like Popper and Eccles' *The Self and Its Brain* (1977), which is a good early example on attempts to change the frameworks and Thibault's (2004) *Brain, Mind, and the Signifying Body* is a more contemporary one and I refer again to the works of Nicolescu. From Chinese research one can mention the work of Wu Kun (Zhou and Brier 2014a, b).

2.7 Evolution and Teleonomy

Going a little back in history then, Jacques Monod highlighted in the famous book *Chance and Necessity* (1971), the apparent epistemological contradiction between the teleonomy of living organisms and the principle of objectivity in science based on the ontological assumption of the natural sciences that there are no intensions or meaning in inanimate nature. Monod (1971) combines scientific realism, positivism and French existentialism in his efforts to show the contingency of human existence opposed to the religious idea of our central importance and ethical obligations in a sort of covenant with the divinely created Cosmos. Monod declares that modern science has broken the old covenant between man and nature. Today man knows that he is alone in the universe's unfeeling immensity. All science can say is that man as an inexplicable fact emerged out of the universe by chance. Neither man's destiny nor his duty is anywhere spelled out in the universe that science knows. Monod – in his very clear and honest book – admits that science cannot explain how human beings can emerge in this meaningless and objective universe, which much of classical physics has claimed to be the whole picture of nature so long ago that we have almost forgotten that it is a metaphysical decision. Thus, we are still stuck with the basic problem of explaining how the inner world of first person experience can arise in the dead deterministic physical and closed world.

In Genesis, it is God who created life, but in the paradigm of evolution, science has to explain life as something, which occurs inside the universe by virtue of the same general principles that science uses to explain the physical and chemical aspects of the universe. Evolution is creative by constantly creating new systems, and these systems become, when they are alive, increasingly creative. However, mechanical systems are not creative[2]. This is also why C. S. Peirce (1892) does not believe that mechanical determinism can be the fundamental paradigm for science.

In the book, *Order out of Chaos* based on Ludwig Boltzmann's probabilistic interpretation of thermodynamics, Prigogine and Isabella Stengers (1984) developed an epistemology and philosophy of science based on a view that took

[2] That is why they fit so well with theistic explanations of divine creation.

complexity and irreversible evolution on the physical level seriously. They, therefore, distanced themselves from the determinism of mechanics and its belief that it is possible to find some abstract and eternally simple universal, natural laws "behind" the complex forms of representations which determine all events in the universe. Laws are not absolute and mechanical but developing forms in the continuum of mind and matter and our ever developing fallibilist knowledge of which symbols is an essential feature. Since mechanical determinism cannot explain the novelty of evolution and the emergence of the laws of nature, Peirce was aware that we needed an alternative ontology to the mechanistic one. As physicist Lee Smolin writes: "The Cosmological questions such as Why these laws? And Why the initial conditions cannot be answered by a method that takes the laws and initial conditions as input." (Smolin 2014, p. 250). But this is what modern classical physics used to do and therefore Smolin's work here is quite revolutionary and he is quite aware that the thought was foundational to Peirce's cosmogony and quotes him several places in the book.

Prigogine and Stengers (1984) accept chance as real – like Peirce and Popper (Miller 1975) – and necessary element of evolution. In their understanding, evolution requires the creation of radial new things, patterns and phenomena that cannot be predicted from a basic physical understanding of the universe. They also realize that the acceptance of the evolutionary idea is in a fundamental paradigmatic conflict with classical physics, but perhaps not with quantum physics. This is certainly a step forward, but we still lack convincing explanations of how self-organizing and self-replicating entities produce life and the ability to experience. Several researchers have continued to try to explain one of the major creative elements in a "self-organizing universe" that could produce life, but the most prominent in the last 20 years has been Stuart Kauffman (1993), whose work never arrives at a theory of consciousness, but at least now to a theory of semiotics (Kauffman 2012).

Since Norbert Wiener established cybernetics and integrated information theory with thermodynamics, information scientists have tried to explain the phenomenon of life using the new concept of information, which Wiener and Schrödinger created. Their starting point was Claude Shannon's mathematics, but they redefined information from being entropy (Shannon's view) to neg-entropy; namely order and structure (Schrödinger 2006). This view has been imported into cognitive science and artificial intelligence research, looking at the human brain as an information processing system in line with the computer. However, such a framework does not give access to theories of qualia a nd first-person consciousness (Brier 2007, 2008a, c, d, 2009a, b) and in my view it needs to integrate them in its foundation before it can explain how experience, qualia, and emotions can arise from computational processes.

2.8 Changing Our Basic Understanding of Physics

It is, therefore, clear for many researchers that an evolutionary theory of information, cognition, meaning, conscious, and communication places certain demands on the ontological presumptions of nature by science if we do not want to bypass the results and methods of science (Küppers 1990). Even if we believe in emergence, it is difficult to take departure in a paradigm of nature based on an ontological materialism that sees nature and the emergence of conscious man as completely determined by absolute and universal natural laws. This would defeat the whole idea of free will and destroy the vision of the human being upon which our culture is built, which is a prerequisite for knowledge as a non -mechanical search for truth. Actually, a theory of emergence is not compatible with mechanical materialist determinism based on a reversible time conception and a belief in a simple ground state of things. Thus, there is no real irreversibility and new levels of complexity as Prigogine managed to describe them in his non-equilibrium thermodynamics (Prigogine and Stengers 1984; Prigogine 1997).

In contrast, Prigogine, like C.S. Peirce (1892), saw the mechanical systems as a special subclass of physical systems, not the foundation for all physical systems. Only a part of nature can be described satisfactorily this way, which the later nonlinear system revolution in mathematics has shown. Most of the physical systems in nature are very complex and dynamic; maybe even hyper-complex, with a stream of energy through them developing in irreversible time into more complex dynamical states in ways not precisely predictable. Furthermore, Einstein's relativity theories told us that matter is not the physical ground state; energy is. Matter is energy stabilized in an interlocked dynamical form, a kind of a causal cybernetic circuit. Information theory's basic definition of information has been developed to be different; form and structure within contexts. Therefore, the ultimate nature of reality is often, these days, answered as being informational and energetic processes and structures.

However, to propose a theory of knowledge, one must dare to say more about the world and its connection to the observer than that it is simply infinitely deep, spontaneous, chaotic, closed, and expending space-time geometry. In modern cosmogony we see energy self-organize and in that process becoming bound up in structures we call matter (atoms) in an uneven way in the field we call gravity, whose drawing force acts like a stabile tendency to produce order in an expanding universe. The uneven distribution of the first particles attract them to each other and when the first Hydrogen atoms are formed their mutual attraction brings them so close together that fusion processes start and form stars wherein heavier elements are created by further fusion, up to iron.

From there on supernova explosions are necessary to create elements heavier than iron and spread the molecules out in space where, they create molecules of a still growing complexity. This matter is collected into planets through gravity and the flow of energy from a star (the Sun for example) creating self-organizing systems far from equilibrium, which get more and more complex and creates the complex macro-molecules necessary for the material foundation of life. These make it possible to build a new system of the same kind by chemical inherence of macromolecular structures like DNA, RNA, and proteins. Membranes and organelles spontaneously self-organize and combine into cells. But it seems highly unlikely that the complicated, static DNA molecule should spontaneously assemble first, and only later it should gather the dynamic, complicated biochemical machinery of a cell around itself. It seems more likely that a proto-cell metabolism emerged first, gradually gaining increasing self-control culminating in DNA and RNA regulation and protein production. In that case, DNA is rather the amazing result of early evolution than its starting point and there was always a cell accompanying DNA – and the structure of that cell, in itself, constitutes inherited information in addition to that of the genes.

However, the incoherent jump in the theory is that now they are suddenly living, while the rest of the objects we have mentioned have been physical or chemical only. Different forms of cells combine into the modern complicated cell with many different organelles like mitochondria and Golgi apparatus. The cells combine into multi-cellular living systems. Later, organs emerge, some of them sense-organs; the combination with a nervous system suddenly makes sense- experience possible. But how? Barbieri attempts to use his code-biology paradigm to solve this problem by referring to new brain-codes. This view has thoroughly been analyzed by Brier and Joslyn (2013) as failing to provide a satisfactory theory of the emergence of experiencer and emotions. Sensing systems can be used in robots to orient the systems related to other structures in the environment with suitable structural couplings as Maturana calls them (Maturana 1983). One can say that these robots, functionally defined, see (if we focus on the visual sensing for a moment), but they do not see in an experiential way. Therefore, the hard evolutionary question is, from where in the received view of physical cosmogony, chemical and then biological evolution does the ability to sense experiences and be aware emerge? As Emmeche (1992) shows, none of the accepted forms of emergence deal with how experiences arise from matter through self-organization. We must further theorize how the processes of cognition and communication develop beyond their basis in the perturbation of and between closed systems to a theory of feeling, awareness, qualia, and meaning.

The German system theorist Niklas Luhmann was inspired by Maturana and Varela's theory of autopoiesis and extended the autopoietic model to the psychological and the socio-communicative level (Maturana and Varela 1987). Luhmann (1990), generalized the theory of autopoiesis and from this abstract model derived a triple autopoiesis model, where both the biologic and psychic systems are silent and only the socio-communicative in its autopoietic form can communicate: (1) Biological autopoiesis functions in the medium of life, and (2) the psychic autopoiesis plus (3) the socio- communicative (mostly different forms of languages) function in the medium of meaning. Thus, communication systems like juridical, political or scientific discourses are autopoietic systems. Luhmann's provoking punch line, "Only communication communicates!" is a meta-biological perspective that processes meaning without intentionality, but from a horizon of subjective expectancies. Loet Leydesdorff writes:

> From the perspective of cultural studies and critical theory, Luhmann's communication-theoretical approach in sociology can still be read as a meta-biology: while biologists take the development of life as a given, Luhmann tends to treat the development of meaning as a cultural given.[.. Meaning is no longer considered as constructed in communication, but meaning processing precedes and controls communication as an independent variable. (Loet Leydesdorff 2012, p. 1)

This gives a totally different view of meaning than the social constructivists or positivists. Meaning is generated through evolution, the social is communication and communication is a way to structure meaning and information as part of communication. Luhmann writes:

> Communication is a completely independent, autonomous, self-referentially closed selections, a mode of constantly changing the forms of meaning material, of reshaping freedom into freedom under changing conditions, whereby (given the premise that the environment is complex enough and not ordered as pure randomness) experiences of reliability gradually accrue and are then re-included in the process. Thus a meaning world emerges through epigenetic evolution that makes possible communication that is less probable. (Luhmann 1995, p. 149)

Science has shown us that reality is very complex and the data we have empirically collected can be interpreted in many ways. There are many competing valid interpretations that are negotiated in the many research communities. We cannot expect a common worldview. We have to negotiate a mutual understanding to find a common working definition. Luhmann's information concept is always integrated with meaning and understanding. He writes:

> If one conceptualizes communication as the synthesis of three selections, as the unity of information, utterance, and understanding, then communication is realized if and to the extent that understanding comes about. Everything else happens 'outside' the unity of an elemental communication and presupposes it. This is especially true for a fourth type of selection: for the acceptance or rejection of the specific meaning that was communicated. (Luhmann 1995, p. 147)

Embodied cognitive science says that some basic part of the common understanding is the prerequisite for the selection resulting from the interaction between body and mind in the process of surviving and preserving the body- mind's organization of the individual living beings. Contrary to dead things, living systems are individuals, and this is the basis for the ability of humans to become a person in a culture.

Habermas (1987) made the argument about this meta-biological foundation of Luhmann's systems theory most forcefully:

> In this way, subject-centered reason is replaced by systems rationality. As a result, the critique of reason carried out as a critique of metaphysics and a critique of power, which we have considered in these lectures, is deprived of its object. To the degree that systems theory does not merely make its specific disciplinary contribution with the system of the sciences but also penetrates the lifeworld with its claim to universality, it replaces metaphysical background convictions with metabiological ones. Hence, the conflict between objectivists and subjectivists loses its point. (Habermas 1987, p. 385)

I agree with Habermas that the problem here is where the first person experiences belong in the intersubjective basis for communication, both social and cultural.

How do systems go from being able to functionally orient themselves in relation to environmental structures and other members of the species – like robots – to having sense-organs giving sense-experiences and constructing an "I"? Moreover, how does the intersubjectivity of communication, language and knowledge emerge? In a way, one can say that intersubjectivity precedes objectivity in the world, because the world is represented as a model within the intersubjective knowledge through language. Luhmann considered human actors as consciousness systems, which are only the environment of the social system, which he saw as communication, and therefore replaced Husserl's concept of 'intersubjectivity' with communication-theoretical concepts like self-organization and horizon of expectations (Leydesdorff and Sander Franse 2009, p. 7).

However, when it comes to the qualia of subjective consciousness, Searle (1989) is not far from Luhmann's views in arguing that the secret must lie in biology. It is true that as far as we know, only biological systems produce the nervous system, and the central nervous systems that create awareness, feeling, sense-experience, and qualia, though we do have some kind of nervous reactions in plants. Nonetheless, biologists insist on describing their subject area in chemical and physiological terms – often supplemented with functionalistic system and cybernetics concepts – and biologists consider molecular biology to be the greatest advantage since Darwin. The vitalism debate has ruled out that there are any differences in the nature of the molecules inside and outside living systems. Thus, the received view in science is that the only difference between pure physical and living biological systems lies in the way the molecules are organized. However, how should that create the difference, which produces consciousness?

In the received view of modern biology it is presumed that over a long period of variation and selection of functional macromolecules, autocatalytic ribosomes develop, which again develops catalytic abilities as templates for the polymerization of polypeptides. This would then, over a long period of time, result in the precise tri-nucleotide "codes",[3] which are used in DNA in all present organisms to determine specific amino acids to be produced by the ribozymes. What is often called "encoding" of information into the DNA through the evolutionary process is actually done by the environment through the processes of "blind variation" and the selective elimination of erroneous variants. Once autopoietic reproduction begins, natural selection becomes possible, and survival knowledge – in the form of structural coupling's readiness to act in an orderly way by certain disturbances from the environment – begins to emerge and grow. These autopoietic structures that are connected to the ability to produce their own macromolecules then create "semantic closure". Solutions to survival problems are kept as a kind of reaction potential within the organism, some of them as molecular structures in the DNA-RNA-protein-synthesis processes. This enables the system to perpetuate its autopoiesis

[3] The term "codes" is (mis)used much in informational biology, but codes is a term that comes from something conscious human beings make to connect two different systems such as the Morse code and the letters in the alphabet. But how can systems that do not have any consciousness, intentions or subjectivity devise 'codes'?

from one instant to the next through generations of self-production as a full-bodied individual. Hoffmeyer and Emmeche (1991) call it code duality. The analogue code is the actual living body as a phenotype and the digital code is the genotype of the genome. These two codes then interchange over time.

Recently, Marcel Barbieri (2009, 2011) has pointed out that there is actually a difference in molecules inside and outside the living systems. This difference is caused due to the fact that many of the proteins, which are constructed by the DNA, RNA and ribosomal protein synthesis machinery, are not found outside living systems at all. They are not spontaneously produced in stardust as so many of the living system's other vital molecules. These special proteins are only produced inside living systems composed of at least one cell. Thus, Barbieri concludes that life is then partly based on artificial molecules, which the living systems' autopoietic machinery has created and keeps reproducing on a regular basis of its own codes. It is on this insight that he develops a code-biology.

Thus, in the beginning, "knowledge" exists only as embodied in the inherent structural dynamics of the autopoietic entity. But is it knowledge defined without life and sense experience? I would not say that a robot has knowledge in itself, as many computer scientists do. We as programmers put it there and is able to appreciate it in relation to the functional goals we have constructing these machines to obtain. So, on one hand, Robots are not alive and do not produce knowledge and on the other hand autopoiesis in itself is not sufficient as a theory for defining experiential life.

I argue here that knowledge needs an experiential component added to the functional, since sense experiences and awareness are usually not part of the standard biological story of the development of life and knowing. Thus, structural couplings in the autopoiesis theory, affordances a la Gibson and Uexküll's tones, are all important parts of a pragmatic evolutionary understanding of cognition, but this is not enough to establish a theory of the emergence of experiential mind in evolution.

Surviving entities in the course of evolution are those where the heritable structures of their DNA molecules contributed to solving survival problems. However, how this should exactly happen as a mechanical process, we do not know. Nonetheless, the general idea is that starting from random noise, the autopoietic functions of the cell make possible the selective filtrate for useful functionality. As such, researchers often say that this process has gradually built knowledge of the world into the DNA sequence; but, how and what kind of knowledge?

Barbieri (2011), in a crystal-clear article, sees the parallel between the problem of the emergence of life from the physico-chemical world and the emergence of experience from the self-organization of the living systems. To him, the production of new codes can solve both. Life is built out of new artificial molecules assembled by the DNA, RNA, and Ribosomal apparatus combining amino acids in new inventive ways. The solution to how the ability to experience emerges from the brains of mammals, is the production of new brain codes, which generates the brains ability in sense experience, emotions, and imaginary abilities. Barbieri, in his most interesting grand theory of code-biology, writes:

> The idea of a deep parallel between life and mind leads in this way to a parallel between proteins and feelings, and in particular to a parallel between the processes that generate them. We already know that the assembly of proteins does not take place spontaneously because no spontaneous process can produce an unlimited number of identical sequences of amino acids. The Code model of mind is the idea that the same is true in the case of feelings, i.e., that feelings are not the spontaneous result of lower level brain processes. They can be generated only by a neural apparatus that assembles them from components according to the rules of a code. According to the Code model, in short, feelings are brain-artifacts that are manufactured by a codemaker according to the rules of the neural code. In the case of proteins, the codemaker is the ribonucleoprotein system of the cell, the system that provides a bridge between genotype and phenotype. It receives information from the genotype in the form of messenger RNAs and assembles the building blocks of the phenotype according to the rules of the genetic code. It must be underlined; however, the codemaking system has a logical and a historical priority over genotype and phenotype, and for this reason it is the third category that has been referred to as the ribotype of the cell.

In the case of feelings, the codemaker is the intermediate brain of an animal, the system that receives information from the sense organs and delivers orders to the motor organs according to Barbieri's theory. The sense-organs provide all of the information that an animal is ever going to have about the world, and represent therefore in an animal what the genotype is in a cell. In a similar way, the motor organs allow a body to act in the world, and have in an animal, the role that the phenotype has in a cell. Finally, the intermediate brain is a processing and a manufacturing system, an apparatus that is in and for an animal what the ribotype is in a cell.

The parallel between life and mind, in conclusion, involves three distinct parallels: one between proteins and feelings, one between genetic code and neural code and one between cell and animal code making systems. The categories that we find in the cell, in other words, are also found in animals, because at both levels we have information, code and codemakers. The details are different, and yet there is the same logic at work, the same strategy of bringing absolute novelties into existence by organic coding (Barbieri 2011, p. 380). It is an interesting and very creative new idea. But I do not think it quite cracks the puzzle of qualia. It is still too functionalistic. In a later section, the article shows that Barbieri thinks of sense experience as modeling. It certainly is, but, in my phenomenologically informed view, it is of a qualitatively different kind. Barbieri writes:

> The results of brain processing are what we normally call feelings, sensations, emotions, perceptions, mental images and so on, but it is useful to have also a more general term that applies to all of them. Here we follow the convention that all products of brain processing can be referred to as brain models. The intermediate brain, in other words, uses the signals from the sense organs to generate distinct models of the world. A visual image, for example is a model of the information delivered by the retina, and a feeling of hunger is a model obtained by processing the signals sent by the sense detectors of the digestive apparatus. (Barbieri 2011, p. 388)

It seems to me to be a hypothesis for new sort of mechanism explaining emergence using the concept of code. Barbieri uses the modeling idea expanded on by Sebeok and Danesi (2000), although not on their basis in Peircean semiotic philosophy. Barbieri's theory is a good "functionalist approach" catching important practical aspects of reality. However, when I make a model of the route I have to follow to get home from a new place in town, I actually visualize the streets. I see them and thereby experience them. I make the images for my "inner eye" and draw on my lifetime's memory of this town, which I have lived in my whole life. It is not just a logical map which directs my way home, it is an embodied conscious experience based on memories of driving through these streets, most of in on bicycle. It is qualitatively different from what such a map is to a robot, not the least because I have the free will to choose not to follow it and change the route. I am not, in any automatic way, determined to follow it. Clayton (2004, p. 601) also argues that the emergence into the quality of experience is different from other emergence theories. Of cause my navigator can change routes, but only when we programmed criteria like: choose the roads that will generate a route that will bring us to our destination with maximum speed and include knowledge of rush hours, etc.

Konrad Lorenz (1971) tried to develop an alternative to behaviorism's mechanical paradigm in the form of a bio- psychological science, which he called ethology. If you follow his work on the theoretical development of this new paradigm, it becomes obvious that after a long struggle with the problem, he failed to integrate the inner phenomenal world with the tradional biological science to create his behavioral science of ethology, a fact also pointed out by Hinde (1970), Brier (1980, 2008a, b, c, d). Biology has not yet been able to produce a concept of qualia or intentionality. Ellis and Newton (1998) and Damasio (2000, 2004) have indicated the importance of emotions for the understanding of cognition, communication and behavior. However, none of them has managed to make a theoretical ontological foundation for a new way to integrate first, second and third-person views on embodied intersubjective linguistically-interacting conscious minds and embodied brains.

Barbieri (2011) attempts bravely to solve this with a new code-semiotic paradigm; building neither on information theory nor on Peirce's semiotic philosophy but on a code-emergence philosophy in some ways close to both system thinking and dialectical materialism. In the regime of computer codes, AI researchers in hard A-life often believe that the agents they create in computers are compatible to living agents (Emmeche (2013) criticizes this). However, many A-life researchers do not see any special abilities in living systems other than complexity. Still, Peircean biosemiotics considers that it is the combination of cells into specialized organs to registries perturbations in the physical chemical environment that gives rise to sense experiences that can make a difference. A difference cannot become knowledge before it has been interpreted to be so meaningful and important that an individual observer/knower in a species or a culture attaches a sign to it and produces an interpretant. Then, it will make that sort of difference that Bateson (1972) writes make a (real) difference. However, neither biology nor computer science has answered the question of how this is possible, and Barbieri seems not to have developed this aspect of his new code-biological paradigm in any explicit way, so far. This is why I still prefer biosemiotics' suggestion that what is transferred in and between living systems are signs, not objective information. Signs have to be interpreted and it has to happen on at least three levels as illustrated in Fig. 2.2.

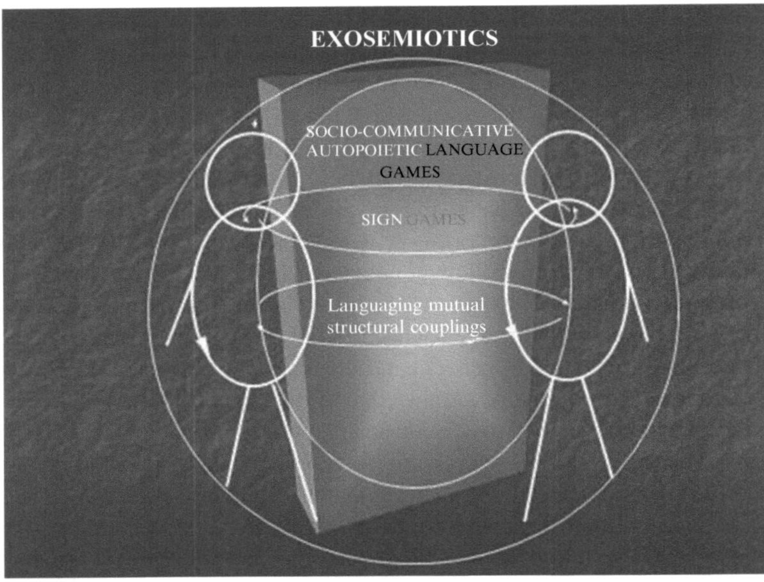

Fig. 2.2 The three levels of systemic semiotic communication in Cybersemiotics

On the most basic level we have the basic coordination between the bodies as a dance of black boxes, like turn taking and mutual orientation and attention, allowing for meaningful exchange. The meaningful exchange then goes on at the next level of instinctual sign plays of drive and emotionally-based communication about meaningful things in life, like mating, hunting, dominating, food seeking, territory, etc. I agree with Barbieri (2011) in his distinguishing between a cybernetic and instinctive aspect of the brain function and his belief that the emotions emerge from the instinctual brain. I agree on this, but cannot see that he solves the problem that Konrad Lorenz (1971) was unable to crack in his creation of the ethological paradigm (Brier 1980, 2008c). Based on these two levels, a new third level of meaning is created so that the socio-communicative system can modulate to conscious linguistic meaning. But this is not understandable from either a pure physicalist or informationalist ontology. Therefore a semiotic one is needed in combination with a systemic and cybernetic one like Luhmann's.

Cognition and communication are socially distributed, bio-physically embodied and culturally embedded. Moreover, there is an integration of the praxis of communication with the praxis of living, of language games with life forms and of the communicative competence with a general socio-cultural competence. An instrumental-pragmatic view of linguistic communication is conceived in linguistic-symbolic behavior and the use of co-evolutionary tools (technology). Donald (1991) and Nelson (1998) believe that it all began with Homo erectus' mimetic mind and culture. It was characterized by representational and reenacted intentionality in the use of fire to cook their food and the institution of the sharing of food among family members. This is assumed to be the start of phatic communion and the development of symbolic codes. Mimesis can be seen as the outgrowth of the primary, proto-semiotic, reflexive stage of languaging, which is securing coordination and community in the general primate episodic mind and culture. Here we go some three million years back (Donald 1991). Mimesis is a precursor to the symbolic stage with its social, communicative, re-enactment side and its individual, cognitive, representational side. Mimesis stages would be from images over diagrams, to metaphors. According to Donald's evolutionary theory (Donald 1991), metaphoricity would include primitive 'narrativity' and develop that further, producing a mythic stage in the Paleolithic epoch of the Stone Age about 35,000 years ago. Narrative skills are a fundamental part of the communicative competence of modern man, Homo sapiens. 'Narrative thinking' in the form of mythos is prior to 'paradigmatic thinking' of analytic thinking, which is the characteristic of the theoretical empirically based scientific type of thinking and explanation. Meaning narratives are a prerequisite for objective science. How are we to integrate that knowledge, if not by including semiotics as biosemiotics in our knowledge foundation?

2.9 Why Brain and Experiential Consciousness Data Do Not Fit

When I think of the problems of interpreting the results of brain research into our social life world of experiential awareness and meaningful language games'existence in their life forms, I think the poem below of the known and unknowns is a fitting metaphor of our problem of formulating the problem:

The Unknown
As we know,
There are known knowns.
There are things we know we know. We also know
There are known unknowns. That is to say
We know there are some things
We do not know.
But there are also unknown unknowns, The ones we don't know
We don't know.
(Donald Rumsfeld Feb. 12, 2002, Department of Defense news briefing poem)

Many researchers think that we just have a "hard problem" of how brains produce awareness, experience and even self- consciousness. Many scientists think that we need to simply find a good chemical or computational description of the processes between that which we can see through brain and neuro-science and that which we can report from our own "inner" experiences. Therefore it is just an unknown that we know that we do not yet know, when we look at it from cognitive brain science. The information and computational cognitive scientists believe they can explain this connection as computation. In the beginning, it was on the basis of the Turing machine's theoretical concept of algorithms. Thus, it was based on an ontology of the world as a sort of Turing machine able to compute by algorithms.

However, in the last couple of years, it has been more widely acknowledged that this foundation is too narrow to be able to explain the emergence of experience and awareness. Researchers, therefore, are now trying to broaden the concepts of computation and information into a theory of natural info-computation partly based on the visions of Gregory Chaitin's (2010) meta- mathematics, where he attempts to view mathematics as a biological process (Dodig-Crnkovic 2010; Dodig-Crnkovic and Muller 2011).

I still fail to see how this paradigm of pan-informational and natural computation can solve the problem, as there is no indication of computational systems producing awareness and qualia (Brier 2010a; Emmeche 2001) or what Peirce calls Firstness. I think that Emmeche (2001) here shows that the epistemology is too simple, and as stipulated by the ontology – which now has a combination of energy/matter, information and computation as its foundational entities – is unable to include a phenomenological and first-person perspective theoretically. Thus, I think our situation is worse than operating strategically with solving a known unknown. My whole argument in this chapter and the rest of my research is that I think the background for the hard and binding problem is an unknown unknown like the dark matter problem in physics and that the problem is not recognized by the ruling paradigms in the field who still believes in a straight -forward manner so that their "normal sciences" can deal with the problem empirically and solve it that way (Edelman and Tononi 2000; Kuhn 1996).

The reason for this unfortunate understanding of the problem is partly based on almost incommensurable communication, because physics and chemistry, on the one hand combined with information and computational-based cognitive brain and linguistic sciences, versus phenomenological and hermeneutic paradigms, on the other hand, which have very different implicit ontologies and epistemologies (Brier 2008a). I think the present attempts to naturalize phenomenology show that many researchers acknowledge this and therefore try to find other connections than the info-computational to the problem (Petitot 1999). However, I think these researchers underestimate the radical nature of the problem if they think there is a simple road from science to phenomenology (see Heelan 1987).

Still, one of the most esteemed philosophers of physicalism, Kim (2007), recognizes the problems of qualia and mental causality to be the two most severe impediments to a physicalist philosopher of science's ability to develop a full-blown physicalism. How are experiences and subjectivity going to be explained by absolute natural laws working on inert matter? One obvious strategy is to invent two independent worlds for mind and matter in a dualism like Descartes'. We have worked with this idea for centuries and it has paved the way for neuroscience. However, as Damasio writes in Descartes Error (1994), most researchers today have realized that Descartes' dualistic solution to this problem does not solve the problem. First of all, because it is impossible to see how any interactions between Res Extensa and Res Cogitans could be possible unless one believes in a pre-stabilized harmony, as Leibniz (1898) did, and combines it with a double aspect theory like Spinoza's. I think this is largely what Chalmers and Damasio have done (Chalmers 1995, 1996; Damasio 2004). However, in this case, mind and matter would be tied together as two aspects of the same reality and governed by absolute deterministic laws or the will of God. That would then leave the experiential domain in the same form of absolute determinism by general laws as the physical domain in the way it was conceived by classical mechanical physics. This would eradicate free will. Consequently many researchers, like for instance Libet (1993), try to show that the concept of free will is an illusion and thereby the independent decision power of the subject will also turn out to be illusory. It would – as I have argued previously – destroy the foundation of the self-same science that was supposed to give the arguments weight. Thus, we would find ourselves in another vicious circle of arguments. As Kant argues:

> It is as impossible for the subtlest philosophy as for the commonest reasoning to argue free will away. Philosophy must therefore assume that no true contradiction will be found between freedom and natural necessity in the same human actions, for it cannot give up the idea of nature any more than that of freedom. (Kant 1909, pp. 75–76)

Kultgen (2009) points out that it is significant that Peirce as well as Whitehead and Griffin (1978) deny Kant's (1909) absolute distinction between nature and freedom in their process philosophy. To Peirce, nature has spontaneity and pure chance at its basis in Firstness, reality in Secondness and reasonability in Thirdness. Peirce furthermore in his semiotic phenomenological-paradigm denies the distinction between the phenomenological and the noumenal (understood as the thing in itself), because this idea of the incognizable appears as a null-term of theoretical and practical thought!

For Peirce, the real is wholly open to our pragmatic observation and thinking, and there is no absolute difference between objects of theoretical and practical thought. Metaphysics is seen as an observable ideal limit of empirical enquiry (Kultgen 2009, p. 288). Thus, Peirce makes a full naturalization of all possible knowing in the universe including the subject and the intersubjective phenomena. This is a philosophical move that modern American philosophers like Sellars, McDowell and Brandom, are known for developing further, but unfortunately not on Peirce's triadic semiotics pragmaticist basis. Robert Brandom (1994) recently declared that Wilfrid Sellars is the greatest American philosopher since Charles Sanders Peirce. Unsurprisingly, Peirce was a great inspiration to Sellars. Like Peirce, Sellars (1991) wanted to move analytic philosophy from its Humean into its Kantian phase. It is a move beyond classical empiricism, or from logical empiricism to logical Kantianism. Peirce, like Sellars, thought the task of philosophy was to provide a 'synoptic' view of how things in the broadest possible sense of the term correlate. How does our common sense outlook fit into our increasingly fine-grained scientific outlook? That is a well-known problem in medicine and make the communication between the patient and the specialized physician difficult as modern medicine is a high specialized scientific endeavor far from the common sense world of most patients no matter how much they have read on the Internet. Another philosophical way of formulating the problem is to ask: How can we make our everyday perception of a blue wall compatible with the same phenomenon described by particle physics? Which is close to the problem of how can we make a patient's experience of for example diabetes 2 compatible with our biomedical knowledge of the causes and processes of the particularly illness.

Both Peirce and Sellars view our non-scientific ways of thinking as being indispensable not only for knowledge but as the very basis for perception and thought. The problem is that empiricist philosophy claims that our ideas arise from direct experience of things. The 'myth of the given'– as Sellars calls the problem that Peirce also was aware of – is the claim that individual pieces of data can be known directly, that is, without any knowledge of associated concepts, paradigms and worldviews. The problem is: how can I say I know what is from the fact that some things look red to me? In both Peirce's and Sellars' view, in order to say anything 'looks blue', we would already have had to require the abstract universal concept of 'being blue'. It is a basic philosophy as well as a philosophy of empirical science's problem that we need universal concepts to distinguish colors from each other, or one taste from another. This means that the current model of the world, which empirically-based science produces, is lacking an integrated reflection on the selfsame consciousness and sign systems – like natural languages – that produced the science by which we attempt to make causal models of the self same consciousness. I am not alluding to a constructivist philosophy of linguistic like Sapir–Whorf–hypothesis of linguistic relativity,[4] but the more realist and logic oriented theory of linguistic supertypes of Durst-Andersen (2011).

My suggestion is that in order to avoid a strange loop in argumentation, we may modify and enlarge our idea of nature, sign and systems. Truth is only mechanical in the formal and abstract world of syllogistic logic, but it is not so in the concrete world as Brenner (2011) argues.[5] Here, the truth of theories with universal aspirations cannot be proved in the ordinary mathematical and logical use of the concept. As mentioned, Penrose (1999) also argues convincingly that important aspects of human consciousness are non-algorithmic. This means that consciousness is not capable of being modeled by a conventional Turing digital computer. Thus, a pan-computational paradigm on this basis will not be able to encompass consciousness. To go from Cartesian dualism to modern pan-computational informationalism does not solve the problem either. There is a weak possibility of this if one changes the concepts of computation and information considerably from the scientific one we have today; indeed Dodig Crnkovic and Mueller have initiated the development of such a new paradigm (Dodig-Crnkovic 2010; Dodig Crnkovic and Mueller 2011). Stephen Wolfram (2002) has announced a new kind of science based on a theory of strong computational universality for complex systems. It was not Wolfram, but Konrad Zuse, who was the first to suggest that the physical universe is being computed on a discrete computer, such as a deterministic cellular automaton. His first paper on this topic dates back to 1967 (Zuse and Raum 1967, vol. 8, pages 336–344). Many develop the computational concept deep into quantum physics in order to reach another sort of computational foundation of reality called qubits like Deutsch, who writes: "Boolean variables, and classical computation are all emergent

[4] http://en.wikipedia.org/wiki/Linguistic_relativity.

[5] This is also the main argumentation of Brenner (2008).

or approximate properties of qubits, manifested mainly when they undergo decoherence" (Deutsch 2013, p. 93). This leads him to the following significant ontological statement:

> The world is made of qubits. Every answer to a question whether something that could be observed in nature is so or not, is in reality a Boolean observable. Each Boolean observable is part of an entity, the qubit, that is fundamental to physical reality but very alien to everyday experience, it is literally not of this universe ... What we perceive to some degree of approximation as a world of single-valued variables is actually part of a larger reality in which the full answer to a yes-no question is not just yes or no, nor even both yes and no in parallel, but a quantum – observable – something that can be represented as a large Hermitian matrix. (Deutsch 2013, p. 100)

Nonetheless, even when trying to go this deep into a quantum-computational foundation for reality as John Archibald Wheeler (1998) also does, I cannot see how any of them can avoid phenomenology and the meaning question of the observer which is so important to quantum physics if they do not at least shift to a biosemiotic approach.

Emmeche (2013), for instance, realizes that a broader idea of ontology is necessary and describes qualitative organicism as one way of making broader ontological stipulations and integrating experience as a part of objective reality. He writes:

> **Qualitative Organicism.** This is a more radical position differing from main stream organicism in its appraisal of teleology and phenomenal qualities. It emphasizes not only the ontological reality of biological higher level entities (such as self-reproducing organisms being parts of historical lineages) but also the existence of qualitative experiential aspects of cognitive behavior. When sensing light or colors, an organism is not merely performing a detection of external signals which then get processed internally (described in terms of neurochemistry or information processing); something more is to be told if we want the full story, namely about the organism's own experience of the light. This experience is seen as real. It may be said to have a subjective mode of existence, yet it is an objectively real phenomenon. (Emmeche 2013, p. 117)

I support the idea that it is a major point of advancement towards a Wissenschaft to realize that the existence of subjective experience is an objective fact. It is real and therefore a part of reality or the real world. My main problem with the standard materialistic scientific evolutionary paradigm is that I cannot see how physics – as an external science – on the basis of the present definitions of matter, energy and deterministic law, can ever alone furnish us with the final understanding of our inner lives and how consciousness arises. When working from an evolutionary view, combining the big bang theory with self-organizing thermodynamics and chemistry, you need to include Darwinism for biological systems. But even when you proceed with a somewhat materialistic theory of the development of the history of language and the culture of man, there still remains the severe problem of explaining consciousness as this inner quality of perception, feeling, volition, and cognition that we all experience. I do not see quantum physics, the general relativity theory or non-equilibrium thermodynamics as being of any help concerning this particular problem, although they may be helpful to explain the physical aspect of consciousness (Penrose 1989). This is my argument of why a bottom up, empirically based physicalism or pan-computationalism is inadequate to solve the gap problem. It is here that Peirce's theory of – what he calls thirdness brings the physical and the mental together in that he sees as the tendency to form habits in both nature and mind. Here is one of those deep quotes of Peirce arguing against the mechanical view of natural law:

> The law of habit exhibits a striking contrast to of forces; but the component motion must actually take place exactly as required by the law of force. On the other hand, no exact conformity is required by the mental law. Nay, exact conformity would be in down-right conflict with the law; since it would instantly crystallize thought and prevent all further formation of habit. The law of mind only makes a given feeling more likely to arise. It thus resembles the 'non-conservative' forces of physics, such as viscosity and the like, which are due to statistical uniformities in the chance encounters of trillions of molecules. (Peirce 1892, CP 6.23)

This is why Thirdness is so important in Peirce's categories and at the same time it is also critical to remember that Thirdness includes Secondness and Firstness. We will return to that below.

The Cybersemiotic transdisciplinarity accepts Peirce's view and sees scientific explanations as going from our present state of socio-linguistically-based conscious semiosis in self-organized autopoietic systems towards a better understanding of the prerequisites of language and the self-conscious being. Science provides us with suitable economic and practically useful understanding of certain processes, often in a way that allows prediction with a desired precision within certain circumstances. However, it does not give universal explanations of the construction of reality, energy, information, life, meaning, mind, and consciousness. Natural science only deals with the outer material aspect of the world and our body, not with experiential consciousness, qualia, meaning and human understanding in its embodiment (Edelman and Tononi 2000, pp. 220–222). Being in the world, in languaging, embodied in a meaningful social context we instead have to start 'in medias res' (center of the Cybersemiotic star). We will always be bound to make metaphysical presumptions based on our present understanding and they will always prove themselves later to be too limited. However, Peirce's semiotics is a very good non-reductionist framework from which to start, since it takes its point of departure in the semiotic mind.

Today, it is widely recognized that what we call a human being is a conscious social being living in language. Terrance Deacon (1998), in his book *The Symbolic Species*, sees language-processing capacity as a major selective force for the human brain in the early stages of human evolution. We speak language, but we are also spoken by language. To a great degree, language carries our cultures as well as our theories of the world and of ourselves (Durst-Andersen 2011). As individuals, we are programmed with language – to learn a language is to learn a culture. As Wittgenstein said in *Tractatus*; the limits of my language are the limits of my world. As such, pre-linguistic children are only potential human beings until they have been linguistically programmed in order to become the linguistic animal cyborgs, which we call humans. However, getting behind language as such is difficult, as already Wittgenstein pointed out. But it may be possible by creating a broader platform than linguistics. Peircean semiotics and its modern development into biosemiotics is such an attempt for a doctrine of cognition and communication, and therefore creating of knowledge in the widest sense.

2.10 Biosemiotics: The Connection Between Meaning, Rationality and Nature

Damasio writes: "Nature appears to have built the apparatus of rationality not just on top of the apparatus of biological regulation, but also *from* it and *with* it" (Damasio 1994, p.128). He agrees with Peirce here. Recently Stjernfelt (2014) has argued that one of the most important lessons to take from Peirce's semiotics is its vast reorientation of the whole domain of sensation, perception, logic, reasoning, thought, language, images, etc. towards the chain of reasoning as its uniting primitive phenomenon. The point of pragmaticism is that this development of reasoning may be formally described, independently of the materials, in which it may be implemented. This view implies that propositions are not primarily entities of language, nor do they presuppose any conscious "propositional stance". Reasoning capacity is developed through evolution in nature. Consciousness and language should rather be seen as scaffolding, serving and increasing reasoning, which is one the most important overall selecting factors during evolution, Stjernfelt (2014) argues. Mind cannot exist or operate at all without a body, which we saw Merleau-Ponty emphasized above. However, something additional, which we still seem to be unable to unravel, is necessary to produce mind. Damasio writes:

> Brains can have many intervening steps in the circuits mediating between stimulus and response, and still have no mind, if they do not meet an essential condition: the ability to display images internally and to order those images in a process called thought. (Damasio 1994, p. 89)

Damasio puts forward an interesting theory of 'somatic markers'. Here, "dispositional representations" set off chains of reaction that reach deep down into the body's accumulated experience and bring forth images of appropriate visceral content intermingled with emotional states, which color everything with moods that regulate our attention and interest. It is a fecund insight, which was already foretold in the work of Konrad Lorenz (1971), in his attempt to build the biological behavioral science of ethology (see Brier 1980, 1999, 2000, 2001). But still, neither Lorenz nor Damasio reveals how the body produces experience as such. The most rudimentary biological cognitive processes of animals with perceptual organs involve the ability to make distinctions.

Any type of distinction must be able to sort out differences that do not make a difference from differences that do in matters of life and death. Thus, the organic is not deterministic or even probabilistic. It is an autopoietically closed individual able to respond to disturbances in a productive way for survival (Maturana 1980; Maturana and Varela 1987).

As such, the life sciences are qualitatively different from the exact sciences like physics and chemistry. Molecular biology and genetics within themselves do not explain the nature and quality of life and how living systems' experiential qualities come about. All we know is that the computational systems we have built thus far are unable to produce an experiential world. Haikonen (2009) gives a convincing analysis of the vast problem the phenomenon of qualia creates in the quest of producing conscious machines (see also his book on conscious machines; Haikonen 2007).

In the literature on biological systems it has for a long time been assumed that the use of the terms "difference", "information", "message", "signal", "communication", "messenger", "cue", "code", "sign", and "meaning" were practical, metaphorical shorthand; but if so, why do they persist and proliferate in scientific articles? One of the points of departure for biosemiotics is to take this "information talk" seriously and develop them into a common framework (El-Hani et al. 2006, 2009). The so-called central dogma in biology postulates a unidirectional flow of "information" from DNA to protein. Scientists hoped that these terms would be effectively reduced to chemical and physical interactions, or at least viewed as computational physical informational processes. Some of these phenomena are instead evaluated in biosemiotics as embodying sign processes, because genetic and biochemical information has shown to be highly context and time dependent. This means that "information" in biological systems is not simple objective "data", but has to be interpreted in a situated context by the cellular or multi-cellular system in order to yield meaning (Kauffman et al. 2007). The simplest such semiotic process is not only the ability of single cells to categorize environmental objects from superficial properties, but also internal exchanges between organelles. E. Coli, for instance, is able to recognize carbohydrates through an active site on the macromolecule. Thus, the active site stands as a code for the whole carbohydrate molecule. This makes it possible for the same kind of active site on another type of molecule – such as artificial sweeteners – to fool the bacteria, just like human beings are fooled by sweeteners in their unhealthy hunt for sugar.

The creative capacity of molecular-biological codes to be interpreted in meaningful ways expresses a generative capacity that is outside the terminology of the molecular-biological language. This example also shows that even at this level of life, a sign is what makes lying possible, as the signs stand for something for someone even though what they stand for need not be present. It is a new level of freedom, indeterminism and risk. Here, context of living becomes vital for interpreting and survival. There is no meaning without a life context and no context determined without meaning. They are bound together by a cybernetic semiotic loop. As the organism is responding to more of the present situation, it reaches deeply into the future and the past and into its construction of its own 'signification sphere'.[6] As anticipation unfolds, variation, plasticity, versatility, and adaptability grow, and semiotic freedom in the form of the enhanced ability to engender new concepts and cognitions develops, which go beyond the genetically determined forms of perception in reflexes and instincts, develops (Hoffmeyer 2008).

The basic reason for developing biosemiotics is thus the ontological postulate that biology is already semiotic. The living world is literally full of organic codes – such as DNA, messenger and transport RNA, ribosomal RNA codes, hormones, transmitters, immunological codes, etc. – and they are associated with all great events of macroevolution from the origin of proteins all the way up to the origin of

[6] 'Signification sphere' is a concept of Cybersemiotics in the form of a Peircean reinterpretation of Jacob von Uexküll's concept of the animal's "Umwelt" (See Brier 1995, 2011).

embryos, the origin of mind and the origin of language (Barbieri 2001, 2006). Not only does life create these semiotic capacities, but it also creates the capacity to create new codes with new semiotic capacities (Brier 2010a, b). Codes and signs cross the old borders between nature and culture, between causality and signification, and between interaction and communication. Biosemiotics suggests developing a reflected semiotic theory of the origin of life, agency, modeling, coding, semiosis, sense perception, conscious awareness and communication. However, it is seldom truly Peircean. As Emmeche (2013, p. 119) – in my view correctly – argues and sees, mainstream biosemiotics has to build on an organismic emergentism like general system theory, if they will not accept Peirce's semiotic pragmaticism. For me, this is a problematic platform from which to use Peirce's semiotics, as its workings are dependent on its paradigmatic triadic phaneroscopic formulation of an ontological framework. Thus organismic emergentism is in my view only defendable on a Peircean semiotic worldview. This includes synechism, which is also close to Whitehead and Griffin's (1978) thinking, namely that the world is a plenum, or a field, where everything is connected to everything else in a hyper-complexity. Much like the one we find in the mathematical line, where a new cut can always be inserted between two points – no matter how refined – they are defined. It also means that all knowledge is fallible – it cannot be proven true. In Peirce's words:

> The principle of continuity is the idea of fallibilism objectified. For fallibilism is the doctrine that our knowledge is never absolute but always swims, as it were, in a continuum of uncertainty and of indeterminacy. Now the doctrine of continuity is that all things so swim in continua... (Peirce, CP 1.171)

In the famous articles "The Fixation of Belief and How to Make Our Ideas Clear", Peirce discusses convergence of different lines of inquiry as a sign, from which inquirers hope to draw near to the truth. Such a sign is inconclusive, though. It involves the belief that the inquiries have been healthy, open, critically examining everything, etc. The interdisciplinary research project of biosemiotics is attempting to re-open the dialogue across the life sciences and the humanities about what terms such as "meaning" and "significance" might refer to in the context of living systems. It does this by treating life as continuous and by discerning semiosis across the realm of nature and culture. It accepts that organisms are agents who co-construct the world and themselves; that they are linking genetic code sequences through intercellular signaling processes and evolving an animal motivated perception with cognition and communicative display of behavior leading to the emergence of language in humans. Communication finally develops by the use of grammar and abstract symbolic thought of representation, meaning and sense into linguistic communication. For the Peircean semioticians, all of this is done in the basic aspect of life that Peirce calls "firstness", or feeling, and which co-occurs with 'possibility' in his phaneroscophy. Peirce defines firstness in this way:

> The idea of the absolutely first must be entirely separated from all conception of or reference to anything else; for what involves a second is itself a second to that second. The first must therefore be present and immediate, so as not to be second to a representation. It must be fresh and new, for if old it is second to its former state. It must be initiative, original, spontaneous, and free; otherwise it is second to a determining cause. It is also something

> vivid and conscious; so only it avoids being the object of some sensation. It precedes all synthesis and all differentiation; it has no unity and no parts. It cannot be articulately thought: assert it, and it has already lost its characteristic innocence; for assertion always implies a denial of something else. Stop to think of it, and it has flown! What the world was to Adam on the day he opened his eyes to it, before he had drawn any distinctions, or had become conscious of his own existence – that is first, present, immediate, fresh, new, initiative, original, spontaneous, free, vivid, conscious, and evanescent. Only, remember that every description of it must be false to it. (Peirce, CP 1.357)

Possibility and potentiality are thus found in Peirce's category of "firstness" as they are in the complexity science behind non-equilibrium thermodynamics and in the vacuum fields behind quantum field theory. The problem is how the modality of possibility – which is so vital for evolutionary thinking in physics, chemistry, biology, and sociology – can function in a non-reductionist and non-scientistic view of a developing cosmos (Deacon 2007, 2008). In contrast, with all other theories of self-organizing evolution, Peirce's view of firstness as both possibility and pure feeling theoretically provides the biological and medical sciences with ontological conditions for felt qualitative experience to emerge in autopoietic systems (Brier 2004, 2007). A non-reductionist view of the cosmos would see it as an infinite being of sheer availability of potential or possible being. In short, ontology is an ongoing process of becoming, as Whitehead and Griffin (1978) also see it in their process philosophy. Thus, Peirce solves Chalmers's problem with determinism in his double aspect theory by introducing a different evolutionary semiotic process ontology.

In Peirce's semiotic category of Firstness, possibility and pure feeling serve as a ground for the disclosure of this infinite potentiality and that pure abstract feeling, which Peirce points out, can be found when inquirers muse freely over nature and the universe in which they are situated. Peirce defines in his paradigm, what he means by his foundational concept of *feeling* as follows:

> By a feeling, I mean an instance of that kind of consciousness which involves no analysis, comparison, or any process whatsoever, nor consist on whole or in part of any act by which one stretch of consciousness is distinguished from another, which has its own positive quality which consist in nothing else, and which is of itself all that it is, however it may have been brought about; so that if this feeling is present during a lapse of time, it is wholly and equally present at every moment of that time. To reduce this description to a simple definition, I will say that by a feeling I mean an instance of that sort of element of consciousness which is all that is positively, in itself, regardless of anything else. (Peirce, CP 1.306)

Peirce does not describe a world of thought or mind other than the material; only the one we are in when having experiences. Like Husserl, Peirce was not a dualist, and therefore did not work with a framework where the distinction between 'subject' and 'object' as well as 'inside' and 'outside' was primary. Thus, his view is compatible with Hans Fink's (2006) suggestion of a new ontology, which he calls an "unrestricted or absolute naturalism". Fink has developed this philosophy from important points in McDowell's (1998) book *Mind, Value* and *Reality*. His view takes the philosophical consequence of realizing that all things and phenomena are developed within the universe in accordance with the evolutionary worldview. We, therefore, do not see culture, mind, meaning, consciousness, and ethics to be outside nature. They are all natural phenomena and therefore inside nature, which is also compatible with the above quotes of Merleau- Ponty. What else can they be, when we do not work with an absolute dualism or any other systems that propose more or less invisible worlds outside nature? Bhaskar (1998) also develops a philosophy much like that, which he calls non-dualism after the Vedic thinker, Shankara. I have suggested categorizing Pierce as such a non-dual thinker (Brier 2014).

The conclusion must be then that we primarily live in a world of signs, the center of the cybersemiotic star (Fig. 2.1), where objects appear when we habitually connect certain differences or 'Secondnesses'– as Peirce calls them – and choose an interpretation by connecting a representamen (a primary sign) with an object into an interpretant in our mind. Some of these objects, which our embodied cognitive experiences show us, turn out to be actual things. Thus, the first impression (immediate objects) is through experiences and communications with other semiotic beings. It is then modified to a true picture of things and processes to what Peirce calls Dynamic Objects. Thus, although Peirce at first blush can appear as a bio-psycho-social constructivist, he is a process realist, believing in universals, but certainly not a physicalist. He calls his stance "Scholastic realism" inspired by Duns Scotus, but adds the important aspect of evolution (Boler 1963). He is thus placing himself somewhere between Plato and Aristotle, but armed with an evolutionary worldview. Susan Haack explains the point very well:

> Though what exists is real, what is 'real' may not exist; existence is reaction, interaction – the characteristic mode of being of particulars, of seconds. This is why Peirce made a distinction between scholastic realism and what he called "nominalistic Platonism" [see CP 5.503 (c.1905); 5.470 (1903); 5.503 (c.1905)]: the thesis that universals like "man" or "horses" refer to abstract particulars, to existents. Peirce objected to nominalism and conceptualism because they deny that generals are real; he objected to nominalistic Platonism because it asserts that generals exist. Peirce's position was that there are real generals, not that generals are real. (Haack 1992, pp. 22–23)

Thus, Peirce's view of reality is very different from a modern physicalistic view combined into a dualism with Platonism in some sort of mathematical variant. The real in Peirce's paradigm is not only comprised of external things! But illnesses could be reals, though it is only individual people that embodies them and thereby becomes ill and experience illness. So, Peircean pragmaticist semiotics does not doubt that the external is real. The existent is that, which reacts against other things. The external world then does not consist merely of existent objects and their reactions; because among the reals Peirce also counts words, signs, general types and would-be's. Peirce writes:

> Thus, for example, the real becomes that which is such as it is regardless of what you or I or any of our folks may think it to be. The external becomes that element which is such as it is regardless of what somebody thinks, feels, or does, whether about that external object or about anything else. Accordingly, the external is necessarily real, while the real may or may not be external; nor is anything absolutely external nor absolutely devoid of externality. Every assertory proposition refers to something external, and even a dream withstands us sufficiently for one description to be true of it and another not. The existent is that which reacts against other things. Consequently, the external world, (that is, the world that is comparatively external) does not consist of existent objects merely, nor merely of these and their

reactions; but on the contrary, its most important reals have the mode of being of what the nominalist calls "mere" words, that is, general types and would-be's. (Peirce, CP 8.191)

It is a fascinating attack on physicalism and, at the same time, Peirce spent his entire life developing and defining scientific knowledge in the belief that it was the highest point of rationality man could attain. His phaneroscopic foundation of qualia is laid down most clearly in the following quote, which repays repeated reading:

No thought in itself, then, no feeling in itself, contains any others, but is absolutely simple and unanalyzable; and to say that it is composed of other thoughts and feelings, is like saying that a movement upon a straight line is composed of the two movements of which it is the resultant; that is to say, it is a metaphor, or fiction, parallel to the truth. ...Whatever is wholly incomparable with anything else is wholly inexplicable, because explanation consists in bringing things under general laws or under natural classes. Hence every thought, in so far as it is a feeling of a peculiar sort, is simply an ultimate, inexplicable fact. Yet this does not conflict with my postulate that that fact should be allowed to stand as inexplicable; for, on the one hand, we never can think, "This is present to me," since, before we have time to make the reflection, the sensation is past, and, on the other hand, when once past, we can never bring back the quality of the feeling as it was in and for itself, or know what it was like in itself, or even discover the existence of this quality except by a corollary from our general theory of ourselves, and then not in its idiosyncrasy, but only as something present. But, as something present, feelings are all alike and require no explanation, since they contain only what is universal.... Finally, no present actual thought (which is a mere feeling) has any meaning, any intellectual value; for this lies not in what is actually thought, but in what this thought may be connected with in representation by subsequent thoughts; so that the meaning of a thought is altogether something virtual. ... At no one instant in my state of mind is there cognition or representation, but in the relation of my states of mind at different instants there is. In short, the Immediate (and therefore in itself unsusceptible of mediation – the Unanalyzable, the Inexplicable, the Unintellectual) runs in a continuous stream through our lives; it is the sum total of consciousness, whose mediation, which is the continuity of it, is brought about by a real effective force behind consciousness. (Peirce, CP5. 289)

It is the subjective and inter-subjectively shared first-person experiential consciousness, as its own first cause, which is for Peirce the basis of his semiotically-based pragmaticist philosophy. Thus, a specific feeling or perceptual experience appears in consciousness as something (secondness). It is compared and identified with another in the present or in the

past (memory), though the theory of thirdness producing symbols and arguments leads into language. This self- representation – the possibility to think and speak of 'me' or 'I' and compare that to 'you'– makes self-consciousness possible. Pure feeling, process, and possibility are connected in Peirce's semiotic philosophy, where signs and cognitive categories are produced, when habits of thirdness mediating between secondness and firstness slowly emerge over time. Kull et al. write about developing biosemiotics from this viewpoint:

Theses on the semiotic study provide a collectively formulated set of statements on what biology needs to be focused on in order to describe life as a process based on semiosis, or sign action. An aim of the biosemiotic approach is to explain how life evolves through all varieties of forms of communication and signification (including cellular adaptive behavior, animal communication, and human intellect) and to provide tools for grounding sign theories. (Kull et al. 2009, p. 1)

Thus, in the biosemiotic paradigm, the primary unit of biosemiotic research is the "sign", not the quark, the atom, the molecule or information. What counts as being true is not a simple given. Knowledge of facts presupposes knowledge of theories (categorizations) and of values, just as knowledge of theories and values presupposes knowledge of facts. Inquiry is never disinterested; questions of what, how and why are always intertwined. But, there are objective and reasonable standards independent of any specific human interest, but not independent of all human interest. Laying out principles of practical reasoning and showing how its universal and contextually relative components work together is the proper task of pragmaticists and was central to the core of Peirce's (1958) endeavor.

2.11 Consciousness as Communicated Lifeworlds

I will here go further into describing why the center of the cybersemiotic star model is social semiotic interaction producing intersubjective knowing, instead of an algorithmic pan-informational and pan-computational impersonal function. We deal with conscious impressions and expressions as the processes of sense experience and thinking, before science has divided the world into subjects and objects – yes, even before we have distinguished and compared our individual feelings. Peirce sees perception as an unlimited continuous stream of experiences and his lifelong analysis leads him to identify three basic categories: firstness, as we have seen, is the stream of felt possibilities that gives rise to semiosis, when the momentary different aspects of consciousness (secondness) interact and are related to one another through self-organization and cognition (thirdness). In the quote below he introduces the three categories on a phenomenological basis:

> First, feeling, the consciousness that can be included with an instant of time, passive consciousness of quality, without recognition or analysis; second, consciousness of an interruption into the field of consciousness, sense of resistance, of an external fact, of another something; third, synthetic consciousness, binding time together, sense of learning, thought. (Peirce, CP 1.377)

Firstness, secondness, and thirdness are thus, to Peirce, the three basic states of consciousness, as well as "outer reality" developing in an evolutionary interplay with each other over time. Thus, there is a deep connection between knowledge and time. This can be connected to the foundational importance of irreversibility in non-equilibrium thermodynamics that defies mechanicism with reversible time as the basis of science. There is also a deep connection between our semiotically guided cognitions and the way the outer world is organized, not in the least because we are connected to it both evolutionarily and ecologically and through the way our culture survives. In a Peircean semiotics, phaneroscophy becomes an intersubjective signification sphere. He writes:

> I use the word phaneron to mean all that is present to the mind in any sense or in any way whatsoever, regardless of whether it be fact or figment. I examine the phaneron and I endeavor to sort out its elements according to the complexity of their structure. (Peirce, CP 8.213)

When we are studying socio-communication and acting from the point of language, we are acting in meaningful language studying other meaningful languages. Knowledge is born within the frame of an unrestricted absolute naturalism. This makes it impossible for any of the other specialized approaches to knowledge (in the four arms of the star) to claim that they encompass a model of all of nature. All perception is embedded in consciousness as signs; from the most rudimentary form as pure feeling in firstness to human linguistic self-consciousness. For a basic trans-disciplinary theory, there is no theoretical interest in looking for something more original (material) "behind" the semiotic sense experience than a reality of potential signs. Materiality and energy are just two of the prerequisites for semiosis that must be present at the same time with experience and language games (on the cultural level), and sign games '(at the level of embodiment) (see Brier 1995). We are, thus, immersed in semiotic webs of communication forms, be they verbal or non-verbal.

We cannot completely escape our life world and language, and thereby culture and power. The cultural-mental universe always informs our knowledge. Nevertheless, that does not leave us in anti-realism and radical constructivism, because we accept the evolution of living systems in an ecological environment as another prerequisite. That, on the other hand, does not make us deny the value of first-person experience in a life world or a "signification sphere" as a prerequisite for sense experience, cognition functions, thinking, and second-person communicative experiences. These two last phenomena are viewed as Thirdness processes. In Peirce's 'Syllabus' of 1903, he introduced the subject of Thirdness and Thought:

> Thirdness is found whenever one thing brings about a Secondness between two things. In all such cases, it will be found that Thought plays a part. By thought is meant something like the meaning of a word, which may be 'embodied in', that is, may govern, this or that, but is not confined to any existent. Thought is often supposed to be something in consciousness; but on the contrary, it is impossible ever actually to be directly conscious of thought. It is something to which consciousness may conform, as a written text may conform to it. Thought is rather of the nature of a habit, which determines the suchness of that which may come into existence, when it does come into existence. Of such a habit one may be conscious of a symptom; but to speak of being directly conscious of a habit, as such, is nonsense. (Peirce 1903, p. 269)

Habits develop meaning by directing attention, not to themselves, but to the real connections between phenomena. Habit and thinking, or thought, consist, in Peirce's semiotic paradigm, of concepts that are far more general than those of just psychology or even sociology, because they are connected to sign interaction and creation in general, or what Peirce calls "the semiotic web". According to Peirce's "Law of Mind" article in *The Monist*, habit is a cosmological principle and not only a psychological one (Peirce 1892). We think in or with thought-signs, but not only in or with brains. Semiosis is meaning-making and as such must have a deep ecological foundation:

> Thought is not necessarily connected with a brain. It appears in the work of bees, of crystals, and throughout the purely physical world; and one can no more deny that it is really there, than that the colors, the shapes, etc., of objects are really there. Consistently adhere to that unwarrantable denial, and you will be driven to some form of idealistic nominalism akin to Fichte's. Not only is thought in the organic world, but it develops there. But as there cannot be a General without Instances embodying it, so there cannot be thought without Signs. ... there can be no isolated sign. (Peirce, CP 4.551)

Peirce's phaneroscophy differs from Husserl's phenomenology as it assumes a monistic hylozoist theory of mind and matter as a continuum. In what physics calls "the beginning", mind is an aspect of matter. Peirce also realizes, as in phenomenology, we have to take seriously the observing and knowing ability of the human animal before it started making science. It is the prerequisite that we have to make clear before we can make any evaluation of scientific knowledge. Deely (2001) argues that Peircean semiotics is a perspective that arises from the attempt to make thematic, a ground common to all methods or, one could say, before all methods. From within this point of view, it becomes clear that Peircean semiotics is the study of the action of signs. It is what he calls a cenoscopic science.

Peirce (see, for instance, CP 1.181) divided the sciences into three types: (1) A science of discovery, (2) A science of review, and (3) Practical sciences. It is within the sciences of discovery that we find the concept of "cenoscopy". In the sciences of discovery, Peirce has the following division: (1) (Pure) Mathematics, understood as that science which draws necessary conclusions about hypothetical objects, (2) cenoscopy, which he also calls primary philosophy, is about all positive perceived phenomena in general (inner or outer), which confront a person at every waking moment. This is where he sees his phaneroscophy placed, and (3) ideoscopic sciences, which is his name for the special or positive sciences. They have the purpose of discovering new phenomena through observation and experiments. This is the typical set up in natural sciences, trying to hold several factors stable in order to focus on one or two variables, but often forgetting the cenoscopic background. Peirce also states:

> Cenoscopic science, with its philosophical reflections, precedes the special or idioscopic sciences and is the place from where their individual contributions to man's knowledge of himself and the world should be evaluated and reflected upon. (Peirce, CP 1.288)

Thus, this article is Cenoscopic or today we would call it a piece of work within philosophy of science. However, in Peirce's phaneroscophy, not all elements in the phaneron are being studied, only the elements that are indecomposable are focused upon. These indecomposable phaneroscopic elements exemplify the most basic universal categories, and therefore become philosophically foundational. According to Peirce, the numbers of categories are three and only three (Peirce, CP 1.418, 1.292), as we have already adumbrated in various ways; he adds:

> Of the three Universes of Experience familiar to us all, the first comprises all mere Ideas, those airy nothings to which the mind of poet, pure mathematician, or another might give local habitation and a name within that mind. Their very airy-nothingness, the fact that their Being consists in mere capability of getting thought, not in anybody's actually thinking them, saves their Reality. The second Universe is that of the Brute Actuality of things and

> facts. I am confident that their Being consists in reactions against Brute forces, notwithstanding objections redoubtable until they are closely and fairly examined. The third Universe comprises everything whose being consists in active power to establish connections between different objects, especially between objects in different Universes. Such is everything which is essentially a Sign – not the mere body of the Sign, which is not essentially such, but, so to speak, the Sign's Soul, which has its Being in its power of serving as intermediary between its Object and a Mind. Such, too, is a living consciousness, and such the life, the power of growth, of a plant. Such is a living constitution – a daily newspaper, a great fortune, a social 'movement'. (Peirce, CP 6. 455)

The dynamic interactions between these three categories make up the triadic sign, where the representamen is firstness, the object is secondness, and the interpretant is thirdness. Together they produce meaning in all living sign-producing beings in the form of primary modeling as a signification sphere (or animal life world) and a secondary modeling in the form of sign games (Cobley 2010). In humans, a grammatically ordered generative system of signs obtains a special social function as the type of modeling system we call "natural language". Language emerged as an evolutionary adaptation over two million years ago. Maybe it started as a mute semiotic modeling system in Homo Habilis. Peirce's semiotics is a kind of double hypothetical realism, since he believes in a reality, which is partly independent from the observer. At the same time, he believes that the embodied observer is a product of this same reality, which thus anchors the result of scientific investigations in a realist evolutionary framework including an ontological place for the phaneroscopic first-person experience (Peirce and Turrisi 1997).

Peirce argues that it is not possible for us to contemplate the immediate immense stream of consciousness that is the 'now' in the 'now'. We can only know the 'now' by attaching signs to it afterwards, and this process is connected to the arrow of time (Brier 2008b). Aristotle wrote that the universe is the place of all things, but it does not have a place of its own. Thus, theories of what the universe is "placed" in and to which degree it is closed is a crucial area of investigation these days. One direction of research focuses on the new theories of multiverses, where billions of universes might exist, but have no empirical contact with each other whatsoever (Carr 2007). The void is not a "something" but a "no-thing". The concepts of nothingness and emptiness are central to Peirce's philosophy, as well as Spencer-Brown's evolutionary theory of how form come into existence in *Laws* of *Form* (1969). Peirce, as well as Spencer Brown's (1979) theories of a dynamic emptiness around and before the material universe are interesting candidates for a new transdisciplinary ontology and may fit well with John Archibald's ontological interpretation of Bohr in a theory of a participatory universe (Brier 2009a; Wheeler 1994, 1998).

Wheeler (1994, 1998) argues that reality exists not on the basis of physical particles alone, but rather because of our acts of observing the universe. In a Peircean framework, observation is based on semiotic interpretation. Where Wheeler formulates his philosophy as "it from bit", a Peircean formulation would be "things from signs". In Wheeler's work and that of other physicists like Stapp's (2007) interpretations of the laws of quantum mechanics, our interactive measurements in experiments at the quantum level influence the universe at such fundamental levels that they might have serious consequences also on a macro level[7]. Based on his interpretation of many delayed choice experiments, Wheeler suggests that the universe could be built like an enormous feedback loop between our conscious mind and pre-physical reality, contributing to the ongoing creation of the present and the future state of reality. He even goes so far as to include the past as well. However, unfortunately – as with most physicists – his philosophy does not deliver a theory of first- person consciousness and its place and emergence in nature and its dependence on signs and semiosis. Therefore, I suggest that it is being replaced by Peircean semiotics.

We are, thus, in this evolutionary ontological theory of Peirce and Wheeler, part of a universe that is still developing and rearranging itself including its own beginning! Nicolescu – who is also a quantum physicist – promotes, like Peirce does, the theory that consciousness is a vital and active part of the wholeness of the universe (Nicolescu 2002, pp. 65–66). The subjective and the objective side of nature make up the whole of reality to an integrated whole based in what Nicolescu calls trans-nature or the zone of non-resistance. Wheeler's view moves the mystery of creation from being not only something in a very distant past upon which we have no influence, to being something that also goes on in the living present. As such, he is close to Peirce's evolutionary concept of hylozoism. In philosophy, "hyle" refers to matter or stuff; the material cause underlying change in Aristotelian philosophy. It is that which remains the same in spite of the changes in forms. In opposition to Democritus' atomistic ontology, hyle in Aristotle's ontology is a plenum or a sort of field, like in quantum field theory. Aristotle's world is an uncreated eternal cosmos, but Peirce used the term in an evolutionary philosophy of a world that has an end and a beginning. Hylozoism – in this context – is the philosophical conjecture that all material things possess life. Hylozoism is different from the panpsychist idea of everything possessing a soul. Instead it attributes some form of sense ability to all matter, very much like Whitehead's pan-experientialism. Hylozoism is not a form of animism either, as the latter tends to view life as taking the form of discrete spirits. Scientific hylozoism is a protest against a mechanical view of the world as dead, but at the same time, synechism upholds the idea of a unity between organic and inorganic nature and derives all actions of both types of matter from natural causes. We are the systems developed in and by the universe that are most highly-developed to make the universe look at itself. As the universe in its fundamental quantum level is still partly undetermined, it is in an ongoing re-arranging process of building itself (even all the way back to the Big Bang). Rugh and Zinkernagel (2009), for instance, doubt the claims of universal time. Nicolescu explains this further when he writes: "Nature seems more like a book in the process of being written: the book of Nature is therefore not so much to be read as experienced, as if we are participating in the writing of it" (Nicolescu 2002, p. 65). That also seems to be Wheeler's view, as well as Peirce's (Davies 2004).

[7] Remember Wheeler considered every measurement-interaction as an "observation". Thus, we are not speaking of the consciousness of the experimenter interacting with the quantum process.

Thus, it simply does not make any sense to ask if the universe would exist if there were no observers, because there would be no knowledge if there were no observers, no language, culture and meaning. This reflective observation also puts an interesting limit to the scope of scientific knowledge. It is a knowledge produced inside the universe in time and space. It is an important part of Wheeler's theory that humans are not only observers but creative participants.

The theory of the participatory universe therefore raises a fundamental problem of whom or what qualifies to be an observer or a thinking agent (Brier 2007, 2009a). New foundational theories of agency and the quality necessary to be an observer have appeared are being developed (Sharov 2010; Arrabales et al. 2009). That problem cannot be solved here, but seems to be related to Peirce's idea of semiosis – the ability to make signs and interpret them meaningfully – as not only being limited to humans, but including all living systems with a blurry border to the precursor systems of life, making thinking something that goes on in an ecological systemic context like Bateson also views it (Brier 2008c).

2.12 The Self-Organizing Universe

I agree with Bateson (1972) and Maturana (1988a, b) that we must start our understanding of information with the process of knowing. Bateson's definition of information as a difference that makes a difference is very fruitful. His problem is that he nearly makes every cybernetic system a communicator and a knower, be it a homeostatic machine, an organism or an ecosystem or organization. However, the big difference between computers and humans is this embodied field of meaning in which human communication operates. The paradox is that the sciences think this domain of awareness, sense experiences and meaning appears later in evolution than energy, matter and information, but we have also shown that it is the prerequisite for the intersubjective knowing process, from which the whole idea of science springs. The irreversible time of evolutionary explanation works one way (outside and in from the arms of the Cybersemiotic Star model) and the explanation of the nature of knowledge and science works the opposite way (from the middle and out in the arms of the model). The production of knowledge seems to be like a kind of breathing in and out; an ongoing process.

As I have argued above, I object to the use of the term "nature", as well as the human body, to define only what physico- chemical sciences can describe. What we can measure inter-subjectively is a part of the reality we call nature; meaning that it has some kind of existence more or less independent of the individual human being, though we are still connected to all other things and bodies by being in the same world or Nature and made by the same "stuff". I see no reason in such a non-reductionist transdisciplinary paradigm to assume that physics has a special privileged position in explaining what this universal "stuff" is. With Peirce, I prefer the concept of hylé, which was fundamental to Aristotle 's philosophy, but which Peirce moved into an evolutionary semiotic process oriented paradigm. On the matter of expanding our ontological basis to construct a transdisciplinary Wissenschaft, I suggest we redefine this "basic stuff" as hylè in a way that can encompass this evolutionary monistic way of understanding the world with quantum field theory. When science reifies this substance we call matter to be inert components in an atomistic thinking (devoid of life and mind, and subject only to mechanical and statistical laws) and creates a worldview, where everything – including life and mind – comes into being through the self-organization of matter through evolution, this move is clearly self-contradictory as it leaves out the observer (Fink 2006; Brier 2010a). The development of organisms is entangled with triadic semiosis, but a semiosis that is broader than life and already at work prior to the emergence of life in bringing about the changes of the physical surroundings, which made the emergence of life possible in the first place. This is what Deely (2001) calls physiosemioitcs.

2.13 Conclusion

Let us return to Kant's quote on nature and free will and continue it a little further. Kant writes about the contradiction between free will and a lawful view of nature:

> It is an indispensable problem of speculative philosophy to show that its illusion respecting the contradiction rests on this, that we think of man in a different sense and relation when we call him free, and when we think of him as subject to the laws of nature ... It must therefore show that not only can both of these very well co -exist, but that both must be thought of as necessary united in the same subject. (Kant 1909, p. 76)

I think it is the kind of work we have pursued here, which moves towards a Wissenschaft of consciousness and the human body, is living up to Kant's formulation of the basic deep problem of human knowing in such ways that it is able to include mental events in an absolute naturalism, which I believe is necessary for the development of an integrative paradigm of psycho-somatic and social medicine. In the Cybersemiotic star model, knowledge is developing in all four 'arms' at the same time. Results from empirical research falsify our theories and force us to theoretically reconfigure our present knowledge into new theories and models to cope with the knowledge and experiences we have now gained. The challenge is now to reintegrate all of the different research paradigms we have developed and specialized into a greater whole. But to make such a shift one needs to develop an ontology that can encompass the ontologies of all four views in a somewhat relativized version and combining them into a transdisciplinary setting. This is what I have sketched here.

I have suggested defining the point of departure in C.S. Peirce's pragmaticist, evolutionary semiotic process philosophy, where semiotic social interactions between embodied – more or less – free minds in nature are viewed as the central process of knowledge production, which is also behind the selfsame "sciences" that attempt to explain meaning production and consciousness. Thus, the view does not deny the necessity of the brain to produce consciousness. However, for a brain to be part of the production of experience, it has to be connected to a feeling body of living flesh and a peripheral nervous system with specialized sense organs. We can model our bodies on animal's behavior, but not its feelings per se. Still, we must accept that first person feelings and perceptions are prerequisites to having consciousness, free will, language, and cultural meaning, which are necessary in order to produce ordinary common sense knowledge, of which scientific knowledge is a culturally developed refinement. This, however, makes it impossible to view mind and brain as two independent entities that have simple, independent and different causal relationships. They are deeply interconnected, which is also shown in Peirce's synechist view of the "basic stuff" of reality as hylé. Thus, we return to a partly Aristotelian view, adding evolution plus modern phaneroscophy (Peircean phenomenology) and biology in the form of biosemiotics. This could be the theoretic foundation for a more semiotic and holistic-based transdisciplinary medical research tradition.

References

Arrabales, R., Ledezma, A., & Sanchis, A. (2009). ConsScale: A pragmatic scale for measuring the level of consciousness in artificial agents. *Journal of Consciousness Studies, 17*(3–4), 131–164.

Barbieri, M. (2001). *The organic codes: The birth of semantic biology*. Cambridge: Cambridge University Press.

Barbieri, M. (2006). Life and semiosis: The real nature of information and meaning. *Semiotica, 2006*(158), 233–254.

Barbieri, M. (2009). Three types of semiosis. *Biosemiotics, 2*(1), 19–30.

Barbieri, M. (2011). Origin and evolution of the brain. *Biosemiotics, 4*(3), 369–399.

Barrow, J. D., Davies, P. C., & Harper, C. L., Jr. (2004). *Science and ultimate reality: Quantum theory, cosmology, and complexity*. Cambridge: Cambridge University Press.

Bateson, G. (1972). *Steps to an ecology of mind: Collected essays in anthropology, psychiatry, evolution, and epistemology*. Chicago: University of Chicago Press.

Bennett, M. R., Hacker, P. M. S., & Bennett, M. (2007). The philosophical foundation of neuroscience. In M. Bennett, D. Dennett, P. Hacker, & J. Searle (Eds.), *Neuroscience and philosophy: Brain, mind, and language*. New York: Columbia University Press.

Bhaskar, R.A. (1997 [1975]). *A realist theory of science*. London: Verso.

Bhaskar, R. (1998). *The possibility of naturalism: A philosophical critique of the contemporary human sciences*. London: Routledge.

Blackmore, S. (2000). *The meme machine*. New York: Oxford University Press.

Boden, M. A. (1990). Escaping from the Chinese room. In M. A. Boden (Ed.), *The philosophy of artificial intelligence*. Oxford: Oxford Readings in Philosophy.

Boler, J. F. (1963). *Charles Peirce and scholastic realism: A study of Peirce's relation to John Duns Scotus*. Seattle: University of Washington Press.

Brandom, R. (1994). *Making it explicit: Reasoning, representing, and discursive commitment*. Cambridge: Harvard University Press.

Brenner, J. E. (2008). *Logic in reality*. Berlin/New York: Springer.

Brenner, J. E. (2011). Information in reality: Logic and metaphysics. *tripleC, 9*(2), 332–341.

Brier, S. (1980). *Der ønskes analyseret (evt. v.h.a. egne undersøgelser), om hierarki og sandsynlighedsbetragtninger i beskrivelsen af adfærd kan anvendes i-og udbygge-een eller flere motivationspsykologiske teorier eller modeller* [In Danish]. Copenhagen/Denmark: Copenhagen University.

Brier, S. (1995). Cyber-semiotics: On autopoesis, code-duality and sign games in biosemiotics. *Cybernetics and Human Knowing, 3*(1), 3–14.
Brier, S. (1999). Biosemiotics and the foundation of cybersemiotics: Reconceptualizing the insights of ethology, second- order cybernetics, and Peirce's semiotics in biosemiotics to create a non-Cartesian information science. *Semiotica, 127*(1–4), 169–198.
Brier, S. (2000). Transdisciplinary frameworks of knowledge. *Systems Research and Behavioral Science, 17*(5), 433–458.
Brier, S. (2001). Cybersemiotics and Umweltlehre. *Semiotica, 134*(1–4), 779–814.
Brier, S. (2004). Cybersemiotics and the problem of the information-processing paradigm as a candidate for a unified science of information behind library and information science. *Library Trends, 52*(3), 629–657.
Brier, S. (2007). Applying Luhmann's system theory as part of a transdisciplinary frame for communication science. *Cybernetics and Human Knowing, 14*(2–3), 29–65.
Brier, S. (2008a). *Cybersemiotics: Why information is not enough*. Toronto: University of Toronto Press.
Brier, S. (2008b). A Peircean panentheist scientific mysticism. *International Journal of Transpersonal Studies, 27*, 20–45.
Brier, S. (2008c). Bateson and Peirce on the pattern that connects and the sacred. In J. Hoffmeyer (Ed.), *A legacy for living systems: Gregory Bateson as precursor to biosemiotics*. Berlin: Springer.
Brier, S. (2008d). The paradigm of Peircean biosemiotics. *Signs, 2*, 30–81.
Brier, S. (2009a). Cybersemiotic pragmaticism and constructivism. *Constructivist Foundations, 5*, 19–38.
Brier, S. (2009b). Levels of cybersemiotics: Possible ontologies of signification. *Cognitive Semiotics, 4*, 28–62.
Brier, S. (2010a). Cybersemiotics and the question of knowledge. In G. Dodig-Cmkovic & M. Burgin (Eds.), *Information and computation*. Hackensack: World Scientific Publishing Co.
Brier, S. (2010b). Cybersemiotics: An evolutionary world view going beyond entropy and information into the question of meaning. *Entropy, 12*(8), 1902–1920.
Brier, S. (2011). Ethology and the Sebeokian way from zoosemiotics to cybersemiotics. In P. Cobley, J. Deely, K. Kull, & S. Petrilli (Eds.), *Semiotics continues to astonish: Thomas A. Sebeok and the doctrine of signs*. Berlin: Walter de Gruyter.
Brier, S. (2014). The riddle of the sphinx answered: On how C. S. Peirce's transdisciplinary semiotic philosophy of knowing links science, spirituality and knowing. In C. Tandy (Ed.), *Death and anti-death: Vol 12. One hundred years after Charles S. Peirce (1839–1914)*. Ann Arbor: Ria University Press.
Brier, S., & Joslyn, C. (2013). What does it take to produce interpretation? Informational, peircean, and code-semiotic views on biosemiotics. *Biosemiotics, 6*(1), 143–159.
Carr, B. (2007). *Universe or multiverse?* Cambridge: Cambridge University Press.
Chaitin, G. (2010). Mathematics as a biological process. In G. Dodig-Cmkovic & M. Burgin (Eds.), *Information and computation*. Hackensack: World Scientific Publishing Co.
Chalmers, D. J. (1995). Facing up to the problem of consciousness. *Journal of Consciousness Studies, 2*(3), 200–219.
Chalmers, D. J. (1996). *The conscious mind: In search of a fundamental theory: In search of a fundamental theory*. Oxford: Oxford University Press.
Churchland, P. (2004). Eliminative materialism and the propositional attitudes. In J. Heil (Ed.), *philosophy of mind: A guide and anthology* (pp. 382–400). Oxford: Oxford University Press.
Clayton, P. D. (2004). Emergence: Us from it. In J. D. Barrow, P. C. Davies, & C. L. Harper (Eds.), *Science and ultimate reality: Quantum theory, cosmology, and complexity* (pp. 577–606). Cambridge: Cambridge University Press.
Cobley, P. (2010). Cybersemiotics and human modeling. *Entropy, 12*(9), 2045–2066.
Cowley, S. J., Major, J. C., Steffensen, S. V., & Dinis, A. (2010). *Signifying bodies: Biosemiosis, interaction and health*. Braga: Faculty of Philosophy of Braga, Portuguese Catholic University.

Damasio, A. R. (1994). *Descartes' error: Emotion, reason, and the human brain paperback*. New York: Penguin.
Damasio, A. R. (2000). *The feeling of what happens: Body and emotion in the making of consciousness*. New York: Harcourt Incorporated.
Damasio, A. R. (2004). *Looking for Spinoza: Joy, sorrow, and the feeling brain*. New York: Vintage.
Davies, P. (2004). John Archibal Wheeler and the clash of ideas. In J. D. Barrow, P. C. W. Davies, & C. L. Harper (Eds.), *Science and ultimate reality: Quantum theory, cosmology and complexity* (pp. 3–26). New York: Cambridge.
Dawkins, R. (2006). *The selfish gene: 30th anniversary edition*. Oxford: Oxford University Press.
Deacon, T. W. (1998). *The symbolic species: The co-evolution of language and the brain*. New York: W. W. Norton & Company.
Deacon, T. W. (2007). Shannon – Boltzmann – Darwin: Redefining information (part I). *Cognitive Semiotics, 1*, 123–148.
Deacon, T. W. (2008). Shannon – Boltzmann – Darwin: Redefining information (part II). *Cognitive Semiotics, 2*, 169–196.
Deacon, T. (2011). *Incomplete nature: How mind emerged from matter*. New York: W.W. Norton & Company.
Deely, J. (2001). Physiosemiosis in the semiotic spiral: A play of musement. *Sign System Studies, 29*(1), 27–48.
Dennett, D. C. (1991). *Consciousness explained*. New York: Little, Brown and Company.
Dennett, D. C. (2007). Philosophy as naïve anthropology. In M. Bennet, D. Dennet, P. Hacker, & J. Searle (Eds.), *Neuroscience and philosophy: Brain, mind, and language*. New York: Columbia University Press.
Deutsch, D. (2013). It from qubit. In J. D. Barrow, P. C. Davies, & C. L. Harper (Eds.), (2004). *Science and ultimate reality: Quantum theory, cosmology, and complexity*. Cambridge: Cambridge University Press.
Dodig-Crnkovic, G. (2010). The cybersemiotics and info-Computationalist research programs as platforms for knowledge production in organisms and machines. *Entropy, 12*(4), 878–901.
Dodig-Crnkovic, G., & Muller, V. (2011). A dialogue concerning two world systems: Info-computational vs. mechanistic. In G. Dodig-Cmkovic & M. Burgin (Eds.), *Information and computation*. Hackensack: World Scientific Publishing Co.
Donald, M. (1991). *Origins of the modern mind: Three stages in the evolution of culture and cognition*. Cambridge: Harvard University Press.
Drummon, J. J. (2003). The structure of intentionality. In D. Welton (Ed.), *The new Husserl: A critical reader* (pp. 65–92). Bloomington: Indiana University Press.
Durst-Andersen, P. (2011). Linguistic supertypes. In *A cognitive-semiotic theory of human communication*. Berlin/New York: De Gruyter Mouton.
Edelman, G. M., & Tononi, G. (2000). *A universe of consciousness: How matter becomes imagination*. New York: Basic Books.
El-Hani, C. N., Queiroz, J., & Emmeche, C. (2006). A semiotic analysis of the genetic information system. *Semiotica, 160*(1–4), 1–68.
El-Hani, C. N., Queiroz, J., & Emmeche, C. (2009). *Genes, information, and semiosis*. Tartu: Tartu University Press.
Ellis, G. F. (2004). True complexity and its associated ontology. In J. D. Barrow, P. C. Davies, & C. L. Harper (Eds.), *Science and ultimate reality: Quantum theory, cosmology, and complexity* (pp. 607–636). Cambridge: Cambridge University Press.
Ellis, R. D., & Newton, N. (1998). Three paradoxes of phenomenal consciousness: Bridging the explanatory gap. *Journal of Consciousness Studies, 5*(4), 419–442.
Emmeche, C. (1992). Modelling life: A note on the semiotics of emergence and computation in artificial and natural living systems. In T. A. Sebeok & J. Umiker-Sebeok (Eds.), *Biosemiotics: The Semiotic Web 1991* (pp. 77–99). Berlin: Mouton de Gruyter.
Emmeche, C. (2001). Does a robot have an Umwelt? Reflections on the qualitative biosemiotics of Jakob von Uexküll. *Semiotica, 134*(1/4), 653–669.

Emmeche, C. (2003). Biosemiotics. In J. Wentzel Vrede van Huyssteen (Ed.), *Encyclopedia of science and religion* (pp. 63–64). New York: Macmillan Reference.

Emmeche, C. (2013). A-life, organism and body: The semiotics of emergent levels. *Proceedings of the 9th international conference on the simulation and synthesis of living systems (Alife IX); 2004 Sep 12*, (pp. 117–24). Boston.

Esposito, J. L. (1980). *Evolutionary metaphysics: The development of peirce's theory of categories*. Athens: Ohio University Press.

Favareau, D. (2010). *Essential readings in biosemiotics: Anthology and commentary*. Berlin: Springer.

Feyerabend, P. (1975). *Against method: Outline of an anarchistic theory of knowledge*. London: NLB.

Fink, H. (2006). Three sorts of naturalism. *European Journal of Philosophy, 14*(2), 202–221.

Firestone, A. R., Scheurer, P. A., & Bürgin, W. B. (1999). Patients' anticipation of pain and pain-related side effects, and their perception of pain as a result of orthodontic treatment with fixed appliances. *European Journal of Orthodontics, 21*(4), 387–396.

Gadamer, H. G. (1989). *Truth and method*. London: Sheed and Ward.

Haack, S. (1992). Extreme scholastic realism: Its relevance to philosophy of science today. *Transactions of the Charles S. Peirce Society, 28*(1), 19.

Habermas, J. (1987). Excursus on Luhmann's appropriation of the philosophy of the subject through systems theory. In J. Habermas (Ed.), *The philosophical discourse of modernity: Twelve lectures* (pp. 368–385). Cambridge: MIT Press.

Haikonen, P. O. (2007). *Robot brains: Circuits and systems for conscious machines*. New Jersey: John Wiley & Sons.

Haikonen, P. O. (2009). Qualia and conscious machines. *International Journal of Machine Consciousness, 1*(2), 225–234.

Heelan, P. A. (1987). Husserl's later philosophy of natural science. *Philosophy of Science, 54*(3), 36–390.

Heelan, P. A. (1988). *Space-perception and the philosophy of science*. Berkeley: University of California Press.

Heil, J. (2004). *Philosophy of mind: A guide and anthology*. Oxford: Oxford University Press.

Hinde, R. A. (1970). *Animal behavior: A synthesis of ethnology and comparative psychology*. Tokyo: McGraw-Hill Kogakusha.

Hoffmeyer, J. (2008). *Biosemiotics: An examination into the signs of life and the life of signs*. Scranton: University of Scranton Press.

Hoffmeyer, J. (2010). A biosemiotic approach to health. In S. J. Cowley, J. C. Major, S. V. Steffensen, & A. Dinis (Eds.), *Signifying bodies: Biosemiosis, interaction and health* (pp. 21–41). Braga: Faculty of Philosophy of Braga, Portuguese Catholic University.

Hoffmeyer, J., & Emmeche, C. (1991). Code-duality and the semiotics of nature. In M. Anderson & F. Ell (Eds.), *On semiotic modeling* (pp. 117–166). Berlin: Mouton de Gruyter.

Hofstadter, D. R. (2007). *I am a strange loop*. New York: Basic Books.

Husserl, E. (1970). *The crisis of European sciences and transcendental phenomenology: An introduction to phenomenological philosophy*. Evanston: Northwestern University Press.

Husserl, E. (1999). *Cartesianske Meditationer: En Indføring i Fænomenologien*. Copenhagen: Hans Reitzel.

Husserl, E., & Bundgard, P. F. (1997). *Fænomenologiens Idé: Fem Forelæsninger*. Copenhagen: Hans Reitzel.

Jackendoff, R. (1987). *Consciousness and the computational mind* (p. 356). Cambridge: MIT Press.

de Jong, A. E., & Gamel, C. J. (2006). Use of a simple relaxation technique in burn care: Literature review. *Advanced Nursing, 54*(6), 710–721.

Kant, E. (1909). *Fundamental principle of the metaphysics of morals* (trans: Abbot, T.K.). London: Forgotten Books.

Kauffman, S. A. (1993). *The origins of order: Self-organization and selection in evolution*. Oxford: Oxford University Press.

Kauffman, S. A. (2012). From physics to semiotics. In S. Rattasepp & T. Bennett (Eds.), *Gatherings in biosemiotics*. Estonia: Tartu University Press.
Kauffman, S. A., Logan, R. K., Este, R., Goebel, R., Hobill, D., & Smulevich, I. (2007). Propagating organization: An inquiry. *Biology and Philosophy, 23*, 27–45.
Kim, J. (2007). *Physicalism, or something near enough*. Princeton: Princeton University Press.
Kuhn, T. S. (1996). *The structure of scientific revolutions*. Chicago: University of Chicago Press.
Kull, K., Deacon, T., Emmeche, C., Hoffmeyer, J., & Stjernflet, F. (2009). Theses on biosemiotics: Prolegomena to a theoretical biology. *Biological Theory, 4*(2), 167–173.
Kultgen, J. H. (2009). The "future metaphysics" of Peirce and whitehead. *Kant-Studien, 51*(1–4), 285–293.
Kuppers, B. O. (1990). *Information and the origin of life*. London: MIT Press.
Latour, B. (1988). *The pasteurization of France*, (trans: Sheridan, A. & Law, J.). London: Harvard University Press. (Original work published 1984).
Latour, B. (1993). *We have never been modern*. Cambridge: Harvard University Press.
Latour, B. (2004). *Politics of nature: How to bring the sciences into democracy*. Cambridge: Harvard University Press.
Latour, B. (2007). *Reassembling the social: An introduction to actor-network-theory*. Oxford: OUP Oxford.
Leibniz, G.W. (1898). *The monadology* (trans: Latta, R.). Retrieved from: http://home.datacomm.ch/kerguelen/monadology/monadology.html.
Levine, J. (1983). Materialism and qualia: The explanatory gap. *Pacific Philosophical Quarterly, 64*, 354–361.
Leydesdorff, L. (2012). Communication-theoretical specification of the 'Phenomena' of Husserl's phenomenology. In E. B. Pires (Ed.), *Public space, power and communication*. Coimbra: University of Coimbra.
Leydesdorff, L., & Franse, S. (2009). The communication of meaning in social systems. *SystemsResearch and Behavioral Science, 26*(1), 109–117.
Libet, B. (1993). The neural time factor in conscious and unconscious events. In G. Bock & J. Marsh (Eds.), *Experimental and theoretical studies of consciousness*. New Jersey: Wiley.
Lorenz, K. (1971). *Studies in animal and human behavior*. Cambridge: Harvard University Press.
Luhmann, N. (1990). *Essays on self-reference*. New York: Columbia University Press.
Luhmann, N. (1995). *Social systems*. Redwood City: Stanford University Press.
Maturana, H. R. (1980). *Autopoiesis and cognition: The realization of the living*. Berlin: Springer.
Maturana, H. R. (1983). What is it to see? *Archivos de Biología y Medicina Experimentales, 16*(3–4), 255–269.
Maturana, H. R. (1988a). Reality: The search for objectivity or the quest for a compelling argument. *The Irish Journal of Psychology, 9*(1), 25–82.
Maturana, H. R. (1988b). Ontology of observing: The biological foundation of self consciousness and the physical domain of existence. *The Irish Journal of Psychology, 9*(1), 25–82.
Maturana, H. R., & Varela, F. J. (1987). *The tree of knowledge: The biological roots of human understanding*. Boston: Shambhala Publications, Incorporated.
McDowell, J. H. (1998). *Mind, value, and reality*. Cambridge: Harvard University Press.
McGinn, C. (2000). *The mysterious flame: Conscious minds in a material world*. New York: Basic Books.
Merleau-Ponty, M. (1962). *Phenomenology of perception*. London: Routledge & Kegan Paul.
Merleau-Ponty, M. (1963). *The structure of behavior*. Pittsburgh: Duquesne University Press.
Merleau-Ponty, M. (2003). *Nature: Course notes from the Collège de France*. Evanston: Northwestern University Press.
Miller, R. W. (1975). Propensity: Popper or Peirce? *British Journal for the Philosophy of Science, 26*(2), 123–132.
Monod, J. (1971). *Chance and necessity: An essay on the national philosophy of modern biology*. New York: Knopf.
Nagel, T. (1974). What is it like to be a bat? *The Philosophical Review, 83*(4), 435–450.

Nelson, K. (1998). *Language in cognitive development: The emergence of the mediated mind*. Cambridge: Cambridge University Press.
Neurath, M. (1983). *Philosophical papers 1913-1946: With a bibliography of Neurath in English*. Berlin: Springer.
Nicolescu, B. (2002). *Manifesto of transdisciplinarity*. Albany: State University of New York Press.
Niiniluoto, I. (1984). *Is science progressive?* (Vol. 177). Berlin/New York: Springer.
Peirce, C. S. (1892). The doctrine of necessity examined. *The Monist, 2*(3), 321–337.
Peirce, C. S. (1903). *A syllabus of certain topics of logic*. Boston: Alfred Mudge & Son.
Peirce, C.S. (1931, 1935, 1958). *The collected papers of Charles Sanders Peirce*. Intelex CD-ROM edition (1994), reproducing vols. I–VI. In C. Hartshorne & Weiss, P. (Eds.), Cambridge: Harvard University Press. 1931–1935 and vols. VII–VIII. In A. W. Burks (Ed.), same publisher, 1958. Citations give volume and paragraph number, separated by a period.
Peirce, C. S. (1958). *Collected papers of Charles sanders Peirce*. Cambridge: Harvard University Press.
Peirce, C. S. (1980). *New elements of mathematics*. Amsterdam: Walter De Gruyter Inc.
Peirce, C. S., & Turrisi, P. A. (1997). *Pragmatism as a principle and method of right thinking: The 1903 harvard lectures on pragmatism*. Albany: State University of New York Pr.
Penrose, R. (1989). *The emperor's new mind: Concerning computers, minds, and the laws of physics*. Oxford: Oxford University Press.
Penrose, R. (1994). *Shadows of the mind: A search for the missing science of consciousness*. London: Oxford University Press.
Penrose, R. (1999). *The emperor's new mind: Concerning computers, minds, and the laws of physics*. Oxford: Oxford University Press.
Petitot, J. (1999). *Naturalizing phenomenology: Issues in contemporary phenomenology and cognitive science*. Redwood City: Stanford University Press.
Popper, K. R. (1972). *Objective knowledge: An evolutionary approach*. Oxford: Oxford at the Clarendon Press.
Popper, K. R. (1979). *Three worlds, the Tanner lecture on human values*. Delivered at the University of Michigan, April 7, 1978. Retrieved from http://tannerlectures.utah.edu/_documents/a-to-z/p/popper80.pdf. (visited 12-26-2014).
Popper, K. (1959). *The Logic of Scientific Discovery*, London: Routledge.
Popper, K. R., & Eccles, J. C. (1977). *The self and its brain: An argument for interactionism*. Berlin: Springer International.
Prigogine, I. (1997). *The end of certainty*. New York: Free Press.
Prigogine, I., & Stengers, I. (1984). *Order out of chaos: Man's new dialogue with nature*. New York: Bantam Doubleday Dell.
Rorty, R. (1980). *Philosophy and the mirror of nature*. Oxford: Basil Blackwell.
Rugh, S. E., & Zinkernagel, H. (2009). On the physical basis of cosmic time. *Studies in History and Philosophy of Modern Physics, 40*(1), 1–19.
Schrödinger, E. (2006). *What is life with mind and matter and autobiographical sketches*. Cambridge: Cambridge University Press.
Searle, J. R. (1980). Minds, brains, and programs. *Behavioral and Brain Sciences, 3*(3), 417–457.
Searle, J. (1989). *Minds, brains and science*. London: Penguin.
Searle, J. (2007). Putting consciousness back in the brain. In M. R. Bennett (Ed.), *Neuroscience and philosophy: Brain, mind, and language*. New York: Columbia University Press.
Searle, J. R., Dennett, D. C., & Chalmers, D. J. (1997). *The mystery of consciousness*. New York: New York Review of Books.
Sebeok, T. A., & Danesi, M. (2000). *The forms of meaning: Modeling systems theory and semiotic analysis*. Berlin: Walter de Gruyter.
Sellars, W. (1991). *Science, perception and reality*. Atascadero: Ridgeview Publishing Company.
Sharov, A. A. (2010). Functional information: Towards synthesis of biosemiotics and cybernetics. *Entropy, 12*(5), 1050–1070.

Smolin, L. (2014). *Time reborn: From the crisis of physics to the future of the universe*. London: Allan Lane.

Sonesson, G. (2009). New considerations on the proper study of man -, and marginally, some other animals. *Cognitive Semiotics, 4*, 133–168.

Spencer Brown, G. (1979). *Laws of Form*. New York: E. P. Dutton.

Spiegelberg, H. (1965). *The phenomenological movement: A historical introduction*. Boston: M. Nijhoff.

Stapp, H. P. (2007). *Mindful universe: Quantum mechanics and the participating observer*. Berlin: Springer.

Steffensen, S. V., & Cowley, S. J. (2010). Signifying bodies and health: A non-local aftermath. In S. J. Cowley, J. C. Major, S. V. Steffense, & A. Dinis (Eds.), *Signifying bodies: Biosemiosis, interaction and health* (pp. 331–355). Braga: The Faculty of Philosophy of Braga Portuguese Catholic University.

Stjernfelt, F. (2014). *Natural propositions: The account of Peirce's doctrine of decisigns*. Boston: Decent Press.

Thibault, P. (2004). *Brain, mind and the signifying body: An ecosocial semiotic theory*. London: Continuum.

Thompson, E. (2003). *The problem of consciousness: New essays in phenomenological philosophy of mind*. Alberta: University of Calgary Press.

Tinbergen, N. (1973). *Animal in its world (explorations of an ethologist, 1933-1972): Vol 1. Field studies*. Crows Nest: Allen & Unwin.

von Uexkül, J. (1957). A stroll through the worlds of animals and men: A picture book of invisible worlds. In C. H. Schiller (Ed.), *Instinctive behavior: The development of a modern concept* (pp. 5–80). Madison: International Universities Press.

von Uexkül, J. (1982). The theory of meaning. *Semiotica, 42*(1), 25–82.

Wheeler, J. A. (1994). *At home in the universe*. New York: American Institute of Physics.

Wheeler, J. A. (1998). *Geons, black holes, and quantum foam: A life in physics*. New York: W. W. Norton & Company Incorporated.

Whitehead, A. N., & Griffin, D. R. (1978). *Process and reality: An essay in cosmology* (2nd ed.). New York: Free Press.

Wilson, E. O. (1999). *Consilience: The unity of knowledge*. New York: Vintage Books.

Wittgenstein, L. (1958). *Philosophical investigations* (1st ed., trans: Anscombe, G.E.M.). New York: Macmillan Publishing Inc.

Wolfram, S. (2002). *New kind of science: Notes from the book*. Champaign: Wolfram Media Incorporated.

Zhou, L., & Brier, S. (2014a). Philosophy of information in Chinese style: Review of Wu Kun's Philosophy of information: Theory, system, method. *Cybernetics and Human Knowing, 21*(4), 83–97.

Zhou, L., & Brier, S. (2014b). Philosophy of information in Chinese Style? An evaluation on Wu Kun's philosophy of information, Chinese title: 中国特色的信息哲学?评邬焜的信息哲学, pp 81–89 in the *proceedings of the international conference Philosophical Spirit of Information Age, at Xi'an Jiaotong University in Xi'an, China, November 2014*.

Zlatev, J. (2009a). The semiotic hierarchy: Life, consciousness, signs and language. *Cognitive Semiotics, 2009*(4), 169–200.

Zlatev, J. (2009b). Levels of meaning embodiment and communication. *Cybernetics and Human Knowing, 16*(3–4), 149–174.

Zuse, K., & Raum R. (1967). *Electroniche Datenverarbeitung* (vol. 8, pp. 336–344). Breunschweig: Vieweg.

Chapter 3
Some Reflections on Non-substance Bound Healing Effects and the Concept of Narrative Medicine

Carl Eduard Scheidt

3.1 The Concept of Narrative Medicine

The term narrative medicine refers to the idea that the construction of meaning is an essential component of medicine. While biomedicine is focusing on the observation of symptoms as indicators of biological dysfunction in the methodological framework of empirical science, narrative medicine is focusing on the interactional processes in which "symptoms" are reframed in the context of the patient's individual history and experience (Charon 2001). Methodologically, narrative medicine advocates a hermeneutic approach in addition to the empirical observational model. The Humanities, from the perspective of narrative medicine, are considered as a valuable contribution adding the dimension of history and the cultural and symbolic mediation of meaning to the understanding of health and disease and the individual patient's suffering. In stressing the relevance of communication in the doctor-patient relationship, narrative medicine is overlapping with the concept of psychosomatic medicine.

It is interesting to note that theories of narrative medicine have only rarely made explicit reference to semiotic theory, but instead have focused on the concept of narration (Labov and Waletzky 1967). The reason might be that narration – in contrast to semiotic theory – is centered around the process of meaning construction in spoken language although narratives emerge in a variety of diverse contexts ranging from esthetically-coded written formats to every day conversation. The doctor–patient interaction can be considered as a form of every day conversation in which narratives are built around experiences of the body and their potential meanings for health and disease. Narratives have been studied also with regard to a social theory of identity and the construction of life scripts and biography (Bruner 1987; Ezzy

C.E. Scheidt (✉)
Department of Psychosomatic Medicine and Psychotherapy, Albert Ludwigs University, University Medical Center, Freiburg, Germany
e-mail: carl.eduard.scheidt@uniklinik-freiburg.de

F. Goli (ed.), *Biosemiotic Medicine*, Studies in Neuroscience, Consciousness and Spirituality 5, DOI 10.1007/978-3-319-35092-9_3

1998; Schiffrin 1996; Mc Adams 1993; Mc Adams et al. 2006). This aspect is of great importance for chronic illnesses, which require life-long processes of coping involving the redefinition of personal goals, norms, and identity. Narration in this context has been increasingly acknowledged as a form of coping independent from the classical problem – or emotion-focused coping strategies (Folkman and Moskowitz 2004). Substantial empirical evidence is supporting the view that the narration of adverse life experiences has strong and lasting (positive) effects on psychological and physical well-being (Pennebaker 1993; Pennebaker et al. 1997; Frattaroli 2006; Pennebaker and Chung 2007).

In the subsequent paragraphs, we will discuss some of the underlying reasons for the phenomenon of non-substance-bound healing effects of language. This will require to discuss differences between narration and placebos. Just as narration, placebo is also a phenomenon associated with non-substance bound healing effects involving meaning and belief systems. However, placebo is involving meaning systems of a rather different kind. While the implicit belief of narration is that there is a world of shared meaning that allows humans to communicate with each other, the basic implicit belief system of placebo is that there is a material world, in which material things can influence other material things. However, despite of these discrepancies in the underlying belief systems, surprisingly narratives and placebos also are sharing some common ground. We hypothesize that the effects of both, narratives and placebos have their origins in developmental trajectories, which can be best understood in the framework of attachment theory in order to elucidate their effects on psychological and physical health.

3.2 Attachment and Physical Health

During the past decades, substantial evidence in the social neurosciences has been accumulated demonstrating the complex effects of social experiences on the structure and function of the brain (Schore 1994). The question how social experiences are transformed into biological processes in the brain is one of the most intriguing areas of neurobiological research. In this context, attachment theory offers a framework in order to conceptualize the developmental trajectories linking early attachment, social behavior, and physiological resilience in later stages of the development. Attachment theory also describes the interactional basis of meaning construction in the early infant-care-giver interaction. In the subsequent paragraphs, we will summarize some findings of attachment theory, which may shed light on the developmental origins of the non-substance bound healing effects of narration and placebo.

3.3 Attachment and Physiological Homeostasis

A milestone of attachment research was the observation that young children in times of threat and distress seek for the reassurance of their primary care givers. The capacity of the primary care givers to relieve the infant's distress is built on the repetitious experience of caring and supportive responses, which are adequately tuned to the child's needs. The notion that MS Ainsworth (1978) has coined for this quality of responsiveness of the primary care giver's reaction to the child's attachment (distress) signals is maternal sensitivity. The continuous experience of prompt and adequate responses of the primary care givers to the child's needs allows the child to anticipate a sensitive response in times of distress in the future. If this anticipation has been integrated into the child's procedural memory, the mother or other primary care givers can be used as a secure base. This implies the expectation that the attachment figure will be responding promptly and adequately to what is needed. Attachment research has extensively investigated infants at 12 months of age and their response to a specific form of threat, namely separation. Separation indeed is a highly ecologically valid form of stress in this age group, due to the relative neurobiological and psychological immaturity, which includes the lack of stable object representations. However, there are all sorts of stress activating the attachment behavioral system in addition to separation. They may all equally activate the attachment behavioral system as one central psychobiological system facilitating the adaptation to danger and threat. Numerous studies have provided evidence that – as Bowlby (1975) had suggested – the attachment behavioral system is a central component in the maintenance of physiological homoeostasis, that is, in protecting the organism against a dysregulation of the stress response. Differences of the quality of the attachment relationship between infant and primary care givers were associated with differences of in the response patterns of the stress axis. This refers to animal studies as well as to research in human infants (Gunnar et al.1991). Meaney and colleagues (Francis and Meaney 1999) used a research paradigm that allowed them to study long-term effects of early deprivation in bush rats. Animals that had experienced less licking behavior – an indicator of maternal caring – during their postnatal development were more vulnerable to stress, showing more anxious behavior and a prolonged stress response in later stages of their development. Current evidence therefore is suggesting a strong role of attachment for the regulation of the physiological homoeostasis, linking individual differences of attachment representation with differences of the physiological response patterns to stress (ibid.).

The concept of attachment in this context is multifaceted. It refers to the observation of patterns of interaction between infants and their primary care givers. This interaction is emerging as the function of motivational systems which have developed during evolution; the attachment behavioral system of the infant and the motivational system of caring behavior on the side of the primary care giver. In a series of studies in the 1980s, Hofer (1984) in the 80s had investigated the psychobiological components of early attachment relationships using an animal model. Hofer described what he termed as "hidden regulators" in the attachment relationship, specific cues of the mother and her caring activity directly regulates the physiological response of the infant animal. One important finding of this research was to demonstrate that attachment interaction is shaping and conditioning biological response patterns. In sum, there is growing evidence supporting the concept that social experiences exert a strong and lasting influence, not only on the psychosocial trajectory of development, but also on individual differences of the stress response. This allows understanding of the influence of social experiences on health and disease and also their potential as non-substance bound healing effects . Social experiences are the basis from which the effects of narration and also of placebo can be derived.

3.4 Attachment and the Doctor-Patient Relationship

From the perspective of narrative medicine, the communication between doctor and patient is the central component of clinical medicine. There are at least three reasons to assume that the doctor-patient relationship should be considered as an attachment relationship. The first reason is its asymmetry, which is due to the fact that one participant of this encounter i s suffering and seeking help, whereas the other is providing skills, experience and knowledge in order to relieve the complaints. This asymmetry of the relationship may foster the forms of transference to which Balint (1964) in his seminal work has drawn attention. In the framework of attachment theory, it can easily be explained how the patient's individual history and experiences with primary care givers are shaping and influencing attitudes towards the doctors and to medical institutions. For instance, a patient with an insecure, dismissing style of attachment might prefer a medical setting or a type of relationship involving less intimacy or closeness and might be expressing his frustration or resentment less open. If this is not recognized, communication might become a source of mutual misunderstanding leading to rejection and anger in the relationship. The influence of attachment patterns on the doctor-patient relationship has been studied recently, but still needs further detailed investigation. Other aspects of the doctor–patient relationship involving attachment issues are the personal involvement; which to a certain degree is inevitable even when technical medical interventions are in the foreground, and also the experience of physical intimacy and closeness during medical assessments, which often is also an unavoidable ingredient component of the medical encounter.

Considering the doctor–patient relationship as a form of attachment contributes to explaining its potentials and also its risks. The potential lies in the beneficial effects of attachment security, which may help to ameliorate anxiety, pain and distress. The risks lie in the possibility that negative experiences in the former attachment context arise in the doctor– patient relationship leading to conflict through enmeshment or rejection. Both sides of the coin, the benefits as well as the risks of the relationship, are rooted in the developmental trajectory of early attachment.

Irrespective of the developmental antecedents, the focus and main goal of the medical encounter is the understanding and removal of physical and or psychological symptoms and complaints. Here the issue of narration comes into play, since before medical treatment begins, the patient has to present his symptoms. This is usually accomplished by giving an account of their history and course. While attachment experiences are the developmental basis providing a matrix of potential meanings with regard to social interaction as well as to bodily experiences, the patient's report of his symptoms is the starting point of a process of meaning construction centered around his perception of the current physical complaints and their cognitive evaluation, for example, as indicators of potential danger.

3.5 Meaning Construction and Somatic Symptoms

There are situations, in which medical assessments have to be performed without the patient giving his own linguistic account of his problems; for example, a patient might be referred to an emergency room after having suffered brain injury associated with loss of consciousness. Symptoms in this situation are not communicated by the patient but rather are observed and interpreted by a medical expert. Symptoms in this context can be conceived of as "signs" in the sense of Peirce since there is an unequivocal link between the sign – loss of consciousness – and a sequence of interrelated events, for example, falling from a roof, hitting of the skull, etc. – its cause or "referent". Therefore in this instance no communication is necessary to establish a narrative linking symptom and cause. For treatments in the emergency room, it is a considerable advantage if the diagnosis can be made based on signs and without communication, because the short cut of a long conversation may facilitate faster decision-makings.

Unfortunately the sign-referent model has been overgeneralized and transposed to other areas of medicine, where it is less appropriate. The much more common situation in clinical medicine is that the explanandum of clinical reasoning – the symptom – is constituted only through communication and through the patient's own narrative of his complaints. It is trivial but important to note that the symptom cannot be registered or assessed independent of the patient's narrative. This is particularly evident for all sorts of pain symptoms. Pain cannot be observed from outside. It is intrinsically linked to the patients' ways of communicating it. The perception of pain may have diverse references (causes). It may be an indicator of an underlying biological dysfunction, or of underlying psychological distress or a combination of both. Even if it could be clarified relatively fast that a biological dysfunction is the only cause of the symptom, future assessment and treatment will heavily depend on the patient's own illness theory and attribution, which is conveyed by his or her narrative.

The process of meaning construction with regard to the patient's symptoms follows three different lines: (1) From the extended narrative of the patient on his/her self-observations, the doctor has to select the features, which according to his knowledge, can be integrated into a pattern or cluster of symptoms, (2) In the second step, this cluster is linked to a variety of potential "causes", either biological or psychosocial, depending on the diagnostic perspective, (3) Bringing the diagnostic hypothesis back to the patient, other aspects of the patient's narrative might come into focus. All of these processes, however, are clearly not rooted primarily in the observation of language-independent biological phenomena, but rather in the listening to the patient's narrative and the construction of meaning in a hermeneutic process.

There may be instances where a patient is cutting short his narration and changing the discourse into a deixis of symptoms, for example, via showing parts of his body in order to communicate his complaints. The deixis of bodily symptoms however cannot be conceived as "observation" in the sense of natural science. It is rather a specific form of meaning construction, in which the patient is using his body in order to express something which otherwise may be difficult to formulate. It also might be an expression of underlying emotions, for example, of worry or anxiety so that the immediacy of the communication is an expression of distress or worry. The deixis of bodily symptoms therefore is a special case of language–based communication and meaning construction in the doctor–patient interaction, not an alternative option.

The hermeneutic understanding of the process yielding a diagnosis in the doctor–patient communication is opposed to the model of subject-independent empirical observation as the central paradigm of biomedicine. Both epistemological approaches have their merits and are essential for modern medicine. Problems arise if one of these approaches is overgeneralized and over-extended at the cost of the other. The assumption, however, that "symptoms" in clinical medicine are accessible primarily by observation independent from the patient's account and narrative is incorrect. Rather, symptoms are constituted by language and communication. The narration of the patient is an over-determined activity. It is not just aiming to provide information to the doctor, but by narrating his symptoms, an attachment relationship is emerging, which may unfold healing effects in itself and independent of the doctor's instrumental actions. We will discuss some of the psychological effects of narration in the subsequent paragraph.

3.6 The Psychological Function of Narratives

Narratives fulfill a variety of psychological functions. They are important elements of every-day communication, conveying personal experiences and their evaluation and describing implicitly how the author sees himself in relation to others. Narratives thus have a strong impact to build up relationships. Another important goal of narratives is to organize experiences in a chronological order and to integrate these experiences into the contextual frame of one's own history. Narratives may range in form from small stories (Bamberg 2004) to life stories (Linde 1993; Maynes et al.

2008). Implicitly, narratives often are addressing issues of identity and probing views of oneself in an intersubjective context.

The aspect of narratives addressing issues of identity becomes even more important if the narrated life events are aversive and threatening. This applies for example to narratives of trauma and bereavement (Waller and Scheidt 2010). Narration in this context can be considered as a form of coping involving the organization of experience in a chronological form and the evaluation of its meaning with regard to personal development and history. From the perspective of memory systems, narratives can be conceived as components of the autobiographic memory system, which allows to perceive of the content of experience as belonging to the own self (Fivush 2001). Narration in coping with trauma experiences may help to reconstitute a sense of self and self-coherence by bringing fragmented memories into a story and by distancing the self from the former experience in the perspective of the here and now of the narration.

In clinical medicine, narratives are an essential part of communication. They are important means to convey the subjective experience of the illness. Narratives about physical and psychological suffering are often incomplete or unclear in structure and in content. The listener then has to encourage the narrator, to enrich, to question or to substitute missing pieces of information, thereby taking the role of a co-constructor (Fivush et al. 1987). The co-construction may be focused around "local experiences" of current symptoms or on extended periods of life for example in the context of coping with chronic illnesses, ambiguities about prognosis of life-threatening disease, etc. The activity of co-constructing the patients' narratives in these instances aims to support the patients' own efforts in constructing meaning in order to better cope with his illness. Sometimes, explanatory models can be integrated into illness narratives, implementing scientific descriptions to organize and interpret physical sensations and processes which otherwise are difficult to comprehend. This applies for example to patients suffering from neuropathic pain, because due to the lack of morphological correlates of the syndrome, patients find the underlying processes difficult to understand. However, explanatory models for example on the pathophysiology of pain, are only tools to transform and organize otherwise incomprehensible experiences of the body into a story which has the primary goal of making sense to the patient. Illness stories may also involve deeper layers of conscious and unconscious phantasy concerning anxieties of bodily harm and distortion associated with illness and treatment. The more comprehensive the patient's narrative on his condition, the better he can communicate his condition and may perhaps find support in the medical setting.

Just as attachment, the co-construction of narratives has a developmental trajectory. The narrative co-construction of experience is an integrated part of the infant's interaction with the primary care givers (Haden et al. 1997; Hirst and Echterhoff 2012). The capacity to narrate in turn is linked to the experience of attachment security. It is the secure base, which in the respective windows of development, supports and encourages the use of language to denominate inner states and their relation to social experience. This is why the co-construction of narratives in the medical setting can help the patient to establish trust and a sense of security. As was outlined above, the experience of attachment security has an impact on the psycho-social development and also on psychobiological adaptation.

3.7 Placebo: Implicit Belief Systems

The reason that makes placebo an interesting topic for narrative medicine is that its effects, like those of attachment, are not substance-bound, but rather are due to conscious and unconscious culturally-transformed stereotyped belief systems (which also can be considered as a form of meaning construction). In pain treatment for instance, a placebo can be analgesic if the patient is convinced that the placebo is an effective analgesic drug. More interesting, however, is the observation that the social context, that is the ways in which the placebo is administered, also plays a crucial role for placebo effectiveness. Current evidence suggests that placebo effects are due to rather different sets of expectations, some of them focusing on the effects of drugs, and some related to communication and psychosocial or attachment experiences. The underlying belief system of the traditional placebo effect is that "material things influence other material things", or to put it more scientifically: that chemical substances are powerful agents influencing biological processes. Such belief systems are obviously rooted in classical bio-medical discourse and its inherent mythology of healing. But on what kind of belief system is the healing power of attachment based? Imagine a patient to whom it is explained that a friendly nurse will comfort him during his tooth extraction instead of getting administering an analgesic injection as part of a trial on placebo effects. Anxiety would probably arise sharply in the attachment condition since there is not a clearly-established culturally-transformed belief system in the healing power of attachment justifying the expectation that a supportive nurse could be equally effective in reducing pain than the application of an analgesic drug. However, although such a belief system in the healing power of attachment and social support is almost non-existent, current empirical evidence demonstrates indeed a strong effect of such "psychosocial agents".

The idea of the doctor and the doctor-patient relationship as a placebo has a long tradition. The concept refers to the fact that the patients' expectations with regard to the doctors' capacity to offer effective help can be a powerful factor in alleviating distress. The underlying belief system of the doctor as a placebo is complex. The beliefs may refer to the power of his instrumental skills or to his capacity to provide comfort and security. As pointed out above, the attachment-related expectations clearly have rather clearly developmental origins. But what is about the belief systems referring to the power of chemical agents? We suggest that these too have developmental origins and are not only based on experiences of relief, for example, but of pain made in the real world. So far, the metaphor of the doctor as a placebo is a good mix integrating traditional placebo, and attachment related to placebo effects. The latter is due to the fact that patients have the capacity to use the doctor as a potential source of security in times of distress. This can be supported if the doctor helps the patient to narrate his symptoms and his suffering in order to make sense of the disturbing experiences associated with his illness.

References

Ainsworth, M. D. S., Blehar, M., Waters, E., & Wall, S. (1978). *Patterns of attachment*. Hillsdale: Lawrence Erlbaum Associates.

Balint, M. (1964). *The doctor, his patient and the illness*. London: Pitman Medical Publishing.

Bamberg, M. (2004). Narrative discourse and identities. In J. C. Meister, T. Kindt, & W. Schernus (Eds.), *Narratology beyond literary criticism: Mediality, disciplinarity* (pp. 213–237). Berlin: de Gruyter.

Bowlby, J. (1975). *Bindung: Eine Analyse der Mutter Kind Beziehung*. München: Kindler.

Bruner, J. (1987). Life as narrative. *Social Research, 54*, 11–32.

Charon, R. (2001). Narrative medicine: A model for empathy, reflection profession and trust. *JAMA, 286*(15), 1897–1902.

Ezzy, D. (1998). Theorizing narrative identity: Symbolic interactionism and hermeneutics. *The Sociological Quarterly, 39*, 239–252.

Fivush, R. (2001). The development of autobiographical memory. *Annual Review Psychology, 62*, 559–582.

Fivush, R., Gray, J. T., & Fromhoff, F. A. (1987). Two year olds talk about the past. *Cognitive Development, 2*, 393–410.

Folkman, S., & Moskowitz, J. T. (2004). Coping: pitfalls and promise. *Annual Review of Psychology, 55*, 745–774.

Francis, D. D., & Meaney, M. J. (1999). Maternal care and the development of stress responses. *Current Opinion in Neurobiology, 9*, 128–134.

Frattaroli, J. (2006). Experimental disclosure and its moderators: A meta-analysis. *Psychological Bulletin, 132*, 823–865.

Gunnar, M. R., Hertsgaard, L., Larson, M. C., & Rigatuso, J. (1991). Cortisol and behavioral responses to repeated stressors in the human newborn. *Developmental Psychobiology, 24*, 487–505.

Haden, C. A., Haine, R. A., & Fivush, R. (1997). Developing narrative structure in parent–child reminiscing across the preschool years. *Developmental Psychology, 33*, 295–307.

Hirst, W., & Echterhoff, G. (2012). Remembering in conversations: The social sharing and reshaping of memories. *Annual Review of Psychology, 63*, 55–79.

Hofer, M. A. (1984). Relationships as regulators. *Psychosomatic Medicine, 46*, 183–187.

Labov, W., & Waletzky, J. (1967). Narrative analysis: Oral versions of personal experience. In J. HELM (Ed.), *Essay on the verbal and visual arts* (pp. 12–44). Seattle: University of Washington Press.

Linde, C. (1993). *Life stories: The creation of coherence*. New York: Oxford University Press.

Maynes, M. J., Pierce, J., & Laslett, B. (2008). *Telling stories: Analysis of personal narratives in the social sciences and history*. New York: Cornell University Press.

Mc Adams, D. (1993). *The stories we live by: Personal myths and the making of the self*. New York: William Morrow & Co.

Mc Adams, D., Josselson, R., & Lieblich, A. (2006). *Identity and story: Creating self in narrative*. Washington, DC: APA.

Pennebaker, J. W. (1993). Putting stress into words: Health, linguistic and therapeutic implications. *Behavior Research and Therapy, 31*, 539–548.

Pennebaker, J. W., & Chung, C. K. (2007). Expressive writing, emotional upheavals, and health. In H. Friedman & R. Silver (Eds.), *Handbook of health psychology* (pp. 263–284). New York: Oxford University Press.

Pennebaker, J. W., Mayne, T. J., & Francis, M. E. (1997). Linguistic predictors of adaptive bereavement. *Journal of Personality and Social Psychology, 72*, 863–871.
Schiffin, D. (1996). Narrative as self-portrait: Sociolinguistic constructions of identity. *Language in Society, 25*, 167–203.
Schore, A. N. (1994). *Affect regulation and the origin of the self.* Hillsdale: Lawrence Erlbaum Associates.
Waller, N., & Scheidt, C. E. (2010). Erzahlen als Prozess der (Wieder-)Herstellung von Selbstkoharenz. Überlegungen zur Verarbeitung traumatischer Erfahrungen. *Zeitschrift für Psychosomatische Medizin und Psychotherapie, 56*(1), 56–73.

Chapter 4
How Can We Reconstruct the Health Anticipation?

Farzad Goli and Reza Johari Fard

Our thoughts are changing day by day, but belief systems are not so flexible and usually resist even positive changes. Various belief systems interact with each other and construct our webs of belief and, consequently, our forms of life. Our beliefs are vastly heterogeneous, from the delusional to the experimental, and the amazing fact is that an irrational or bizarre belief may overcome plenty of concordant rational beliefs even from our responses in a maladaptive manner. Various health beliefs such as "depression is a disease", "my illness is due to evil eye", "my spouse's behavior is the cause of my anxiety", "smoking is an unhealthy behavior", "energy enhancement of kidney meridian cooks ameliorate glomerohephritis", and "I can control my pain" could be categorized in sociocultural, individual, and healing belief systems. The examples mentioned demonstrate some of the different beliefs which are aroused from various healing systems (biomedicine, acupuncture), subcultures (transitional, modern) and personal belief systems. Each of these beliefs, separately and/or in interaction with other beliefs, can determine our locus of control, self-efficacy, coping strategies and expectations, ultimately changing our illness behavior and psychoneuroimmonologic responses.

Quine and Ullian (1978), in their influential book *Web of Beliefs*, defined believing as "a disposition to respond in certain ways when the appropriate issue arises" (p. 4). To believe in something this way means the tendency to interpret the relevant stimuli and the effect of the relevant objects in a particular manner. Thus, a belief is a symbolic sign in itself, which can conduct a special flow of psychophysical signs to/in/from our body. Let us draw on one of previously mentioned examples: When

F. Goli (✉)
Head of Danesh-e Tandorosti Institute, Isfahan, Iran

Energy Medicine University, Mill Valley, CA, USA
e-mail: Dr.fgoli@yahoo.com

R. Johari Fard
Department of Psychology, Ahvaz Branch, Islamic Azad University, Ahvaz, Iran
e-mail: rjoharifard@gmail.com

F. Goli (ed.), *Biosemiotic Medicine*, Studies in Neuroscience, Consciousness and Spirituality 5, DOI 10.1007/978-3-319-35092-9_4

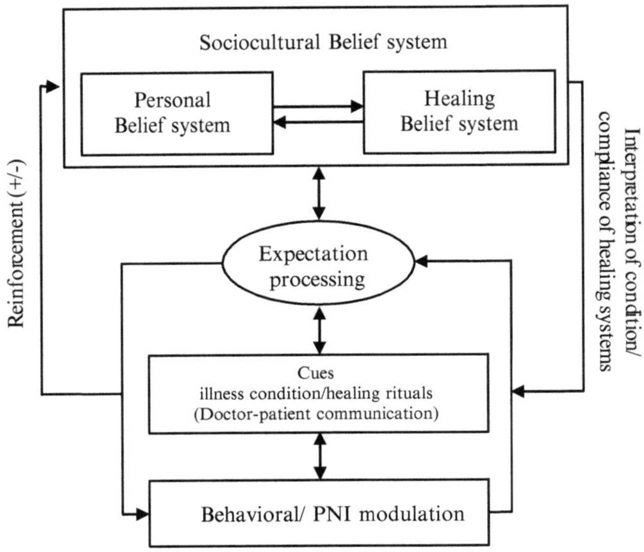

Fig. 4.1 Through the webs of belief

I believe in my pain control ability (sign), it can change my expectations and consequently, my somatic (neuroimmune modulation), psychological (calmness/anxiety, hopelessness/hopefulness), and social (seeking behavior/aggression) responses (interpretant). The flow of signs through intra/intercorporeal fields is determined via biopsychosocial interpretations in the cultural, social (healing systems), psychological and physical contexts. Cultural beliefs such as "pain is a divine examination" can change the patient's attitude towards an assertive and generative manner. Not only can this interpretation make a hopeful and positive meaning but it can also moderate our illness behavior, relationships, and even the interpretation of our T cells.

In Fig. 4.1, we have summarized the hypothesis of this chapter, and displayed the relationships between personal and healing belief systems in the frame of sociocultural beliefs and the role of expectations in healing response. Each person, due to his/her schemas, attachment, history and narrations, constructs a belief network which addresses causes, consequences, control, and anticipations around his/her life, and, finally, health and illness.

Healing systems have their own worldviews and interpretations of health and illness conditions. You can imagine that a psychoanalyst, a neurologist, an acupuncturist, and a physical therapist each have their specific narration of a particular health condition. They then induce their indirect suggestions to the client. The interaction between the personal and healing belief systems are configured in a sociocultural belief system, which can facilitate or disturb compliance and meaning responses of clients. These interactions shape expectations and are embodied in the form of clinical rituals and tasks. The expectation-ritual interactions mediate the belief network as well as psychoneuroimunologic and behavioral responses.

The implicated and manifest meanings of procedures and remedies are formed in the personal and social context and actualized through performing clinical rituals. These psychosomatic dynamisms can be activated even when we intend to perform a particular prescription. It is not uncommon to hear reports such as: "Just when I decided to go to the doctor, I felt better" or "I got better right after I had visited the physician."

Through the expectation-rituals interactions, biosemiotic procedures not only can anticipate illness behavior, but also modulate vital systems. In this chapter we will develop this model and present documents which can explain how meaning formation in the context of belief network leads social and physiological behavior to a special way of healing.

4.1 Belief System and Expectation

Believing, whether defined as disposition of action in a certain manner, judgment of reality (Freud 1895, p. 333), or cognitive-emotional acceptance of an idea as being a truth (See Thagard 2007), could be considered as a pattern of bioenergy distribution and bodily attunement which is configured to a linguistic pattern. This pattern systemically forms certain sign processes and determines our cognitive, emotional, and behavioral responses. Belief is closer than knowledge, and have darer but more stable relation with action in comparison with emotion (See Frijda et al. 2000, p. 46). Therefore, dysfunctional beliefs can evidently distort our lifeworld and bring about complexities and facilitate health. The renunciation of belief is then an educational task and a psychological struggle, both liable to encounter great resistance. The idea has been proposed that many faith-based beliefs are actually delusional beliefs (Dawkins 2001; Harris 2004). Myths and irrational beliefs are also responsible for behaviors which could be harmful to one's health. Albert Ellis (2004), in his rational emotive behavior therapy theory (REBT), showed the role of irrational beliefs in illnesses, their formation, and aggravation.

Belief system is an organized way of trying to explain the world around us. It is something that distinguishes human beings and becomes an integral part of our culture. As belief systems grow in complexity beyond simple common-sense generalizations, these systems attempt to also explain and understand. There are two forms of belief systems: evidence-based and faith-based (Siegel 2004). Science is used to build an evidence-based belief system under the premise that the world is ultimately understandable through observation, experiment, and prediction. The key element of science is the recognition that human beings possess individual beliefs, and are consequently capable of introducing biases into their interpretation of the world. As a result, science attempts to militate against such biases by requiring strict definitions of terms and conditions, as well as demanding that any evidence be capable of independent verification by others. This ensures that accepted results have been subjected to trials. These may not be free from bias, but by strict adherence to procedure, such biases will cancel each other out and produce conclusions that are largely objective.

Faith-based belief systems are mental constructs, which are not necessarily based on evidence. This is not meant to disparage them or to diminish their value, but rather to define an important difference. In short, a faith-based belief system does not have a foundation based on evidence. It is usually defined by the properties such as archetypal and subconscious conceptions, and/or evidence, which may be impossible to collect (ibid.).

Based on these descriptions, it would be easy to consider faith-based beliefs as somehow lesser in value, but this would be incorrect. We tend to draw these conclusions because we are all convinced that our particular beliefs are correct, regardless of what we individually believe. It is this fascination with being "correct" that leads to such discrepancies. Thus, we tend to defend our particular belief systems vigorously as being the only means by which one can experience "truth" or "reality". However, it is important to note that not all beliefs are subject to verification, and that this is precisely where these two forms of belief may often collide.

Therefore, beliefs coming from either evidence or faith can introduce our personal sense (or narration) of "reality" (Johari-Fard 2012). Both interpersonal and scientific beliefs were often seen as the outcomes of emotional responses to issues or persons (Frijda et al. 2000). As human beings, we tend to use all of these belief systems to varying degrees in order to cope with events in our lives. One of these events is to patent. When we get sick, our belief system about disease, doctors, healing, etc. can affect our behaviors and what we then plan to do. But there is a mediating variable between belief systems and behavior called "expectation" (Stewart-Williams and Podd 2004).

An expectation is a belief about the probabilities associated with a future state of affairs (Olson et al. 1996). The expectancy approach holds that the placebo effect is driven by anticipation that a treatment will result in a particular outcome (Bootzin 1985; Kirsch 1999; Stewart-Williams and Podd 2004). Expectation is what is considered to be the most likely outcome. An expectation, which is a belief that is centered on the future, may or may not be realistic.

Thus, simply put, expecting the suggested reaction is said to lead to the generation of that reaction. The expectancy framework that has received the most attention in the placebo literature is that of response expectancy theory (Kirsch 1997, 1999). According to this framework, response expectancy is one's anticipated automatic reaction to situational and behavioral cues. Response expectancies are believed to be directly self-confirming (Kirsch 1997). Figure 4.2 displays the circular relationships between the belief system, expectation, and experience.

Glover (2011) suggested that beliefs have to be considered holistically and that no belief exists in isolation in the mind of the believer. They always implicate and relate to other beliefs. Glover provided the example of a patient with an illness who returned to the doctor and the doctor said that the prescribed medicine was not working. At that point, the patient had a great deal of flexibility in choosing what

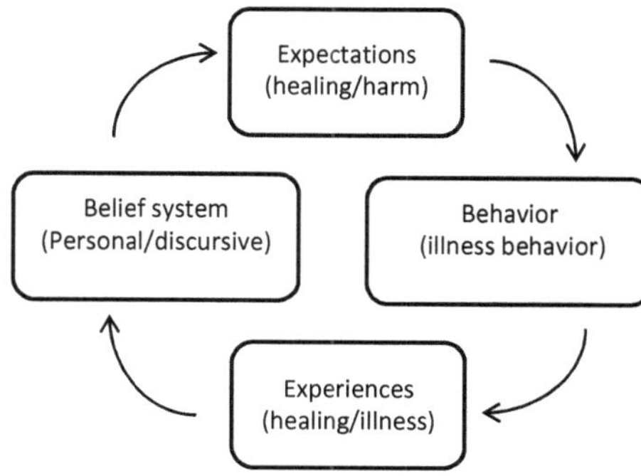

Fig. 4.2 Cycle of belief, expectation, behavior, and experience (adapted from Kirsch 1997)

beliefs to keep or reject – the patient could believe that the doctor is incompetent, that the doctor's assistants made a mistake, that the patient's own body is unique in some unexpected way, that western medicine is ineffective, or even that western science is entirely unable to discover truths about ailments.

Thus, the associative network around a simple fact-based belief (the treatment does not work) can switch the patient's mind to the other fact or faith-based beliefs which determine his/her illness experience and behavior. These biopsychosocial responses are formed in a complex network of webs of beliefs on the personal, healing system, and sociocultural levels. The role of the webs of belief in the sign processing of symptom formation, and especially healing response, is being discussed in this chapter.

4.2 Personal Belief System

A personal belief system is built upon our life experiences. Every word that has been spoken to us, every personal success or failure, our family, education, friends, work, dating experiences and even physical trauma and illness will lead us to build our personal belief system, for good or bad. These are the things that make up who we are, and how we feel and act. In this part, we discuss more about the interactions of these factors in personal belief system formation.

John Bowlby (1969), an English psychiatrist who pioneered attachment theory, argued that the primary motivation of the infant was to attach itself to a stable caretaker. From the perspective of developmental neuropsychology, the goal of attachment is to promote maturation of the brain regions responsible for configuring a progressive hierarchy of behavioral organization (Main 1995). This is achieved by progressively bringing lower levels of primitive reactivity, such as the spinal reflexes, under the influence of higher cortical brain areas via top-down regulation (Toates 1998). This organization of the early self is integrally linked to and motivated by the brain's affect centers. Psychologists' term of developmental stages are, in reality, new categories of dynamic skills that emerge with the progressive maturation of the nervous system.

There is general agreement that secure attachment protects against psychopathology (Fonagy 2001). In fact, as Michael Ainsworth, an expert on attachment noted, secure attachment may be "the primary defense against trauma-induced psychopathology" (Kumin 1996). As previously noted, secure attachment develops through reciprocal mutually attuned preverbal interactions between mother and infant. When separated from the mother, the child engages in exploratory behavior and eventually shows signs of missing her, subsequently returns to reestablish physical contact with her. Under good enough circumstances, the child develops the ability to self-soothe and a cohesive sense of self (Kradin 2011). From the developmental neuropsychological standpoints, the function of attachment is the progressive organization of brain and behavior. In other words, the pattern of attachment coordinates cortical and subcortical activities such as neuroimmune responses (See Toates 1998; Main 1995). On the other hand, a secure attachment facilitates optimal doctor- patient communication (Balint 1972) and, subsequently, response to treatment (Siegel 2003).

Beliefs refer to assumptions about reality that form how one interprets events, and can thus be considered as determinants of appraisal. For example, pain beliefs develop during the lifetime as a result of an individual's learning history and cover all aspects of the pain experience (e.g., the causes of pain, its prognosis and suitable treatments). Appraisal and beliefs about pain can have a strong impact on an individual's affective and behavioral response to pain. If a pain signal is interpreted as harmful (threat appraisal) and is believed to be associated with actual or potential tissue damage, it may be perceived more intensely and may evoke more escape or avoidance behavior. For instance, pain associated with cancer is rated as more unpleasant than labor pain, even though the intensity is rated equally (Price et al. 1987). Similarly, Smith et al. (1998) demonstrated that cancer patients, who attributed pain sensation after physiotherapy directly to cancer, reported more intense pain than patients who attributed this pain to other causes. Perception of danger of an experimental pain stimulus (cold-presser test) may also lead to avoidance of this stimulus (Cipher and Fernandez 1997). Arntz and Claassen (2004) experimentally manipulated the appraisal of a mildly painful stimulus (a very cold metal bar against the neck), by suggesting that it was either very hot or very cold. It was assumed that it would be stronger with tissue damage (hot) than cold. As expected, participants rated the stimulus as more painful in the condition where they were informed that it was hot. In addition, the effect appeared to be mediated by the belief that the stimulus would be harmful. These studies demonstrated the important role of people's interpretations regarding the meaning of pain.

Pain appraisal and pain beliefs are also prominent determinants of adjustment to chronic pain (Jensen et al. 1999; Turner et al. 2000; Turner and Aaron 2001). The following pain beliefs have been identified as particularly maladaptive in dealing with pain: pain is a signal of damage, activity should be avoided when one has pain, pain leads to disability, pain is uncontrollable, and pain is a permanent condition (Jensen et al. 1995; Turner et al. 2000). The belief that pain is a signal of damage and the belief that activity should be avoided in order to recover fro m pain appear to be widespread (Balderson et al. 2004; Ihlebaek et al. 2003). Two months after seeking treatment, a large majority of back pain patients believed that a single wrong movement could have serious negative consequences. Moreover, this belief was associated with reduced activity levels and increased disability (Balderson et al. 2004; Thorn et al. 1999).

Health-related self-efficacy, as a cognitive factor, may be one of the mechanisms of the placebo effect. Perceived self- efficacy refers to a psychological construct concerning the belief that one's abilities organize and execute behaviors with experimental manipulation of self-efficacy impacting stress, autonomic nervous system activation and neuroendocrine changes (Bandura 1997; Buckalew and Ross 1981).

Treatment regimens that actively engage the patient to have some sense of control over their disease process may produce better outcomes than those that are less actively engaging to the patient. Studies with adequate control groups that can clearly differentiate positive expectancy from self-management are, however, lacking (Crow et al. 1999). High-success biofeedback that improves one's sense of control may improve clinical outcomes, independent of the accuracy of the biofeedback (Holroyd et al. 1984). Adherence to a drug regimen may relate to an expectancy of the drug working in this sense of control. Subjects more adherent to a placebo intervention do better than those less adherent to the placebo regimen even with gross major medical outcomes (Hoewitz et al. 1990; Simpson et al. 2006).

However, other than expectancy, subjects more adherent to a prescribed medical regimen may have different characteristics, such as personality or mood (Osterberg and Blaschke 2005; Flegal et al. 2007), which may correlate with compliance and other aspects of medical intervention or health-promoting behavior. A systematic review found that positive expectations towards the outcome were associated with indeed better results (Mondloch et al. 2001).

Personality traits have a tendency to influence selection and reinforcement in beliefs. Plenty of studies have investigated the correlation between personality and placebo response. Currently, the most popular such model emphasizes that the search for a placebo personality factor must be combined with the measurement of situational expectancy. Expectancy is widely considered the central mechanism of placebo phenomena (Price et al. 2008; Tracey 2010), with variability in expectations influencing the variability of the response (Vase et al. 2005; Flatten et al. 2006). The most promising personality traits interacting with the mechanism of expectancy are optimism or pessimism, defined as a generalized and relatively stable expectancy for positive or negative future outcomes (Solbergnes and Segerstorm 2006). A considerable amount of research indicated that optimism is related to the flexible use of adaptive mental and behavioral coping strategies when faced with stressful life situations (Solbergnes and Segerstorm 2006). More importantly, optimists tend to exhibit attentional bias for positive information (Isaacowitz 2005; Geers et al. 2003). Therefore, optimism (and even pessimism) might serve as a moderator of placebo that responds by influencing the strength and/or the direction of the relation between expectancy and specific placebo effects.

Another example of the situation-personality interaction has been demonstrated by the traits of extroversion, as well as agreeableness to a lesser extent (Kelley et al. 2009). Extroversion is closely aligned with the temperament of positive emotionality/affect, referring to people who are described as sociable, talkative, energetic and assertive. With agreeableness, this manifests itself in individual characteristics perceived as kind, sympathetic and cooperative (John and Srivastava 1999). Kelley et al. (2009) examined the relationships between personalities of patients with irritable bowel syndrome and response to placebo acupuncture in different therapeutic settings, such as warm emphatic interaction, neutral interaction or waitlist control. Several personality dimensions were significantly associated with a placebo response, but extroversion was the only independent predictor, holding true for the warm emphatic therapeutic setting. The authors suggested that extroverted and agreeable patients responded in a better way to the efforts of emphatic clinicians; thus, facilitating the warm therapeutic relationship further. At the psychological level, this caring interaction could have reduced anxiety and increased positive expectancies. Conversely, when placebo effects are a consequence of medication with a minimal or neutral patient-clinician interaction, these personality traits will not have such a relevant moderating role.

There have been inconsistent results from studies evaluating whether certain personality traits predispose some to experience improvements from placebo administration more than others. While some studies have been negative (Fercund et al. 1972; Buckalew et al. 1981), other studies suggest there may be contributions to the placebo response from factors such as social acquiescence (McNair and Barrett 1979), suggestibility, or hypnotizability and absorption– which is the degree to which one can focus on a single theme (Evans 1985; Challis and Stam 1999; Raz 2007). This lack of consensus on individual differences to placebo administration may be related to an interaction between personality factors and the specific experimental condition.

The individual response to placebos also differed based on optimism–pessimism scale in the 100 % deceptive, but not in the 50–50 % conditional expectancy of receiving an active drug (Geers et al. 2005). The effect of personality traits such as optimism on placebo response may be dependent on the specific treatment and context (Geers et al. 2007; Hyland et al. 2007). Uncertainty in diagnosis and prognosis produces expectancy effects on health outcomes (Thomas 1987), possibly through some mechanism related to stress or anxiety.

High levels of neuroticism, along with depression and anxiety, helped to predict analgesic responses in patients with discogenic back pain (Wasan et al. 2006). Individual differences may contribute to variations in placebo effects in other ways. The individual experience of actual pain contributes significantly to neurotransmitter activity during placebo analgesia (Zubieta et al. 2006). Personality may relate to placebo responses either through the neurotransmitter systems, which were thought to be related to these traits, or to interactions with these traits. The mechanisms of expectance nausea and vomiting associated with chemotherapy overlap with the placebo effect. Higher anticipatory nausea and vomiting were not related to measures of absorption and autonomic perception (Challis and Stam 1999).

4.3 Healing Belief System

Each healing culture has its belief system, jargon, metaphors, and rituals. Healers and those being healed of each system are living in common cultures all over the world. For example, homeopaths and their clients in Germany, China and Iran all implicitly or explicitly, more or less, believe in a monistic worldview; a bioenergetic explanation of health and illness, the effective role of vibrational medicines in moderation of energetic balance and of course, the law of "like cures like". Many medical-anthropologic studies show the cultural diversity of healing systems and how this symbolic aspect of medical practice can induce physiological, psychological and social effects (Kirmayer 2004; Kleinman 1978). Even in the biomedicine's "culture of no culture", behind its positivistic claims are many recognizable mythical, metaphorical and rituals elements (Lock and Nguyen 2010; Taylor 2003; Coulehan 2003).

At the moment, the majority of people, especially because of worldwide media, live in a cultural marginality, and the nationality borders cannot determine cultural identities (See, e.g., Horback and Rothery- Jackson 2007). We are virtual nomads who travel through the cultures and, in addition to that, live simultaneously in several cultures. Thus, healing systems are a set of global belief systems, which could be more or less compatible with the various personal and cultural belief systems.

A study for the Institute of Noetic Sciences by Paul Ray (1996) had similar findings regarding different health beliefs and behaviors among different subcultures of a society. Ray identified three subsets of American culture that are influencing the demand for health services:

1. Heartlanders preserve traditional or rural values, tend to resist change, are somewhat isolationist, and are most often among middle to lower-income populations.
2. Cultural moderns are found in the mainstream in all income categories.
3. Cultural creatives are most often found in upper income levels, and tend to be leaders of cultural change and see a desirable future.

Cultural creatives have nontraditional values that require a different paradigm of health; they are willing to try a variety of approaches to health care. These consumers believe in holistic health through a unified approach to the body, mind, and spirit. Although this group tends to be fairly healthy, some members have also been described as the worried. They are more prevention–oriented than the two other groups that make up the U.S. population.

Thus, even in a certain society, health care systems should distinguish different subcultures, their belief systems, and their preferences. To minimize the chaotic response to the therapeutic programs and optimize placebo responses, we should mention the compatibility of the healing belief system with the personal and sociocultural systems. The next chapter is focused on the healing belief systems and their biopsychosocial effects. Afterwards, we will discuss further the healing power of the forms of clinical settings, and healing rituals.

4.4 Sociocultural Context

A cultural system is the interaction of different elements of culture. Cultural system is quite different from social system, however, sometimes both are referred to as the socio-cultural system. We live our lives in both the social systems of relationships and institutions, and also in the cultural systems of beliefs and discourses. Wilber (2000) illustrates the human's nest in a window by four fields –I (intentional), IT (behavioral), WE (cultural), and ITS (social) (Fig. 4.3).

He explains how our lifeworlds are formed through these symbolic-intentional and physical-behavioral worlds. Therefore, the sociocultural context indicates a matrix of material, energic, symbolic, and intentional elements of the collective aspect of our life. Healing occurs in such a complex network.

Healing involves a basic logic of transformation from sickness to wellness that is enacted through cultural salient metaphorical actions. Kirmayer (2004) applied the notion of a hierarchy of metaphoric spaces, through which multiple levels of meaning are generated to the range of symbolic healing practices. Figure 4.4 depicts some of the many processes involved in symbolic healing. In the central column of the figure, these are arranged as a hierarchy of organizational levels: first within the central nervous system (CNS), then on to family and to the larger levels of community, the physical and social environment, and the spiritual world. Each of these levels has its own metaphorical logic and dynamics corresponding to specific neuropsychological, interpersonal, social, political, or ecological processes.

Gottman et al. (2002) emphasized that beyond the processes associated with levels of information processing within the CNS, it is useful to distinguish at least two further levels in biological organization: the social groupings of family and community. Families and other groups of people living together develop implicit rules of interaction that may give rise to problems that are not reducible to psychological

Fig. 4.3 Four quadrants (adapted from Ken Wilber 2000)

I Interior-Individual (Intentional) -Subjective-	IT Exterior-Individual (Behavioral) -Objective-
WE Interior-Collective (Cultural) -Intersubjective-	ITS Exterior- Collective (Social) -Interobjective-

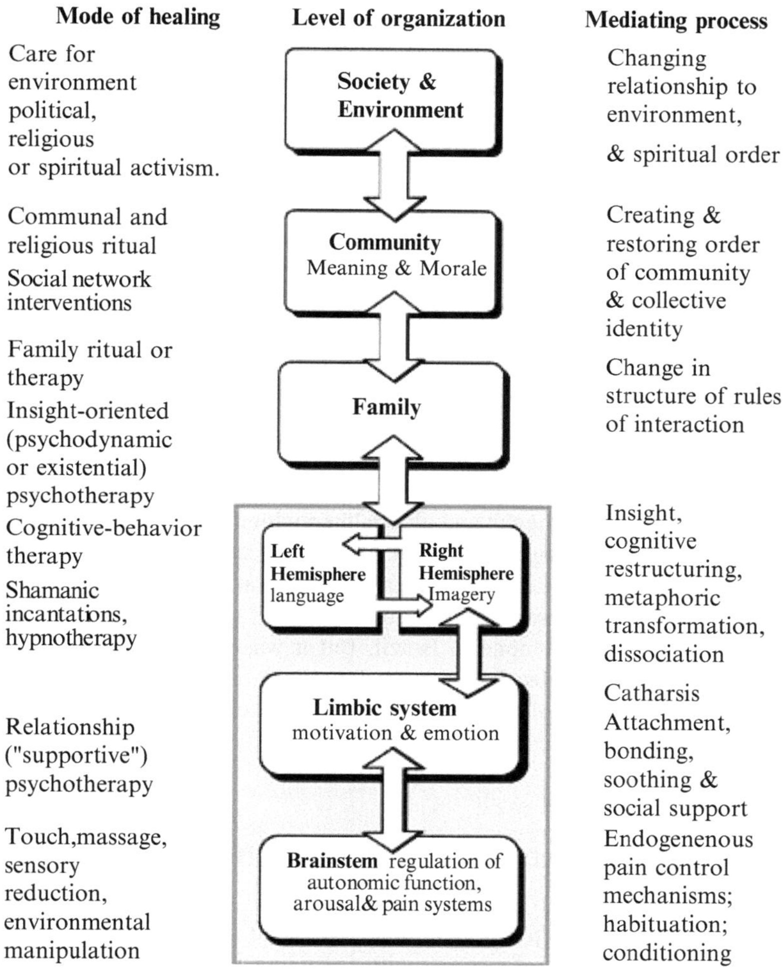

Fig. 4.4 A hierarchy of healing mechanisms (reprinted from Kirmayer 2004, by permission of Oxford University Press)

conflicts within one individual. Consequently, the unraveling of such interpersonal conflicts demands a change in family structure or the rules of interaction. Family rules are rarely articulated as such. Instead, family members conceive their group in terms of family myths and metaphors. Therefore, a change in metaphor prescribed by a therapist and subscribed to even one family member can change the pattern of interaction in widespread ways. The family unit is embedded in a community or larger social grouping with a collective history and way of life. At this sociocultural level of organization, people participate in the construction of institutions and shared symbolic meanings that confer an order, beauty and diversity that surpass individual experiences. Psychological healing at this level employs the extended metaphors of secular and religious rituals to create and restore the order of the community and the relationship to the environment, the larger cosmos and with it, the sufferer's experience of meaning and morale (Turner 1974).

For example, cultural aspects common to Native Americans usually include being oriented in the present and valuing cooperation. Native Americans also place great value on family and spiritual beliefs. They believe that a state of health exists when a person lives in total harmony with nature. Illness is viewed not as an alteration in a person's physiological state, but as an imbalance between the ill person and natural or supernatural forces (McLaughlin and Braun 1998).

As mentioned before, all cultures have their especial health belief systems, which explain the causes of illness, the consequences, the meaning and how to control them. The consistency of a healing model or even technique in a sociocultural context has a crucial role in compliance and psychoneuroimmunologic response (See McLaughlin and Braun 1998; Roth 2003).

We should know the meaning of therapeutic concepts, tools, and procedures in the various cultural frameworks, otherwise, we cannot recognize their expectations and interpretations of their health and illness, therewith rapport, compliance, illness behavior, and even physiological responses will be affected.

Some cross-cultural studies unfold the psychocultural aspects of physiology. For example, Klein (2003) showed that in Germany 60 % of stomach ulcers were healed by placebos but practically none in Brazil. But it was nearly impossible to treat German hypertension by placebo, whereas Brazilians reacted quite well. It seems that the symbolic meaning of illness and treatment in each sociocultural context can change the pharmacokinetic responses.

4.5 Changing Expectation and Placebo Effect

A ritual is a set of actions performed mainly for their symbolic value. It may be prescribed by the traditions of a community, including a religious community. The term usually refers to actions which are stylized, excluding actions which are arbitrarily chosen by the performers. In psychology, the term ritual is sometimes used in a technical sense for a repetitive behavior systematically used by a person to neutralize or prevent anxiety; it is a symptom of obsessive– compulsive disorder. Among anthropologists, and other ethnographers, who have contributed to ritual theory are Victor Turner, Ronald Grimes, Mary Douglas and the biogenetic structuralists. Anthropologists from Émile Durkheim through Turner and contemporary theorists like Michael Silverstein treat rituals as social actions aimed at particular transformations, often conceived in cosmic terms. Although the transformations can also be thought of as personal (e.g. the fertility and healing rituals Turner describes), they become a sort of cosmic event– one stretching into "eternity" (McNeill 1995).

To reiterate, the term placebo effect will be used to refer to a physiological state anticipating and contributing to the occurrence of some future health-related outcome through learning, conditioning, or other related processes. Other terms used to describe these effects include expectancy effects (Crow et al. 1999), context effects (Di Blasii et al. 2001) and meaning response (Brody and Brody 2000; Moerman 2002). Expectancy is what people predict will happen, rather than what they desire. Expectancies are primarily based upon social norms and specific characteristics of the communicators. Some placebo researchers have used the term expectancy narrowly to mean placebo effects due to anticipation that has been verbally or consciously mediated. Also, as mentioned in Chap. 1, meaning response evidently includes expectancy effects that impact health aside from the placebo effects such as cultural effects (Moerman 2002), uncertainty in diagnosis and prognosis (Thomas 1987), the impact of pessimism and hopelessness on disease and function (Anda et al. 1993; Maruta et al. 2002), and the nocebo or negative placebo effect (Hahn 1997; Barsky et al. 2002). Placebo effects also encompass neural systems not only simply related to anticipation or expectancy, but also to the desire to achieve a particular goal (Price et al. 2008).

A placebo can be in the form of any clinical intervention including words, gestures, pills, devices and surgery (see Chap. 1). The term "sham" or "trick" is sometimes used to describe a placebo intervention, such as in the context of surgery. Placebo effects do not include methodological factors resulting in improvement that are unrelated and active alterations of outcome measures, for example, natural history, regression to the mean (McDonald and Mazzuca 1983), the Hawthorne effect (Bouchet et al. 1996), or poor experimental designs such as subject biases (Clayden et al. 1974) or the purported inert control condition not being inert (Kienle and Kiene 1997; Ader 2000; Miller et al. 2004).

The natural history is particularly problematic, because it is impossible to infer anything about the frequency or size of placebo effects without a control for the placebo condition. Unfortunately, it is rare in modern clinical trials to have untreated control groups. A recent systematic review of placebo effects found only 114 clinical trials out of all clinical trials spanning several decades that had both a placebo treatment arm as well as a non-treatment arm in a clinical trial (Horbjartsson and Gotzsche 2001). Subject biases resulting from non-blinding, especially in a crossover design, may confound placebo research (Ader 2000).

Meaning effects presumably have different mediators depending on the specific learned association and affiliation to acquisitions or contexts of practice. The CNS is the primary location and mediator of the physiological basis of the placebo effects through its role in learning and memory, and its outputs on sensory, motor and autonomic pathways, as well as the immune and endocrine system. People have individual traits that predispose them to be more or less responsive to certain stimuli; the interaction between the learned associations of the clinical situation and the person's particular biology produces a response. The response could be a basic physiological process, such as modulation of sensory processing, release of neurotransmitters, alterations in the Hypothalamic-Pituitary-Adrenal (HPA) axis, or immune system activity. The placebo response could also be a more complex physiological process including change in mood, change in motivation/effort, or cognitive set-shifting.

Learned associations producing placebo effects can be acquired through conditioning (Vodouris et al. 1989; Price et al. 1999; Wickramasekera 2000; Siegel 2002).

The conditioned pharmacotherapeutic effects have been characterized in animal models (Ader and Cohen 1982; Jones et al. 2008). While the relative contribution of conditioning to placebo effects remains uncertain (Montgomery and Kirsch 1997; Benedetti 2003; Kirsch 2004; Stewart-Williams and Podd 2004), non-conscious mechanisms such as conditioning may be particularly important for immune or endocrine placebo effects (Kirsch 2004). Animal models are useful models of some components of placebo effects but are intrinsically limited placebo effect models because there are no verbally mediated expectancy changes.

Conditioning in placebo research studies has consisted of exposure(s) prior to administration of placebo of either the active drug itself (Laska and Sunshine 1973; Amanzio and Benedetti, 1999), or of an apparent effect of a placebo, for example, due to surreptitiously turning down the pain intensity at the same time as the placebo is administered (Vodouris et al. 1989). Since most adults have had previous exposures to clinical experiences such as taking oral analgesics, clear separation of conditioning from other aspects of the placebo response in human experiments is difficult.

Conditioning is only one aspect of the placebo effect which can form in sociocultural context. Many aspects of placebo effects, including verbal communication, encompass more top-down and cortically mediated change in behavior than the term conditioning usually implies. Some learned anticipations acquired over longer periods of time than are usually studied in the conditioning experiments may be related to: interaction between person and health care provider (Brody and Brody 2000), health care setting and practitioner characteristics (Di Blasii et al. 2001), physical characteristics of a pill (Buckalew and Coffield 1982), type of treatment (e.g., pill versus injection versus surgical) (Kaptchuck et al. 2000), and pill administration frequency (de Craen et al. 1999). Additionally, anticipation or expectancy can refer to a response expectancy or self-efficacy expectancy, which is one's sense of being able to achieve an outcome (Caspi and Bootzin 2002). Desire or motivation for improvement is another aspect of the placebo effect (Hyland et al. 2007; Price et al. 2008). All of these variables can form in the sociocultural context.

There is some data to suggest that placebo effects are greater for psychological and self-rated measures than other objective measures of disease activity (Horbjartsson and Gotzsche 2001). A study that evaluated patients in placebo arms of rheumatoid arthritis drug trials found essentially no change over 6 months on the erythrocyte sedimentation rate, but there was a significant improvement in articular index and morning stiffness (Porter and Capell 1993).

While placebo responses may be generally greater for self-ratings, there are many studies that demonstrate the changes in more objective outcome measures including C-reactive protein (Hashis et al. 1988), elevation of liver enzymes (Merz et al. 1997), changes in pulmonary function (Luparello et al. 1970; Butler and Steptoe 1986; Kemeny et al. 2007), postprandial glucose (Sievenpiper et al. 2007) and the neurobiology studies.

Even though there is no control over the placebo condition, the balanced placebo has shed light on expectancy effects. For example, in a balanced placebo design among cocaine abusers, administration of methylphenidate when the client expects to receive methylphenidate produces significantly increased brain glucose metabolism compared to the administration of methylphenidate with the expectation of simply receiving placebo (Volkow et al. 2003). Expectation of receiving caffeine produced dopamine release in the thalamus, measured by raclopride positron emission tomography (PET) following administration of placebo (Kaasinen et al. 2004). Therefore, as we displayed in Fig. 4.1, expectation has an effect on the illness condition and healing rituals, and these expectations/rituals change the placebo effect.

4.6 Behavioral-PNI Modulation

The two-way communication between the brain and the immune system (Ader et al. 2001) contributes to aspects of the placebo response, both in its potential relationship to conditioning and in relationships mediated by stress and HPA axis activity (Ader 2000). A beneficial immunosuppressive effect was obtained with placebo through conditioning of administration of cyclophosphamide with saccharine in a murine systemic erythematous model (Ader and Cohen 1982). Even a commonly-used clinical immune marker, the tuberculin reaction, can be significantly diminished through conditioning (Smith and McDaniel 1983).

There are factors related to a clinical interaction that may produce improvement in patient outcomes without directly affecting the underlying pathophysiology of a disease. Methodological artifacts have contributed to confusion about these factors. However, there are clearly effects on outcomes that are dependent on patient expectations, whether these expectations are related to patients, culture, previous interactions with the clinical setting, verbal communication, conditioning, or some combination of factors. Figure 4.5 shows these behavioral-PNI modulation pathways.

In accordance with this model, these meaning effects are mediated through change in neocortical and subcortical systems. It is likely that some therapies and therapists have been successful in improving people's health because of their utilization of these beneficial effects and affective coordination of the healing narration within the psychosocial context. Sustaining these effects is important, and many current placebo effect studies actually serve to extinguish the beneficial placebo response through lack of reinforcement of the response (Oken 2008).

In conclusion, a certain therapy or clinical experience, as a set of signs, is interpreted by different sets of cognitive, emotional and behavioral responses through the various sociocultural contexts. On the other hand, each sociocultural context has its own especial normative attitudes, belief systems, and healing rituals. Health professionals have to be familiar with both the clients' reaction to the scientific management and their own ethnic healing methods. They should also have a non-judgmental attitude, systemic approach, and deep attention to the verbal bodily messages. Reframing, adjustment and reinforcement of both professional and traditional health beliefs should be based on our knowledge and awareness of the cultural belief system.

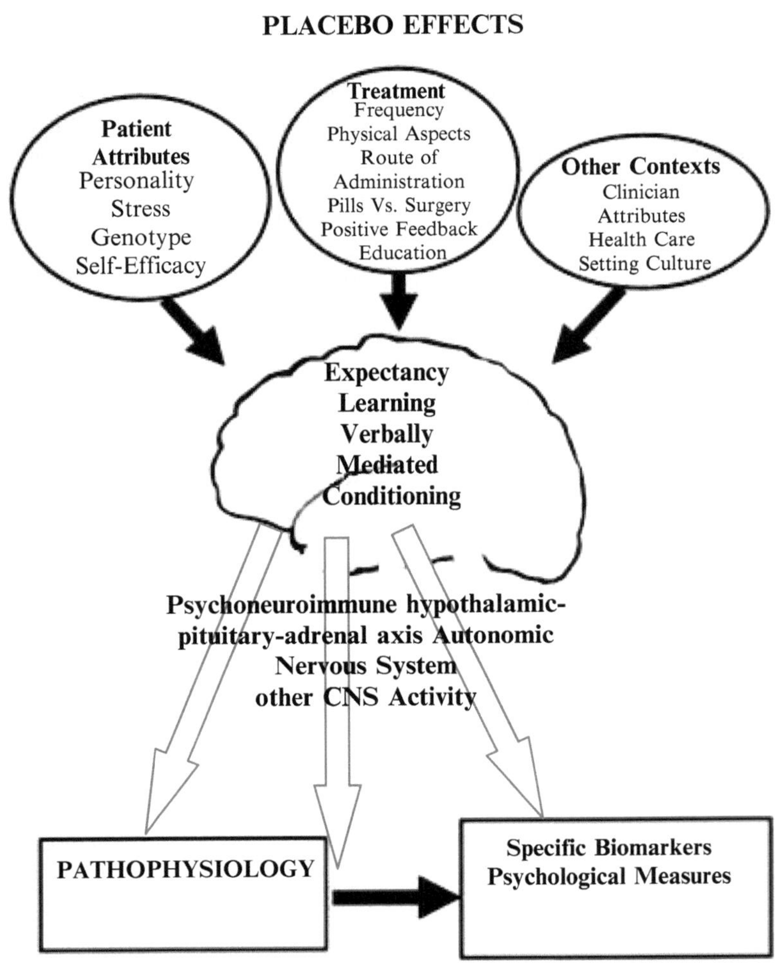

Fig. 4.5 Theoretical model of issues impacting development of expectancy and how brain outputs may produce a placebo effect (reprinted from Oken 2008, by permission of Oxford University Press)

References

Ader, R. (2000). The placebo effect: If it's all in your head, dose that mean you only think you feel better? *Advances In Mind-Body Medicine, 16*(1), 7–11.

Ader, R., & Cohen, N. (1982). Behaviorally conditioned immunosuppression and murine systemic lupus erythematosus. *Science, 512*, 1534–1536.

Ader, R., Felten, D. L., & Cohen, N. (Eds.). (2001). *Psychoneuroimmunology*. San Diego: Academic Press.

Amanzio, M., & Benedetti, F. (1999). Neuropharmacological dissection of placebo analgesia: Expectation-activated opioid systems versus conditioning–activated specific subsystems. *The Journal of Neuroscience, 19*(1), 484–494.

Anda, R., Williamson, D., Jones, D., Moacera, C., Eaker, E., & Glassman, A. (1993). Depressed affect, hopelessness, and the risk of ischemic heart disease in a cohort of U.S. adults. *Epidemiology, 4*(4), 258–294.
Arntz, A., & Claassens, L. (2004). The meaning of pain influences its experienced intensity. *Pain, 109*(1–2), 20–25.
Balderson, B. H. K., Lin, E. H. B., & von Korff, M. (2004). The management of pain-related fear in primary care. In G. J. Asmundson, J. W. S. Vlaeyen, & G. Crobez (Eds.), *Understanding and treating fear of pain* (pp. 267–292). England: Oxford University Press.
Balint, M. (1972). *The doctor, his patient and the illness*. New York: International University Press.
Bandura, A. (1997). *Self-efficacy; the exercise of control*. New York: W. H. Freeman and Company.
Barsky, A. J., Saintfort, R., Rogers, M. P., & Borus, J. F. (2002). Nonspecific medication side effects and the nocebo phenomenon. *JAMA, 287*(5), 622–627.
BenedettI, F., Pollo, A., Lopiano, L., Lanotte, M., Vighetti, S., & Rainero, I. (2003). Conscious expectation and unconscious conditioning in analgesic, motor, and hormonal placebo/nocebo responses. *Journal of Neuroscience, 23*, 4315–4323.
Bootzin, R. R. (1985). The role of expectancy in behavior change. In L. White, B. Tursky, & G. E. Schwartz (Eds.), *Placebo: Theory, research, and mechanisms* (pp. 196–210). New York: Guilford Press.
Bouchet, C., Guillemin, F., & Briancon, S. (1996). Nonspecific effects in longitudinal studies: Impact on quality of life measures. *Journal of Clinical Epidemiology, 49*, 15–20.
Bowlby, J. (1969). *Attachment and loss* (Vol. 1), Attachment. New York: Basic Books.
Brody, H., & Brody, D. (2000). Their perspectives on the placebo response: Expectancy, conditioning, and meaning. *Advances in Mind-Body Medicine, 16*, 216–232.
Buckalew, L., & Coffield, K. H. (1982). An investigation of drug expectancy as a function of capsule color and size and preparation from. *Journal of Clinical Psychopharmacology, 2*, 245–248.
Buckalew, L., & Ross, S. (1981). Relationship of perceptual characteristics to efficacy of placebo. *Psychological Reports, 49*, 955–961.
Buckalew, L., Ross, S., & Starr, J. B. (1981). Nonspecific factors in drug effects: Placebo personality. *Psychological Reports, 48*, 3–8.
Butler, C., & Steptoe, A. (1986). Placebo responses: An experimental study of psychophysiological processes in asthmatic volunteers. *British Journal of Clinical Psychology, 25*, 173–183.
Caspi, O., & Bootzin, R. R. (2002). Evaluating how placebo produce change: Logical and causal traps understanding cognitive explanatory mechanisms. *Evaluation & the Health Professions, 25*, 436–464.
Challis, G. B., & Stam, H. J. (1999). A longitudinal study of the development of anticipatory nausea and vomiting in cancer chemotherapy patients: The role of absorption and autonomic perception. *Health Psychology, 11*, 181–189.
Cipher, D. J., & Fernandez, E. (1997). Expectancy variables predicting the tolerance and avoidance of pain in chronic pain patients. *Behaviour Research and Therapy, 35*, 437–444.
Clayden, J. R., Bell, J. W., & Pollard, P. (1974). Menopausal flushing: Double-blind trial of a non-hormonal. *British Medical Journal, 1*, 409–412.
Coulehan, J. (2003). Metaphor and medicine: Narrative in clinical practice. *Yale Journal of Biology and Medicine, 76*, 87–95.
Crow, R., Gega, H., Hampson, S., Hart, J., Kimber, A., & Thomas, H. (1999). The role of expectancies in the placebo effect and their use in the delivery of health care: A systematic review. *Health Technology Assessment, 3*, 1–48.
Dawkins, R. (2001). Has the world changed? *The Guardian*,10–11.
de Craen, A. J., Moerman, D. E., Heisterkamp, S. H., Tytgat, G. N., Tijssen, J. G., & Kleijnen, J. (1999). Placebo effect in the treatment of duodenal ulcer. *Journal of Clinical Pharmacology, 48*, 853–860.
Di Blasii, Z., Harlness, E., Erns, E., Georgiou, A., & Kleijnen, J. (2001). Influence of context effects on health outcomes: A systematic review. *Lancet, 357*, 757–762.

Ellis, A. (2004). *Rational emotive behavior therapy: It works for Me: It can work for you by*. Amherst/New York: Prometheus Books.
Evans, F. J. (1985). Expectancy, therapeutic instructions, and the placebo response. In L. White, B. Tursky, & G. E. Schwartz (Eds.), *Placebo: Theory, research, and meachnisms* (pp. 215–228). New York: The Guilford press.
Fercund, J., Krupp, G. g., Goodenough, D., & Preston, L. W. (1972). The doctor-patient relationship and drug effect. *Clinical Pharmacology and Therapeutics, 13*, 172–180.
Flatten, M. A., Aslaksen, P. M., Finset, A., Simonsen, T., & Johansen, O. (2006). Cognitive and emotional factors in placebo analgesia. *Journal of Psychosomatic Research, 61*, 81–89.
Flegal, K. E., Kishiyama, S., Zajdel, D., Hass, M., & Oken, B. S. (2007). Adherence to yoga and exercise interventions in a 6-month clinical trial. *BMC Complementary and Alternative Medicine, 7*, 37.
Freud, S. (1895). Project for a scientific psychology (1950) [1895]. In J. Strachey (Ed.), *The standard edition of the complete psychological works of Sigmund Freud* (Vol. 1, pp. 283–398). London: Hogarth Press (Originally written 1895).
Frijda, N. H., Manstead, A. S. R., & Bem, S. (2000) *Emotions and beliefs: How feelings influence thoughts*. Cambridge: Cambridge University Press.
Geers, A. L., Handley, I. M., & McLarney, A. R. (2003). Discerning the role of optimism in persuasion: The valence- enhancement hypothesis. *Journal of Personality and Social Psychology, 85*, 544–565.
Geers, A. L., Helfer, S. G., Kosbab, K., Weiland, P. E., & Landry, S. J. (2005). Reconsidering the role of personality in placebo effects: Dispositional optimism, situational expectations, and the placebo response. *Journal of Psychosomatic Research, 58*, 121–127.
Geers, A. L., Kosbab, K., Helfer, S. G., Weiland, P. E., & Wellman, J. A. (2007). Further evidence for individual differences in placebo responding: An interactionist perspective. *Journal of Psychosomatic Research, 62*, 563–570.
Glover, J. (2011). Systems of belief. *Philosophy Bites Podcast*, October 9, 2011. [Online] Available from: http://philosophybites.com/2011/10/jonathan-glover-on-systems-of-belief.html. 12 Nov 2012.
Gottman, J. M., Murrary, J. D., Swanson, C. C., Tyson, R., & Swanson, K. R. (2002). *The mathematics of marriage: Dynamic nonlinear models*. Cambridge: MIT Press.
Hahn, R. A. (1997). The nocebo phenomenon: Concept, evidence, and implications for public health. *Preventive Medicine, 26*, 607–611.
Harris, S. (2004). *The end of faith: Religion, terror, and the future of reason*. New York: W.W. Norton and Company.
Hashis, I., Hai, H. K., Harvey, W., Feinmann, C., & Harris, M. (1988). Reduction of postoperative pain and swelling by ultrasound treatment: A placebo effect. *Pain, 33*, 303–311.
Hoewitz, R. I., Viscoli, C. M., Berkman, L., Donaldson, R. M., Horwitz, S. M., & Murray, C. J. (1990). Treatment adherence and risk of death after a myocardial infarction. *Lancet, 336*, 542–545.
Holroyd, K. A., Penzien, D. B., Hursey, K. G., Yobin, D. L., Rogers, L., & Holm, J. E. (1984). Change mechanisms in EMG biofeedback training: Cognitive change underlying improvements in tension headache. *Journal of Consulting and Clinical Psychology, 52*, 1039–1053.
Horback, S., & Rothery- Jackson, C. (2007). Cultural marginality: Exploration of self- esteem and cross cultural adaption of the marginalized individual: An investigation of the second generation Hare Krishnas. *Journal of Intercultural Communication, 14*. Retrieved from:http://immi.se/intercultural/nr14/horback.htm
Horbjartsson, A., & Gotzsche, P. C. (2001). Is the placebo poweress? An analysis of clinical trials comparing placebo with no treatment. *New England Journal of Medicine, 344*, 1594–1620.
Hyland, M. E., Whalley, B., & Geraghty, A. W. A. (2007). Dispositional predictor of placebo responding: A motivational interpretation of flower essence and gratitude therapy. *Psychosom Res, 62*, 331–340.
Ihlebaek, C., Love, T., Eilertsen, D. E., & Magnussen, S. (2003). Memory for a staged criminal event witnessed live and on video. *Memory, 11*, 319.

Isaacowitz, D. M. (2005). The gaze of the optimist. *Personality and Social Psychology Bulletin, 31*, 407–415.
Jensen, M. P., Turner, J. A., Romano, J. M., & Strom, S. E. (1995). The chronic pain coping inventory: Development and preliminary validation. *Pain, 60*, 203–216.
Jensen, M. P., Romano, J. M., Turner, J. A., Good, A. B., & Wald, L. H. (1999). Patient beliefs predict patient functioning: Further support for a cognitive-behavioral model of chronic pain. *Pain, 81*, 95–104.
Johari-Fard, R. (2012). *Introduction to psychosomatic medicine*. Aahvaz: IAU publication.
John, O. P., & Srivastava, S. (1999). The big five trait taxonomy: History, measurement, and theoretical perspectives. In L. A. Pervin & O. P. John (Eds.), *Handbook of personality: Theory and research* (2nd ed., pp. 102–138). Guilford.
Jones, R. E., Moes, N., Zwickey, H., Cunningham, C. L., Gregory, W. L., & Oken, B. (2008). Treatment of experimental encephalomyelitis with alpha lipoic acid and associative conditioning. *Behavior and Immunity, 22*, 538–543.
Kaasinen, V., Aalto, S., Nagren, K., & Rinne, J. O. (2004). Expectation of caffeine induces dopaminergic responses in humans. *European Journal of Neuroscience, 19*, 2352–2356.
Kaptchuck, T. J., Goldman, P., Stone, D. A., & Stason, W. B. (2000). Do medical devices have enhanced placebo effects? *Journal of Clinical Epidemiology, 53*, 786–792.
Kelley, J. M., Lembo, A. J., Ablon, J. S., Villanueva, J. J., Conboy, L. A., & Levy, R. (2009). Patient and practitioner influences on the placebo effect in irritable bowel syndrome. *Psychosomatic Medicine, 71*, 789–797.
Kemeny, M. E., Rosenwassre, L. J., Panettieri, R. A., Rose, R. M., Berg-smith, S. M., & Kline, J. N. (2007). Placebo in asthma: A robust and objective phenomenon. *Journal of Allergy and Clinical Immunology, 119*, 1375–1381.
Kienle, G. S., & Kiene, H. (1997). The powerful placebo effect: Fact or fiction? *Journal of Clinical Epidemiology, 50*, 1311–1318.
Kirmayer, L. J. (2004). The cultural diversity of healing: Meaning, metaphor and mechanism. *British Medical Bulletin, 69*(1), 33–48.
Kirsch, I. (1997). Specifying nonspecifics: Psychological mechanisms of placebo effects. In A. Harrington (Ed.), *The placebo effect: An interdisciplinary exploration* (pp. 166–186). Cambridge, MA: Harvard University Press.
Kirsch, I. (1999). *How expectancies shape experience*. Washington, DC: American Psychological Association.
Kirsch, I. (2004). Conditioning, expectancy, and the placebo effect: Comment on Stewart-Williams and Podd. *Psychological Bulletin, 130*, 341–343.
Klein, S. (2003). Ohne rislken und nebenwirkugrn. *GEO, 10*, 48–64.
Kleinman, A. (1978). Concepts and a model for the comparison of medical systems as cultural systems. *Social science & medical anthropology, 12*, 85–93.
Kradin, R. (2011). The placebo response: An attachment strategy that counteracts the effects of stressrelated dysfunction. *Perspectives in Biology and Medicine, 54*(4), 438–454.
Kumin, I. (1996). *Preobject relatedness*. New York: Guilford Press.
Laska, E., & Sunshine, A. (1973). Anticipation of analgesia: A placebo effect. *Headache, 1*, 1–11.
Lock, M., & Nguyen, V. K. (2010). *An anthropology of biomedicine*. New York: Wiley- Blackwell.
Luparello, T. J., Leist, N., Lourie, C. H., & Sweet, P. (1970). The interaction of psychologic stimuli and pharmacologic agents on airway reactivity in asthmatic subjects. *Psychosomatic Medicine, 32*, 509–513.
Main, M. (1995). Attachment: Overview with implications for clinical work. In S. C. Goldberg, R. Muir, & J. Kerr (Eds.), *Attachment theory: Social, developmental and clinical perspectives*. Hillsdale: Analytic press.
Maruta, T., Colligan, R. C., & Malinchoc, M. (2002). Optimism-pessimism assessed in the 1960s and self-reported health status 30 years later. *Mayo Clinic Proceedings, 77*, 748–753.
McNair, D. M., & Barrett, J. E. (1979). Two bass scale factors and response to placebo and anxiolytic drugs. *Psychopharmacology, 65*, 163–170.

McDonald, C. J., & Mazzuca, S. A. (1983). How much of the placebo effect is really statistical regression? *Statistics in Medicine, 2*, 417–427.
McLaughlin, A. L., & Braun, K. (1998). Asian and Pacific Islander values: Consideration for health care decision-making. *Health and Social Work, 23*(2), 116–126.
McNeill, W. H. (1995). *Keeping together in time: Dance and drill in human history*. Cambridge, MA: Harvard University Press.
Merz, M., Seiberling, M., Hoxter, G., Holting, M., & Worthan, H. (1997). Elevation of liver enzymes in multiple dose trials during placebo treatment: Are they predictable? *Journal of Clinical Pharmacology, 37*, 791–978.
Miller, F. G., Emanuel, E. J., Rosenstein, D. L., & Straus, S. E. (2004). Ethical issues concerning research in complementary and alternative medicine. *JAMA, 291*, 599–604.
Moerman, D. E. (2002). Explanatory mechanisms for placebo effects: Cultural influences and the meaning response. In H. A. Guess, A. Kleinamn, J. W. kusek, & L. W. Engle (Eds.), *The science of the placebo*. London: BMJ Books.
Mondloch, M. V., Cole, D. C., & Frank, J. W. (2001). Does how you do depend on how you think you'll do? A systematic review of the evidence of a relation between patient's recovery expectations and health outcomes. *Canadian Medical Association Journal, 165*, 174–179.
Oken, B. S. (2008). Placebo effects: Clinical aspects and neurobiology. *Brain, 131*, 2812–2823.
Olson, J. M., Roese, N. J., & Zanna, M. P. (1996). Expectancies. In E. T. Higgins & A. W. Kruglanski (Eds.), *Social psychology: Handbook of basic principles* (pp. 211–238). New York: Guilford Press.
Osterberg, I., & Blaschke, T. (2005). Adherence to medication. *New England Journal of Medicine, 353*, 487–497.
Porter, D. R., & Capell, H. A. (1993). The natural history of active rheumatoid arthritis over 3-6 Months – an analysis of patients enrolled into trials of potential disease-modifying antirheumatic drugs, and treated with placebo. *British Journal of Rheumatology, 32*, 463–466.
Price, D. D., Harkins, S. W., & Barker, C. (1987). Sensory-affective relationships among different types of clinical and experimental pain. *Pain, 28*, 297–307.
Price, D. D., Milling, L. S., Kirsch, I., Duff, A., Montgomery, G. H., & Nicholls, S. S. (1999). An analysis of factors that contribute to the magnitude of placebo analgesia in an experimental paradigm. *Pain, 83*, 147–156.
Price, D. D., Finniss, D. G., & Benedetti, F. (2008). A comprehensive review of the placebo effect: Recent advances and current thought. *Annual Review of Psychology , 592*, 1–226.
Quine, W. V., & Ullian, J. S. (1978). *The web of belief* (2nd ed., p. 10). New York: McGraw-Hill.
Ray, P. (1996). *The integral culture survey: A study of the emergence of transformational values in America*. Sausalito: Institute of Noetic Sciences.
Raz, A. (2007). Hypnobo: Perspectives on hypnosis and placebo. *American Journal of Clinical Hypnosis, 50*, 29–36.
Roth, R. S. (2003). A biopsychosocial perspective on the placebo effects: Comment on Benedetti et al. (2003). *Prevention and Treatment, 6*(1), 8–17.
Siegel, S. (2002). Explanatory mechanisms for placebo effects: Pavlovian conditioning. In H. A. Guess, A. Kleinman, J. W. Kusek, & L. W. Engle (Eds.), *The science of the placebo*. London: BMJ Books.
Siegel, D. J. (2003). An interpersonal neurobiology of psychotherapy: The developing mind and the resolution of trauma. In M. Solomon & D. J. Siegel (Eds.), *Healing trauma: Attachment, mind, body, and brain* (pp. 1–56). New York: Norton.
Siegel, H. (2004). Faith, knowledge and introduction: A friendly response to hand. *Theory and Research in Education, 2*, 343–353.
Sievenpiper, J. L., Eztagha, A., Dascalu, A., & Vuksan, V. (2007). When a placebo is not a 'placebo': A placebo effect on postprandial glycaemia. *British Journal of Clinical Pharmacology, 64*, 546–549.
Simpson, S. H., Eurich, D. T., Majumder, S. R., Padwal, R. S., Tsuyuki, R. T., & Varney, J. (2006). A meta-analysis of the association between adherence to drug therapy and mortality. *British Medical Journal, 333*, 15–20.

Smith, G. R., & McDaniel, S. M. (1983). Psychologically mediated effect on the delayed hypersensitivity reaction to tuberculin in humans. *Psychosomatic Medicine, 45*, 65–70.
Smith, W. B., Gracely, R. H., & Safer, M. A. (1998). The meaning of pain: Cancer patient's rating and recall of pain intensity and affect. *Pain, 78*, 123–129.
Solbergnes, L., & Segerstorm, S. C. (2006). Dispositianal optimism and coping: A meta-analytic review. *Personality and Social Psychology Review, 10*, 235–251.
Stewart-Williams, S., & Podd, J. (2004). The placebo effect: Dissolving the expectancy versus conditioning debate. *Psychological Bulletin, 130*, 324–340.
Taylor, J. S. (2003). Confrounting "culture" in medicine's "culture of no culture". *Academic Medicine, 78*(6), 555–559.
Thagard, P. (2007). How cognition meets emotion: Beliefs, desires, and feeling as neural activity. In G. Braun, V. Doguoglu, & D. Kuenzle (Eds.), *Epistemology and emotions*. Aldershot: Ashgate.
Thomas, K. B. (1987). General practice consultations: Is there any point in being positive? *British Medical Journal, 294*, 1200–1202.
Thorn, B. E., Rich, M. A., & Boody, J. L. (1999). Pain beliefs and coping attempts. *Pain Forum, 8*, 169–171.
Toates, F. (1998). The interaction of cognitive and stimulus response processes in the control of behavior. *Neuroscience and behavioral Reviews, 22*, 59–83.
Tracey, I. (2010). Getting the pain you expect: Mechanisms of placebo, nocebo and reappraisal effects in humans. *Nature Medicine, 16*, 1277–1283.
Turner, V. (1974). *Dramas, fields, and metaphors: Symbolic action in human society*. Ithaca: Cornell University Press.
Turner, J. A., & Aaron, L. A. (2001). Pain-related catastrophizing: What is it? *Clinical Journal of Pain, 17*, 65–71.
Turner, J. A., Jensen, M. P., & Romano, J. M. (2000). Do beliefs, coping, and catastrophizing independently predict functioning in patients with chronic pain? *Pain, 85*, 115–125.
Vase, L., Robinson, M. E., Verne, G. N., & Price, D. D. (2005). Increased placebo analgesia over time in irritable bowel syndrome (IBS) patients is associated with desire and expectation but not endogenous opioid mechanisms. *Pain, 115*, 335–347.
Vodouris, N. J., Peck, C. L., & Coleman, G. (1989). Conditioned response models of placebo phenomena: Further support. *Pain, 38*, 109–116.
Volkow, N. D., Wang, G. J., Ma, Y., Fowler, J. S., Zhu, W., & Maynard, L. (2003). Expectation enhances the regional brain metabolic and the reinforcing effects of stimulants in cocaine abusers. *Journal of Neuroscience, 23*, 11461–11468.
Wasan, A. D., Kaptchuk, T. J., Daver, G., & Jamison, R. N. (2006). The association between psychopathology and placebo analgesia in patients with discogenic back pain. *Pain Medicine, 7*, 217–228.
Wickramasekera, I. (2000). How to produce not only powerful but, more importantly, reliable placebo healing and analgesia. *Advances in Mind-Body Medicine, 16*, 211–216.
Wilber, K. (2000). *Integral psychology: Consciousness, spirit, psychology, therapy*. Boston: Shambhala.
Zubieta, J. K., Yau, W. Y., Scott, D. J., & Stoher, C. S. (2006). Belief or nedd? Accounting for individual variations in the neurochemistry of the placebo effect. *Brain Behave Immun, 20*(1), 15–26.

Chapter 5
The Ritual Effect: The Healing Response to Forms and Performs

Farzad Goli and Mahboubeh Farzanegan

> This story is about a tribesman who gets sick and, despite all rituals and natural remedies, his illness becomes worse. He goes to a clinic in a big city for the first time. When he arrives at the clinic, he is confronted with a myriad of technical and showy words, which he has never heard before and does not know the exact meaning of: "Doctor…, Specialist in…, Faculty member of…, Fellowship in…".
>
> After entering the clinic, he sees the secretaries and patients in a suspended mysterious waiting room as a special space for registration. Finally, it is his turn and the secretary leads him in to the practice room. He sees the doctor with a white coat behind a desk sitting with a more or less dignified, superior posture. His gaze is felt-penetrating and deep. Evidently, it seems that he can see the underlying events beyond the physical and mental boundaries and interprets even patient's meaningless signs; finally, somebody who knows the problem and how to fix it. After several questions and examinations with mysterious instruments and additional diagnostic rituals in a laboratory, involving radiology, the wise man decodes the natural signs, usually in a deep silence. After some short comments, he writes some special jargon on a piece of paper and gives the tribesman a coded treasure map, he remembers that their witchdoctor writes such spells when they get sick. It is the same ancient way for healing.
>
> In this moment, the tribesman feels relaxed and relieved of his symptoms without any reason. Full of hope at the prospect of finding a solution, he goes to a pharmacy to have his prescription decoded. Then he receives his *pharmacon*, a special set of tablets, ampoules, syrups, etc. The tribesman has to use these drugs at determined times every day and he should perform these rituals carefully…

From this viewpoint, a surgical operation, psychotherapy or an acupuncture session could be seen as healing rituals. A tribesman can consider all of them as different types of neoshamanistic ceremonies for healing. He may not care about their unknown contents and chemophysical elements.

At first sight, it may seem meaningless and illogical, but this thinking model could be acknowledged as a structuralistic analysis of medical practice, which is

---

F. Goli (✉) • M. Farzanegan
Head of Danesh-e Tandorosti Institute, Isfahan, Iran

Energy Medicine University, Mill Valley, CA, USA
e-mail: Dr.fgoli@yahoo.com

F. Goli (ed.), *Biosemiotic Medicine*, Studies in Neuroscience, Consciousness and Spirituality 5, DOI 10.1007/978-3-319-35092-9_5

focused on the effect of the forms of healing rituals and contexts instead of ingredients and contents. It is not so common in biomedicine, but it is a very well -known approach in the humanities. Various clinical procedures can be analyzed as communicative forms and rituals, which can change the energy-information flow and biopsychosocial responses.

Many experimental studies show the significant effect of healing rituals on physiological procedures such as wound healing (Hall 2011). Thus, the rituals, by their bodily -spatial and verbal suggestions, can change our psychosomatic responses (Krimayer 2004; Boudewijnse 2006).

On the other hand, plenty of open -label placebo studies uphold the pure for mal and ritual aspect of the therapies without any direct or indirect suggestion. In this method, there is no chemophysical effective agent, and no expectation to receive such therapies, because patients are informed that they are using a placebo (see, e.g., Sedgwick 2014; Kaptchuk et al. 2010; Day and Williams 2007). It seems that this pure ritual effect could be seen as an estimated ingredient of the placebo effect, and of course medical practice. To understand these amazing phenomena, we should first know a bit more about rites as perfomative metaphors and the elements and dynamisms of them. After wards, we can consider the role of rites in healing procedures.

5.1 The Function of Rituals

Rituals are a feature of all human societies, large and small, modern and traditional. They are an important part of the way that any social group celebrates, maintains and renews the world in which it lives, and the way it deals with the dangers and uncertainties that threaten the world (Helman 2000).

Rituals represent symbolic actions that restructure meaning and create situations which participants aim to control (Lévi-Strauss 1969; Choi 2003). Turner (1968, 1969) explains two basic functions of rituals: an expressive function and a creative function. In its expressive aspect, a ritual portrays certain key values and cultural orientations in a symbolic form, that is, it expresses these basic values in a dramatic form, and communicates them to both participants and spectators. In its creative aspects, rituals actually create or recreate the categories through which man perceives reality – the axioms underlying the structure of society and the laws of the natural and moral orders. It therefore restates, on a regular basis, certain values and principles of a society and how its members should act vis-à-vis other men, gods and the natural world. It helps to recreate the collective view of the world in the minds of the participants.

Both expressive and creative aspects indicate a transmission from the inside out, or from idea to matter. As Turner (1982) indicates, rituals of birth, death, marriage, seasonal changes, initiation, and healing are all indeed the vitas of passage or transition. In a similar way, rituals are transitional objects in a potential space (Winnicott 1971) which connect internal mental processes to external social processes

(Boudewijnse 2006), the ways in which we can control our drives (Gay 1975), develop our individuality in relation with the "other" (See Jung 1968) and a symbolic, adaptive response to reality.

In other words, rituals are spatial -bodily metaphors in which intentions and emotions are translated into the form of gestures, movements, postures clothing, makeup, icons, chemical (e.g. smoking, eating), and physical (e.g. drumming, dancing, singing, biting, whispering) procedures. According to Parkin (1992), each ritual is a bodily journeying and a symbolic passage to a new desirable condition which connotes.

Each part of the body, each movement in space, body orientation, and the relation with others ' bodies is a metaphor which can express emotions and ideas, and communication with the self, the other and the whole (see Halprin 2003). These bodily disciplines can change our mental activities. Rossano (2011, p. 40) defines a ritual as an attention -getting, formalized and invariantly ordered sequence of behaviors designed to convey particular meaning (see also Bell 1997; Rappaport 1999).

From a developmental standpoint, modern cognition emerged as a byproduct of the mental requirements for successful ritual performance -sustaining attentional focus, increasing working memory, inhibiting pre -potent responses, and retaining calm equanimity in the face of distracting, even threatening signals. In short, rituals made us human (Rossano 2011, p. 51).

Indeed, the ritual is a performative, stereotypic, and metaphoric language, which tries to translate mind to body, self to other, past to present, and heaven to earth. The power of rituals may come from these symbolic functions and the way it can integrate our intra/inter/transpersonal fields. Many great anthropologists and semioticians explain the semiotic, synthetic, linguistic, and pragmatic views of rituals.

Kreinath (2006) introduces a set of characteristics of sign processing among rituals. The combination of different types and amounts of each variable in a particular context determines the uniqueness of dynamics and effects of rituals. He states:

> The signs in ritual have to show by their usage that they follow their own logic and composition and thereby unfold their own dynamic and efficacy, which can be ascribed to them… with regard to a set of seven distinctive features:
>
> 1. *Sequentiality*, that is, how ritual acts and utterances are related to one another in a particular way and function therefore as specific vectors and not as abstract variables;
> 2. *Regularity*, that is, how the rules that inherently regulate the performance of ritual acts and utterances configure the respective pattern in the ritual performance, in terms of self-similarity;
> 3. *Referentiality*, that is, how ritual acts and utterances constantly indicate themselves by referring back to their respective contexts;
> 4. *Formality*, that is, how ritual performances indicate that they are based on particular modes of action and utterance by embodying themselves and becoming similar to themselves and sensitive of, and dependent upon, the contexts that they generate;
> 5. *Temporality*, that is, how ritual acts and utterances exist only in the present moment of their performance by mirroring their actual presence in that they create their own frame of reference;
> 6. *Dynamics*, that is, how every interplay among participants, which presupposes their agency to choose intentionally between options, configures reciprocal patterns of interaction and relation among them (as those who act and on whom is acted), which

change over the course of ritual performance and have irreversible consequences for the outcome of the ritual performance; and

7. *Efficacy*, that is, how the performance of ritual acts and utterances establishes and transforms the (symmetrical and asymmetrical) relations among the participants by determining the differences and similarities between them in charging or discharging their agency. (pp. 429–470)

Now, it is more clear that when we talk about, for example a healing ritual, it is not exclusively about a superstitious, ethnic, compulsive, and/or historical matter, but a current pervasive phenomenon by a vast variety of types and effects. We are still communicating with others and changing ourselves by the embodied language of rituals in birth, death, marriage, and healing. These goal-directed, interactive, stereotypic behaviors by any origin – historical or natural events, and/or science and technology – could induce plenty of biopsychological responses. We should be aware of the ritual effect, and we can use this metaphoric language systematically to optimize its functions.

By reviewing the different aspects, elements, and effects, we can conclude the functions of rituals as follows:

1. Developing working memory
2. Sustaining attentional focus
3. Bodily expression of intentions and emotions
4. Inhibiting pre-potent responses
5. Altering state of consciousness
6. Inducing a cathartic transition
7. Increasing suggestibility
8. Including verbal suggestions
9. Spatial-bodily metaphors
10. Role playing
11. Establishing psychosocial identity
12. Integrating intra/inter/transpersonal fields

A glance at the list discloses the biopsychosocial dynamisms of rituals. By contemplating these functions, we can imagine how a combination of these dynamisms among a certain ritual can change our mood, beliefs, behaviors and/or psychophysical procedures. It is not hard to understand why we perform rites to form our life.

5.2 Rites of Healing

Healing is a lifeworld recreating process. In the course of the healing process we should reframe our illness, our future and our self. Turner (1969) has examined the forms and meanings of ritual symbols, particularly those used in healing rituals. Each symbol has a wide range of associations for those taking part in the ritual. It tells them something about the values of their society, how it is organized and how it views the natural and supernatural worlds. This restatement of basic values is particularly important in times of danger and uncertainty when people feel that their world is threatened by misfortunes such as an accident, famine, war, death, severe interpersonal conflicts, or illness (Helman 2000). Therefore, a disaster like an illness pushes the afflicted person to draw a new lifeworld by performing healing rituals. Many anthropologists and psychologists find a great deal of similarities between rituals and therapies, specifically psychotherapies. In this part, we will focus more on the mutual relation of ritual and healing, rituals as healing procedures, and healing procedures as rituals.

Boudewijnse (2006, pp. 129–130) reviews the relation between ritual and treatment in three fields: first, the usefulness of a "ritual" in psychotherapeutic treatments. In the case of problematic mourning processes, for instance, the wholesome effects of designing and practicing personal rituals of transition are emphasized (Aune and Demarinis 1996). Secondly, the similarities between (religious) rituals and therapy are addressed; the therapeutic session as a ritual setting on one hand and the religious ritual as a therapeutic setting on the other. Thirdly, the importance of family rituals in patient management is emphasized (Imber-Black et al. 1988). Therefore, rituals could be held as biopsychosocial therapies and all different types of the therapies could be studied as healing rituals.

Rituals also create an opportunity for important work to be done independent of a therapist. They create a process for growth and healing that depends on the clients' energy and commitment to growth. This is a resource of a therapeutic process that empowers the client and gives him a tool that can be used throughout life, perhaps instead of returning to the therapist (McMillan 2006, p. 31).

As we discussed before, the cathartic and transformative effects are common in the various types of rituals, but these properties are more tangible in the healing rituals. The forms and meanings of the healing rituals and symbols are arranged around a journey from "being ill" to "being healthy". It is a metaphoric voyage through mind, body, space, time, and of course others.

Helman (2000, p. 164) explains healing rituals as social transitions by which an "ill person" is transformed into a "healing person". For instance, a patient admitted to hospital, leaves his/her normal life behind and enters a state of limbo characterized by a sense of vulnerability and danger. Their clothing is removed and replaced by a uniform of pajamas or a nightdress. In the ward, they are allocated a number, and transformed into a "case" for diagnosis and treatment. Later, when they have recovered, they regain their own clothes and rejoin their community in the new social identity of either a "cured" or a "healthier" person. Van Gennep (1960) explains three stages of separation, transition and incorporation, as illustrated in Fig. 5.1.

In the biomedical frame of reference, these procedures are introduced as formal and pragmatic measures relied o n scientific axioms and rules, but from an anthropological point of view, these healing rituals are based upon the biomedicine's web of belief and work as performative, metaphoric suggestions which determine responses of both the healers and those being healed.

Individual doctors employ the potent symbols of medical science (such as a white coat or a stethoscope) in their rituals of healing in the same way that non-

Fig. 5.1 Hospitalization as a ritual of social transition (Form Van Gennep 1960)

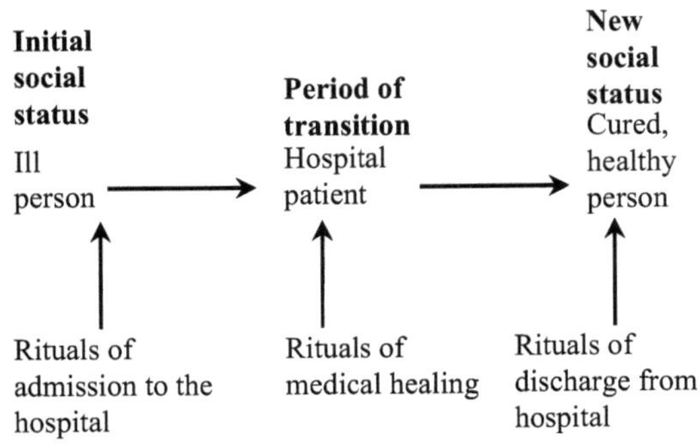

western healers employ certain religious symbols or artifacts (such as certain plants, a talisman, divination stories, holy tents, or statuettes) that also symbolize powerful healing forces (such as gods, spirits or ancestors). In this way, the use of these symbols brings the wider values of the society directly into the doctor -patient interaction. Ritual symbols can be "decoded" only by looking at the context in which they appear. For example, a white coat worn in a hospital setting has a different range of associations from one worn by a supermarket employee. Some of these associations are a license to practice medicine, authority to take a patient's history, examine their body, prescribe treatments, also cleanliness, emotional and sexual detachment, reliability efficiency, and a familiarity with suffering and death (Helman 2000, pp. 157–159). Thus, each element should be interpreted in the whole story of healing.

Medicine is largely about storytelling in which interpretations, narratives, metaphors, and symbols are fundamental tools of the trade (Charon 2001; Montgomery 1991). Ill persons experience meaning in their illnesses; they see themselves as characters in a life narrative, and they approach medicine as a vast network of healing symbols (Coulehan 2003).

The meaning of a healing technique is related to the clinical context; like a word to a sentence, the clinical setting to the medical model. We need a clinical narratology in order to realize the systemic properties of each therapeutic element. From this structuralistic viewpoint, the rationales and documents behind the efficacy of each therapeutic measure are less important than how each element can work in a certain set of therapeutic elements.

In the study of literary works, structuralism and its science of signs is distinguished by its rejection of those traditional notions, according to which literature "expresses" an author 's meaning or "reflects" reality. Instead, the "text" is seen as an objective structure activating various codes and conventions independent of author, reader, and external reality. Structuralist criticism is less interested in interpreting what literary work s mean than in explaining "how" they can mean what they mean, that is, in showing what implicit rules and conventions are operating in a given work (see Baldick 2008). Thus, a simple sign such as a white coat or a diagnostic procedure in the clinical context can obviously induce plenty of healing suggestions and there is no need to explain that such healing signs and therapeutic metaphors saturate the text of therapy.

5.3 Medical Systems as Embedded Therapeutic Metaphors

The metaphor in itself is figurative and bodily, and in fact is the base of the language; the hieroglyph of mind which forms and changes our lifeworlds (see Lakoff and Johnson 2003). Obviously, rituals as verbal-bodily-spatial therapeutic metaphors can change attention, memory, and healing expectation, and create a deep impression of the possibility of being healthy. The explanatory model, preparation phase and performance of a ritual can semiotically change health belief, behavior, and psychoneuroimmunologic responses by their direct and indirect suggestions. Krimayer (2004) emphasized that healing rituals and other symbolic actions can thus have effects on physiology, experience, interpersonal, interaction and social positioning. The comparative study of healing systems has shed light on the universal elements of healing as well as culture-specific features.

The great diversity of systems of medicine is reflected in the comparable diversity of models and metaphors for healing. As mentioned, healing involves a basic logic of transformation from sickness to wellness that is enacted through culturally-salient metaphorical actions. At the heart of any healing practice are metaphorical transformations of the quality of experience (from feeling ill to wellness) and identity of the person (from afflicted to healed). The metaphoric logic of specific modalities of healing often follows the associated model of affliction (Ibid). Where illness is understood as the result of mechanical of physical injury, specific physical measures may be taken. When the spirit comes to dwell within or possess the afflicted, it must be exorcized (Goodman 1988).

Healing rituals are associated with old traditions and ethnomedicine but biomedicine, despite its powerful experimental support and its no time-no space narrative (Wilce 2007), is also linked to non-experimental beliefs and some culture-bounded rituals. Many theoretical and medical anthropologic studies properly explore the myths and rituals of biomedicine (see, e.g., Focault 1975; Kleinman 1989).

Krimayer (2004) sums up that different traditional and modern healing systems rely on their core beliefs, explanatory models, and the interventions (see Table 5.1). We can add biomedicine to this list as a widespread western system which is based on two paradoxical western theories; reductionism and mind-body dualism. In this system, the disease is basically a chemophysical malfunction and the related psychosocial phenomena are considered as epiphenomena (Wulff et al. 1986). It can be predicted that the biomedical practice would be seen as a chemophysical intervention. Thus, in the next part we will focus more on the metaphors and myths of the modern medical discourse.

Table 5.1 Some common systems of healing

	Region	Theory of affliction	Healing practices
Aurveda	Indian subcontinent	Imbalance of elements or humours (dosas)	Diet, Purification, medicines
Chiropractic	Europe and north America	Misalignment of spinal column	Physical manipulations
Christian healing	Americas	Moral error, sin; demonic possession	Prayer, restitution; demonic exorcism
Divination		Offending spirits or ancestors	Offering or propitiation
Homeopathy	Widespread (orig. Northern Europe)	Life force out of balance	Administration of homeopathic remedies
Islamic medicine	Widespread (orig. possession by spirit allows propitiation Middle East)	Disturbance of heart as centre of spiritual, emotional and physical	Recitation of Quran
Naturopathy	Europe and North America	Experience Weakened state of body	Strengthen body through diet, cleansing, 'natural' remedies
'New age' (e.g. aromatherapy, Crystal healing, light therapy, Polarity therapy, Reiki)	Europe and North America	Energy imbalance	Use of materials and manipulations to 'rebalance' energy possession by spirit allows propitiation
Possession Cults (e.g. Candomble, Zar)	Widespread (Africa, Asia, South America)	Offending spirits or ancestors	
Psychotherapies	Widespread (orig. Europe)	Psychological conflict or maladaptive learning (behavioural, cognitive)	Corrective relationship, re-learning (through exposure, and cognitive or behaviour modification), insight shamanic healer travels to spirit world and with aid of spirit helper (usually an animal), redresses wrong
Shamanism	Hunter-gatherer peoples	Offending spirits, magical attack, accident	
Traditional Chinese Medicine	East Asia	Imbalance in energy (chi'i, Ying/yang) or in five phases (air, earth, wind, fire, water)	Herbal and other medicines, diet, moxibusion, acupuncture
Unani medicine	Indian subcontinent Middle East	Imbalance in humours or life force	Herbal or mineral medicines

Reprinted from Kirmayer 2004, by permission of Oxford University Press

5.4 Cultural Contexts and Healing Metaphors

It seems that each medical model could be considered as a sign system organized around one or more core metaphors. All the concepts, relations, actions and roles are generated by these metaphors. For example, "fighting the disease" is a common metaphor in biomedicine which has its roots in nineteenth century and inspired by dealing with microbial factors and these days it involves non-communicable diseases. Some medical expressions like "struggle with cancer" or "fighting with depression" are seen in the topics of several articles, books, and also as title of social institutions. Even the more diplomatic expressions such as "coping with diabetes" are based on the war metaphor but with less hope on winning and of course more emphasis on surviving. All these metaphoric descriptions show that there is an animistic image of disease in modern medicine. We imagine diseases as negative entities instead of the name of categories, and consider them as enemies instead of types of natural disorders. We do not have anything named cancer, depression or diabetes, but we have adaptive or maladaptive responses to cancerous, depressive and diabetic states. Obviously, the shamanistic mentality is still working in the core of modern medicine.

In the "fighting the disease" metaphor, treatment gives power to the warrior (e.g., serum therapy, vitamins, immune enhancer drugs, and rehabilitation), kills the enemy (e.g., antibiotics and cytotoxins), suppresses enemy (e.g., sedatives and immunosupressores), or drives the enemy out (e.g., chelation therapy and surgery).

There are also some metaphors in the popular health discourse which can affect the healing process. For example, "illness as punishment" is very common, especially in chronic and life threatening conditions. Many educated patients never confess that they have such a mythic-religious belief, but this metaphor is still much more influential than has been admitted.

It can be a reason for resisting recovery, because the patient feels that they deserve punishment. Each metaphor evokes its associated network and emotions. For instance, cancer is an obscene, unspeakable and shameful condition; the disease closely related to sin or guilt. Because of their metaphors, especially the first, people who suffer from cancer experience isolation and shame. They do not talk about their illness. They delay seeking medical care and their friends and family shy away from them. In *Illness as Metaphor*, Sontag (1978) also examined the nineteenth century cultural beliefs about tuberculosis and found that they, too, detracted from a "true", that is scientific understanding, of the disease. 10 years later, with *AIDS as Metaphor* (Sontag 1988), the author extended her anti-metaphorical analysis to HIV/AIDS, which she claimed had largely replaced cancer as the unspeakable disease in our society because it was associated with homophobia and believed to be a punishment from God (Coulehan 2003).

Had western medicine been incorporated into the patients' cultural expectations, they would need to arrange a "Sing" in order to address the more narrative dimensions of the illness, that is, to re-experience themselves as part of a meaningful story. A number of writers have looked beyond the day-to-day language to discover

Table 5.2 Medical metaphors

War metaphor Disease is the enemy. Physician is a warrior captain. Patient is a battleground.	**War statements** "I treat all my patients aggressively…" "He's a good fighter." "The war on cancer."
Parental metaphor Disease is a threat or danger. Physician is a loving parent. Patient is a child.	**Parental statements** "She's too sick to know the truth…" "We don't want him to lose hope."
Engineering metaphor The disease is malfunction. Physician is an engineer or technician. Patient is a machine.	**Engineering statements** "He's in for a tune-up." "Something's wrong, doc… you fix it." "We need to ream out your plumbing."

Reprinted from Coulehan 2003, by permission of Yale Journal of Biology and Medicine

the basic models or metaphors we use when thinking about medicine (Sontag 1988; May 1983). There are several such metaphors that to a large extent generate our vocabulary of the patient-physician relationship. Table 5.2 lists three of the most prominent and some of their implications.

It is clear that metaphors and even mythic thinking are still alive and influential in the folk, popular and professional health sectors. We must be aware of their functions and we should sometimes enlighten, reframe or deconstruct them, and sometimes we can use them consciously in our explanations and healing rituals. The relative demise of paternalism (which at least implied a human, caring interaction) has accompanied the rapid advance of engineering and war metaphors, both of which tend to objectify and dehumanize the patient (Coulehan 2003).

To clarify the ritual aspects of the diversity of medical systems, we will present different cases which are engaged in different types of therapies, one modern (a psychiatric visit), traditional (an acupuncture session), and/or shamanistic. In the case reports, we will highlight the direct and metaphoric suggestions of each therapeutic technique by neglecting the mechanisms and effectiveness of them. Therefore, we continue our narratological approach and focus on the healing power of the forms of therapies:

A friend of yours who has been suffering from migraine headaches tells you that he no longer experienced headaches after 3–4 sessions of acupuncture, and he strongly urges you to try it. After that, you go to an acupuncture practitioner. In the office, you see some Chinese icons showing Meridian's cobweb network. From the traditional viewpoint of Chinese medicine, this network provides bioenergy (Qi) for feeding organs and each obstruction in them can cause disturbances in the related organ's function. You also observe diagrams of hands, feet and the whole body with the place of each organ depicted on them. These plans show you the acupoints of the organs that, when motivated, re-stimulate the flow of bioenergy restoring healthy function of the organ. When healthy, our inner energy circulation is in coordination with universe, and female– male energies (Yin and Yang) are in equilibrium (see Wiseman et al. 1993).

When you meet the practitioner, he greets you and smiles. After taking down your medical history and performing a physical exam, he tells you about the points on the axioms of Acupuncture and prepares you for the start of the treatment. If you accept trying this treatment, it means that you, to some degree, believe these suggestions:

- There is an existential and energic connection between human and galaxy.
- There is a living matrix that relays energy and information, and forms natural functions of the body.
- The cause of your disorder is an obstruction of one or more energic flows.
- Injecting a needle in your acupoints eliminates these obstructions.
- By these stimulations, energetic equilibrium and also the natural function of organs will be restored.

By administering needle injections and the feeling of subtle electric currents in your body, the aforementioned suggestions become activated and, in addition to bioenergetics pulsations, these psychosomatic inductions can change gate-control of pain, vasomotor reactivity and the other automatic responses (ibid).

Although the surgical processes are clear and of course mostly irreplaceable, it is not exempt of mysterious inductions and therapeutic metaphors. The efficacy of sham surgeries (Moseley et al. 2002) could be due to implicated suggestive beliefs such as the cause, control, consequences and meaning of disorders. For instance, the metaphor of "removing the matter of illness from the body" is an ancient feature of healing and very easy to believe. All of the points stated above used to explain the curative process and healing belief systems can be written in the form of a hypno-therapeutic script.

We now come to our second example:

> I have identified your suffering. I have diagnosed it frequently before…Your depression is a chemical disorder in your brain. The drug you will consume will manipulate your brain in order to increase electrochemical coordination. At this time the light will return to your brain and life again. You will be happy and all things will find meaning again.

These are just a part of the inductions which someone receives after visiting a psychiatrist with a biological approach; a short visit, the prescription of a drug and finally the daily rituals of taking prescribed drug. Most of these inductions will be received after accepting a clinical setting. If the personal web of belief is compatible with the healing belief system, subsequently the compliance, illness behavior, and psychoneuroimmunologic response will be more facilitated and synergetic.

A person for whom energy flow and energy equilibrium metaphor is more believable than the chemical metaphor may prefer to go to an energy healer and will have more compliance and more effective meaning response.

As to how the responses to these therapeutic metaphors differ among peoples, it can be different among various cultures, because each culture has its own web of beliefs and each web of belief resonates the relevant health and illness behaviors, and therapeutic metaphors. Each culture and subculture, as a set of narrative beliefs and behaviors, can facilitate certain habits, epigenetic functions (see e.g. Wallace and Wallace 2011, Cheruch 2009, Rossi 2002) and psychological responses (see e.g. Rebhun 2004). Therefore, even being a member of a cultural group can facilitate related culture-bound responses. Performing rituals is the active feature of membership. Thus, performing a healing ritual would actualize our membership, and at the same time facilitate the transformation to a healthy person.

Levi-Strauss (1967) argued that the transformations of healing involve a symbolic mapping of bodily experience onto a metaphoric space represented in myths and rituals. The narrative structure of the ritual then carries the participants into a new representational space, and with this movement, transforms their bodily experience and social position. Dow (1986) builds on Levi-Strauss's account to suggest that symbolic healing involves mapping a personal problem onto a collective rhythmic world through emotionally charged symbols. The emotion evoked by the symbols then insures that manipulating the symbols within that mythic world will lead to corresponding transformations of patients ' illness experiences (as cited in Krimayer 2004).

As mentioned before, changes in cognition, emotion, and behavior occur during the healing, initiation, and action. In this period, we perceive – directly and indirectly – suggestions from explanation, perception, and performance. These inductions, like hypnotherapeutic interventions, can prompt modulations in the mind and body. This pattern could be more or less considered as an archetypical model for healing. When we introduce ourselves as a member of a culture and/or subculture, it means that some of our responses are epigenetically facilitated to some extent. We accept an explanatory and management model, and our compliance, health behavior, expectations and bodily responses will be subsequently modified.

For example, among the Navajo, all serious illnesses are thought to result from disharmony. To become sick, a person has somehow fallen out of harmony with himself, his family, his clan, and the network of relationships that constitute the Navajo way. To be healed is to have that harmony restored (Explanatory suggestions). In order to accomplish this, the patient first has to consult a diagnostician who, by means of hand trembling or other forms of divination, establishes the cause of the illness (Preparation suggestions). The diagnostician then prescribes an appropriate ceremony or "Sing" which consists of storytelling, chanting, sand painting and other elaborate rituals that may go on for three to nine days (Performance suggestions).

It implicates that by accepting a membership, we receive the explanatory suggestions; that is, we receipt a set of beliefs related to various situations and modes. When we encounter the mentioned situation or mode, the associated beliefs are initiated; the hypnotic procedure which will be accomplished by preparation and performance trials.

The therapeutic metaphors can change the semiosis through the molecular, cellular, personal, familiar and social fields.

The compatibility of these membership, preparation and performance metaphors with the personal factors such as temperament, attributional style and coping strategies can resonate placebo responses. For example, I suggest you to take this case scenario into consideration:

> Naser was a 19 year-old boy and a freshman. He had been suffering from tonic clonic attacks since he was 16. He lived on *Qeshm* Island in the Persian Gulf. In the manner of the ethnic beliefs of his local culture, he had suffered from a kind of possession by non-organic creatures called *Zar*. In line with the neighbors' advice, his parents took him to two witch doctors (Babazar and Mamazar) and they held a *Zar* therapy ceremony for him. The boy and his family reported that after visiting the witch doctor, the intensity and frequency of his attacks had decreased for 2–3 weeks but they came back again and after a few weeks the frequency had increased to every day. The boy went to a neurologist following the advice of his classmate and after taking a physical exam and examining his medical history and EEG, the doctor made the diagnosis that he had juvenile myoclonic epilepsy. The doctor explained the disease and assured him that his disease is caused by his abnormal electrochemical brain discharge that led to abnormal motional and cognitive functions. The physician prescribed him *Lamotrigine*, which controls and stabilizes neural cell activity by sodium channel inhibitions. Because he did not have enough money to buy his drugs, it took some weeks before he was able to purchase them. Interestingly, his anxiety and troubled thoughts about the illness were quelled and his attacks even stopped. Afterwards, his seizures were controlled completely by using the drug.

As we see here, parallel and consecutive flows of signs in different levels of organization changed the course of the patient's symptoms; the signs such as: "Explaining seizures as possession" or "electrochemical irregularity", "belief in hypnotic rites with drumming and praying by bizarre jabbers" which are supposed to drive the unknown creature out, and/or "belief in the antiepilepsy medicine" which is supposed to promote electrical and neurological balance, and finally "*Lamotrigine* molecules" which can inhibit stimulation by voltage-sensitive-sodium-channel and give more stability to neuron membrane.

In accordance with the history, Naser belongs to at least two subcultures: the traditional (Qeshm culture) and the modern (academic culture). Each of them facilitated compliance and psychophysical response to a medical model, but it seems that the modern belief system was more compatible for him. If he had believed more in the shamanistic model, he might have benefitted from it more like many from his traditional culture.

These days, cultural identity or feeling of belonging to social groups is much more complex than what an authentic Navajo feels. We can recognize several subcultures in a society, and even in a family. This is very important knowledge because each subsystem induces its special set of expectations and prescriptions. Ray (1996) identifies three subsets of American culture that are influencing the demand for health services: Heartlanders, who preserve traditional values; cultural moderns; and cultural creatives, who tend to favor holistic health and health-oriented services. Interestingly enough, one can find different cultural identities even within the same person. We can belong to various cultures of nationalities, media, globalization, professions, religions, and political systems at the same time, and each system – by its special semiosis – could be switched in a certain situation. Usually, in overwhelming and/or chronic conditions, we tend to regress to older and mythic modes of thinking. These phenomena are distinguishable both in individuals (see, e.g., Reich 2013; Sparrow 2014; Cassirer 1946) and societies (see, e.g., Mercer 2011; Corradi 1983). Thus, it is crucial for health delivery systems to consider these cultural subsystems and their related myths, metaphors, and rites in order to optimize the semiotic effect of the services. The third case study shows the mobility of a person through the different subcultures and how a proper narrative by its accurate therapeutic metaphors can moderate the signs and quality of life.

5.5 Conclusion

Each clinical procedure is first and foremost a healing ritual by a formulation of verbal–spatial–bodily metaphors. Even before beginning the behavioral and chemophysical affects, these semiotic agents start working. The healing rituals could be considered as contextual meaning effect in comparison to the placebo studies, which are focused more on the main objects of therapy.

In fact, the healing rituals are sets of therapeutic techniques. A number of healing methods are used including those based on logic, hypnosis, intuition, telepathy, autosuggestion, the interpretation of dreams, and a kind of psychotherapy by encouraging the patient and providing hope of recovery (Stutley 2003). Thus, it is not so difficult to find similarities between ancient healing rituals and modern psychotherapies and psychosomatic techniques. It seems that healing rituals have been gradually purified from their mythic and supernatural assumptions and transformed to more parsimonic and experimental forms.

The metaphors and rites implicate our limitations in the tolerance of reality. A fundamental mechanism in ritual behavior, apparently caused by an innate inability to directly cope with reality, gives rise to an unconscious process of misrecognition (Boudewijnse 2006, p. 126). So we ought to use these metaphors and perform the healing rituals very carefully in order to avoid misleading signs. Sometimes, we must reframe these metaphors to perform them in a more adaptive way. As a result, we need to construct our own realities in order to stand the unbearable lightness of being (see Kundera 1984), but sometimes we must dare to accept the essential meaninglessness of our being and perform our autogenic role based on our hermeneutic liberty.

This is the way of an authentic human being, as Heideggger (1977) addressed, creating our story by recreating our own life's author and vice versa, a way out of a priori myths and rites. By switching to a proactive meaning making, the sign flow can change to a more integrative and adaptive manner.

It seems that we can use the metaphors and rituals in our health system; it would be a shortcut for activating unconscious healing responses. But we also have an enlightening mission to unfold the deception of regressive ways of narrating and healing illness. By accepting our existential responsibility to create our worlds and ourselves, we can create our own healing narratives, metaphors, and rites; an automatic way of formulating semiotic healing cocktails.

References

Aune, M. B., & DeMarinis, V. (Eds.). (1996). *Religious and social rituals: Interdisciplinary explanations*. Albany: SUNY Press.

Baldick, C. (2008). *Oxford dictionary of literary terms*. NewYork: Oxford University Press.

Bell, C. (1997). *Ritual: Perspectives and dimensions*. New York: Oxford University Press.

Boudewijnse, B. (2006). Rituals and psyche. In J. Krenath, J. Snock, & M. Stausberg (Eds.), *Theorizing rituals: Issues, topics, approaches, concepts*. Leiden/Boston: Brill academic Publishers.

Cassirer, E. (1946). *The Myth of the state*. London: Yale University Press.

Charon, R. (2001). Narrative medicine. A model for empathy, reflection, profession and trust. *JAMA, 286*, 1897–1902.

Cheruch, D. (2009). *The genie in your genes: Epigenetic medicine and the new*. Santa Rosa: Energy Psychology Press.

Choi, C. (2003). The artistry and ritual aesthetics of urban Korean shamans. In G. Harvey (Ed.), *Shamanism: A reader* (pp. 170–185). London: Routledge.

Corradi, R. B. (1983). Psychological regression with illness. *Psychosomatics, 34*(12), 1139–1145.

Coulehan, J. (2003). Metaphor and medicine: Narrative in clinical practice. *Yale Journal of Biology and Medicine, 76*, 87–95.

Day, R. O., & Williams, K. M. (2007). Open-label extension studies: Do they provide meaningful information on the safety of new drugs? *Drug safety: An International Journal of Medical Toxicology and Drug Experience, 30*(2), 93–105.

Dow, J. (1986). Universal aspects of symbolic healing: A theoretical synthesis. *American Anthropologist, 88*, 56–69.

Focault, M. (1975). *The birth of the clinic: An archeology of medical perception*. New York: Vintage.

Gay, V. P. (1975). Psychopathology and ritual. *Psychoanalytic Review, 62*, 493–507.

Goodman, F. D. (1988). *How about demons? Possession and exorcism in the modern world*. Bloomington: Indiana University Press.

Hall, H. (2011). Sufism and healing. In H. Walach, S. Schmidt, & W. B. Jones (Eds.), *Neuroscience, consciousness and spirituality*. Netherlands: Springer.

Halprin, D. (2003). *The expressive body in life, art and therapy: Working with movement, metaphor, and meaning*. London: Jessica Kingsly.

Heideggger, M. (1977). The question concerning technology. In D. F. Krell (Ed.), *Basic writings* (pp. 283–317). San Francisco: HarperCollins Publishers.

Helman, C. G. (2000). *Culture, health and illness*. Oxford: Butterworth Heinemann.

Imber-Black, E., Roberts, J., & Whiting, R. (1988). *Rituals in families and family therapy*. New York: Norton.

Jung, C. G. (1968). *Man and his symbols*. New York: Dell Publication.

Kaptchuk, T. J., Friedlander, E., Kelley, J. M., Sanchez, M. N., Kokkotou, E., Singer, J. P., Kowalczykowski, M., Miller, F. G., Kirsch, I., & Lembo, A. J. (2010). Placebos without deception: A randomized controlled trial in irritable bowel syndrome. *PloS One, 5*, e15591.

Kleinman, A. (1989). *The illness narrative: Suffering, healing, and the human condition*. New York: Basic Books.

Kreinath, J. (2006). Semiotics. In J. Kreinath, J. Snock, & M. Stausberg (Eds.), *Theorizing rituals: Issues, topics, approaches, concepts*. Leiden/Boston : Brill academic Publishers.

Krimayer, L. J. (2004). The cultural diversity of healing: Meaning, metaphor and mechanism. *British Medical Bulletin, 69*, 33–48.

Kundera, M. (1984). *The unbearable lightness of being*. New York: Harper & Row.

Lakoff, G., & Johnson, M. (2003). *Metaphors we live by*. Chicago: University of Chicago Press.

Levi-Strauss, C. (1967). *Structural Anthropology*. New York: Basic Books.

Lévi-Strauss, C. (1969). *The Raw and the Cooked* (J. Weightman, & D. Weightman, Trans.). New York: Harper and Row.

May, W. F. (1983). *The physician's covenant: Images of the healer in medical ethics*. Philadelphia: The Westminster Press.

Mcmillan, D. W. (2006). *Emotion rituals (A resource for therapists and clients)*. New York: Routledge.

Mercer, J. (2011). The concept of psychological regression: Metaphors, mapping, Quin square, and Tavistock square. *History of Psychology, 14*(2), 174–196.

Montgomery, K. (1991). *Doctors' Stories: The narrative structure of medical knowledge*. Princeton: Princeton University Press.

Moseley, J. B., O'Malley, K., Petersen, N. J., Menke, T. J., Brody, B. A., Kuykendall, D. H., Hollingsworth, J. C., Ashton, C. M., & Wray, N. P. (2002). A controlled trial of arthroscopic surgery for osteoarthritis of the knee. *The New England Journal of Medicine, 347*, 81–88.

Parkin, D. (1992). Ritual as spatial direction and bodily division. In D. D. Coppet (Ed.), *Understanding rituals*. London/New York: Routledge.

Ray, P. H. (1996). *The integral culture survey: A study of the emergence of transformational values in America*. Sausalito: Institute of Noetic Sciences.

Rebhun, L. A. (2004). Culture-bound Syndromes. In C. R. Ember & M. Ember (Eds.), *Encyclopedia of Medical Anthropology: Health and illness in the world's cultures* (pp. 319–327). New York: Klower Academic/Plenum Publishers.

Reich, W. (2013). *The mass psychology of fascism*. New York: Farrar, Straus and Giroux.

Rossano, M. J. (2011). Setting our own terms: How we used ritual to become human. In H. Walach, S. Schmidt, & W. B. Jonas (Eds.), *Neuroscience, consciousness and spirituality*. New York : Springer.

Rossi, E. (2002). *The psychobiology of gene expression*. New York: Norton.

Sedgwick, P. (2014). Placebo controlled trials. *BMJ, 348*, g1635.

Sontag, S. (1978). *Illness as metaphor*. New York: Farrar, Straus and Giroux.

Sontag, S. (1988). *AIDS as metaphor*. New York: Farrar, Straus and Giroux.

Sparrow, L. M. (2014). Beyond multicultural man: Complexities of identity. In M. K. Asante, Y. Miike, & J. Yin (Eds.), *The global intercultural communication reader*. New York: Routledge.

Stutley, M. (2003). *Shamanism: An introduction*. London: Routledge.

Turner, V. W. (1968). *The drums of affliction*. Oxford: Clarendon.

Turner, V. W. (1969). *The ritual process: Structure and antistructure*. Middlesex: Penguin.

Turner, V. W. (1982). *From ritual to theatre: The human seriousness of play*. New York: Performing arts Journal Publication.

van Gennep, A. (1960). *The rites of passage* (trans: Vizedom, M. D. & Cafee, G. L.). London: Rutledge and Kegan Paul.

Wallace, R., & W., D. (2011). Cultural epigenetics: On the heritability of complex disease. In C. Priami et al. (Eds.), *Transactions on computational systems biology XIII*. Berlin: Springer.

Winnicott, D. W. (1971). *Playing and reality*. London: Psychology Press.

Wiseman, N., Ellis, A., Zmiewski, P., & Li, C. (1993). *Fundamentals of Chinese medicine*. Taipei: SMC Publishing.

Wulff, H. R., Pedersen, S. A., & Rosenberg, R. (1986). *Philosophy of medicine: An introduction*. Oxford: Blackwell.

Chapter 6
Hypnosis, Placebo, and Performance: Recovering the Relational Aspects of Medicine

Shahram Rafieian and Howard Davis

Consultation between a physician and a care seeker is a frequent and essential event in health care systems and its outcome depends not only on the medical knowledge and technical skills of the doctor, but also on their ability to communicate effectively. Changes in the paradigm of medicine affect the quality of this encounter, and the dominance of biomedicine as the working paradigm of modern medicine has resulted in a change to the pattern of doctor–patient interaction (Morgan 2008). In biomedicine, the emphasis is on the human as a biological and physiological machine, and as a result, sciences such as biology, physiology, and anatomy are considered to be the "basic sciences" (Pauli et al. 2000; Greaves 2002). The model of interaction which has emerged in the biomedical approach is the paternalistic (guidance–cooperation) relationship, in which the doctor is the expert and source of knowledge who gives advice that the patient should passively accept. Although this model functions well in some situations such as emergencies and in the management of acute infectious diseases, it is insufficient in many other contexts such as the care and management of chronic diseases (Morgan 2008). In spite of many efforts to eradicate chronic health problems such as high blood pressure and diabetes, there is still a long way to go to achieve this goal (Wagner et al. 2001). In chronic health conditions, after a while, the patient becomes familiar with the basic medical and technical knowledge about their problem and clinical consultations mainly focus on issues related to long term management of the disease. As a result, it has been argued that a patient-centered approach, in which the patient's views, feelings, thoughts and needs are respected, is especially necessary in the care of

S. Rafieian (✉)
Danesh-e Tandorsti Institute, Isfahan, Iran

Bangor University, Bangor, UK
e-mail: rafieiansh@yahoo.com

H. Davis
School of Social Sciences, Bangor University, Bangor, UK
e-mail: h.h.davis@bangor.ac.uk

F. Goli (ed.), *Biosemiotic Medicine*, Studies in Neuroscience, Consciousness and Spirituality 5, DOI 10.1007/978-3-319-35092-9_6

chronic conditions. Such an approach improves the quality of decision making for patient care, results in adherence to treatment and in more effective ways of addressing the patient's concerns (Mead and Bower 2000). Furthermore, as will be discussed in this chapter, this relationship has a healing effect per se.

Considering these facts, the aim of this chapter is to explore the power and potentials of the relational space as an important but insufficiently explored area of research in medicine. For this purpose, two fields of research, hypnosis and placebo, are briefly considered and the place of performance in the development of healing response in these contexts will be examined. Finally, we will try to show the consequences of these understandings, such as the need for the use of qualitative research methods.

6.1 Hypnosis: Imagination and Role-Taking

Although the astonishing power of suggestion has been described in texts dating back to ancient times, the first reports of hypnosis being used as a therapeutic modality go back to the eighteenth century and Austrian physician Franz Anton Mesmer (1734–1815). In his system of treatment, it was believed that there is a special fluid present everywhere in the body and any disturbance in the distribution of this fluid results in the formation of diseases. The idea of the presence of such a fluid was not verified but certain characteristics of Mesmer's healing method have been regarded as important factors in his success. In a treatment session, a crisis was an essential episode in which patients experienced a trance-like state, after which the healing occurred. To induce the crisis, the dramaturgical nature of Mesmer 's practice was crucial. He was a charismatic character, dressed in a showy manner and the room in which treatment was performed was elaborately decorated (Ellenberger 1970; Whorwell 2005).

Since that time, extensive research has been done in the field of hypnosis in medicine and the effectiveness of hypnotherapy for the treatment of different health conditions has been shown. Hypnotherapy helps patients with a wide range of problems such as chronic pain, asthma, irritable bowel syndrome, warts, migraine, and anxiety (Olness 2008; Heap and Aravind 2002; Kroger 2008). In spite of this proven effectiveness, hypnotherapy has not been incorporated into mainstream treatments in modern medicine. Upshaw (2006) argues that negative perceptions among medical educators, practitioners, patients and even the general public are barriers to appropriate use of hypnosis in medicine. As Olness (2008) explains, there are certain misperceptions that make the general public fearful of hypnosis, such as the incorrect idea that hypnosis is sleep or the notion that the hypnotist takes control of the subject during hypnosis.

In terms of the underpinning mechanisms of hypnotherapy, there is a controversy among researchers. Some believe that hypnosis happens when the person goes into a special state of consciousness different from ordinary consciousness in everyday life. Others believe that although alteration of consciousness happens under hypnosis, an extraordinary change in the state of consciousness is not necessary for experiencing hypnotic phenomena. The debate between these two groups of researchers has resulted in the formation of so-called state and non-state theories of hypnosis. The non-state theories of hypnosis, also called sociocognitive theories of hypnosis, emphasize the psychosocial context in which the hypnosis happens (Lynn et al. 2008; Kallio and Revonsuo 2003).

The sociocognitive theorists of hypnosis believe that hypnotic phenomena are the result of suggestions and can be experienced without hypnotic trance. Researchers in this field have empirically shown that all hypnotic phenomena such as hallucinations, changes in perceptions, and feeling of automaticity and involuntariness can be experienced out of the trance state as well. They conclude that the trance state is not the essential component for experiencing hypnotic phenomena and there are other factors important for these to occur (Lynn et al. 2008; Coe and Sarbin 1991). Kirsch (2001) explains:

> If the effects of suggestion are not produced by magnetism or by trance states, then how are they produced? The data point to two factors. One is a talent or ability to experience an imaginary state of affairs as if it were real. The second is the person's beliefs and expectations. The effects of suggestion, in or out of hypnosis, may be due to a tendency to experience the world as one expects to experience it. (p. 800)

Hence, in the view of the sociocognitive theorists of hypnosis, imagination, belief, and expectations of hypnotic phenomena are the most important components necessary for their development. As a result, they define hypnosis as a process of "believed-in imaginings" (Sarbin 1998).

Considering these facts, it is obvious that defining the situation has an important role in the success of the hypnotist and the quality of the experience of the hypnosis. For example, the knowledge of the subject about the nature of hypnosis and the assumptions that they have about what is experienced during hypnosis can determine the quality of the experience. Apart from that, environmental factors such as the space, smells, sounds, etc. are also factors involved (Coe and Sarbin 1991; Lynn et al. 2008). Furthermore, although it is generally assumed that the subject under hypnosis has a passive role, this is not in reality the case. During hypnosis, especially at the beginning, the subject has to cooperate and actively participate and use their imagination to experience the hypnotic phenomena. In fact, it is after a while and only in so called "deep hypnosis" that the feeling of automaticity or involuntariness is experienced (Miller 1994; Straus 1978). Consequently, the general belief in people's minds that the relationship in hypnosis is unidirectional and the hypnotist applies their power and controls the subject by suggestions is wrong. Hypnosis is impossible without the cooperation of the subject and in fact they need to actively participate and use their imagination and creativity in order to experience the hypnotic phenomena. The hypnosis will not have a good outcome if the beliefs, interests, feelings, and emotions of the subject are not considered when formulating the suggestions (Laurens 2007). In other words, as Coe and Sarbin state (1991), by performing the rituals of hypnosis induction, the hypnotist implicitly or explicitly conveys this message to the subject: "Please participate in a miniature drama" (p. 317), and it is the choice of the subject whether to take on this role or not.

To clarify the meaning of believed-in imaginings, a distinction should be made between role playing and role taking. Hypnosis is a role taking process similar to other social roles that people take in their life. For example, when teachers perform their job in the classroom, they believe that they are teachers and because of that, embody this role with the totality of their mind-body system. It means that they are not faking the role of a teacher. In a similar way, the person under hypnosis takes the role of a hypnotic subject. Defining hypnosis as role taking has actually been a source of misunderstanding for many researchers, but sociocognitive theorists do not say that hypnosis is role playing or faking. Instead, they define it as role taking similar which resembles social roles that people embody in their daily lives (Lynn et al. 2008; Coe and Sarbin 1991; Kirsch 1998).

Even in the case of role-playing, it has been shown that deep engagement with a role can alter the state of consciousness. In an empirical study by Scheiffele (2001), a scale measuring different dimensions of altered states of consciousness was given immediately after an improvisational acting exercise to a group of drama students and the results show that a significant alteration occurred in their state of consciousness. The change in the state of consciousness after acting can be so profound that a director may sometimes use de-roling techniques to bring the actor back to the ordinary consciousness (Scheiffele 2001, p. 187; Scheiffele 2003, p. 16). Similarly, in his contribution, Zarrilli (2011) has explored the effects of performance on altering consciousness in various modes of role playing and theatre in West and East. In a historical approach, he points to the fact that, the modern aesthetic theatre is rooted in the early forms of ritual/shamanistic practice. These practices have various functions such as pleasing gods, predicting the future and getting power for diagnosis and treatment of diseases. In the context of non-western traditions of performance, he states that, in these practices, direct experiential knowledge has always been central. He brings different examples of these psychophysical methods of altering consciousness in eastern traditions such as Yoga, Zen meditation, martial arts, and genres of performance such as kutiyattam and kathakali in India, and noh in Japan. In the West, he notes that under the influence of Russian director Konstantin Stanislavsky (1863–1938), an innovative system of actor training began in which various modalities and techniques were applied to actualize the altered states of consciousness necessary for performance.

In the context of medicine, the corollary of the above discussion is that the healing effects of hypnosis might also be achieved outside the hypnotic trance, a consequence which connects hypnosis to the placebo effect (Kirsch 1985, 1999). Here, placebo effects will be introduced briefly and then the connection between these two will be explored in more detail.

6.2 Placebo, Meaning, and Interpersonality

Placebo is an inert medicine or intervention, such as a sugar pill which is prescribed for the patient condition. Usually the doctor knows that the placebo does not have any therapeutic effect in that condition but tells the patient that it will help and in many cases it is effective (Miller and Kaptchuk 2008, p. 222). Studies have shown that many doctors prescribe placebos in their practice (Meissner et al. 2012; Fässler et al. 2010; Howick et al. 2013). There is an ongoing debate about the ethical concerns related to prescribing placebos because especially in modern medical ethics, there is an emphasis on the autonomy and rights of the patients to know about the treatments and the interventions used in the management of their problems. In spite of that, many clinicians believe that prescribing placebos is not unethical and their benefits outweigh the disadvantages (Bensing and Verheul 2010).

American Anthropologist Daniel Moerman (2002a, b, c, 2006, 2011) emphasizes the fact that we already know the placebo itself – in the form of a pill or any other intervention – is inert and the mechanisms of healing involved are not related to the action of placebo. Then if the placebo has no role here, how does the healing happen? Moerman (2011) believes that the meaning of the placebo for the patient is the key factor and describes the placebo response as a "meaning response". Consequently, any change in the intervention that could alter its meaning could affect the outcome of the treatment. For example, injecting the drug may be more effective that prescribing it orally. Also any information given about the intervention could change its effects as well. Even minor characteristics of the intervention such as the color, shape, or smell of the pills and medicines used are important. In summary, the context in which the placebo is prescribed assigns meaning to it and this meaning is the basis for activation of the healing response involved in the emergence of placebo effects.

As it might be imagined, the results of suggestions coming from the environment are not always positive; they can create negative consequences as well. In fact, this notion has been identified and the term nocebo has been coined to describe it. Here this concept will be briefly reviewed.

6.3 The Nocebo Effect

Nocebo effects occur when negative expectations or affective states result in the development of sickness, symptoms, or exacerbation of an existing health condition (Hahn 1997; Cohen 2014). The classical form of nocebo effect is "Voodoo death" which has been described by American physiologist, Walter B. Cannon (1942). His work brought to light anthropological evidence from the natives of various regions such as South America, Africa, Australia, and New Zealand that sudden and inexplicable death happened in response o being cursed. When these people were informed that they had been cursed, their health condition deteriorated suddenly and died within a short period of time. Cannon attributed this phenomenon to psychological shock and emotional stress resulting from the bad news and their impact on the internal physiological balance of the victim's body.

Hahn (1997) considers four categories for nocebo effect. The first originates in the internal world of the individual. Negative mood and certain psychological conditions are more frequently associated with negative expectations. For example, depressed people usually lack hope and more frequently expect pathology. The second is related to nosological categories and self-scrutiny. For instance, the people who are always concerned about their heart condition are more frequently diagnosed with non-cardiac chest pain and evidence shows that the belief in being susceptible to heart problems is a risk factor in developing real heart attack. The third type is the sociogenic illness. In this case, observing and learning illness or symptoms in others results in the development of similar conditions in the person. Outbreaks of "epidemic hysteria" and an increase in suicide rate after the release of suicide news of celebrities or in response to media stories about suicide are examples of this type. Lastly, the nocebo effect can be sickness or symptom induced. In this form, negative suggestion can negatively affect the medical or surgical intervention. For example, asthmatic patients more frequently react to nebulised normal saline when they are told that they are inhaling irritants or allergens. Clearly, these different forms of nocebo effect could have diverse consequences in various area of health care. One consequence is related to the notion of informed consent and the ethical concerns about the nocebo effect resulting from giving information regarding the possible negative outcomes. Calling this the nocebo effect of informed consent (NEIC), Cohen (2014) argues that the ethical dilemma raised by this, is important and needs to be addressed properly. On the one hand, according to the need respect patient autonomy and the right of patients to know about the possible side effects of interventions, it is necessary to inform the patients about the possible negative consequences and complications of the interventions they are receiving. On the other hand, there is compelling evidence available that shows that informing the patient about side effects and complications can result in their actual formation, which conflicts with the rule of "do no harm" in medical ethics. In addressing the question of optimal balance between disclosure of information to the patient and nonmaleficence, the dilemma of the NEIC should be considered.

Accordingly, the effects of suggestions in the matrix of relationships can be harmful as the suggestions and the meaning of the events and interactions with others and the environment are not always positive. These facts reveal the complexity of the effects of psychosocial interactions on the health states of the people. In next section, the connections between the above findings in the fields of placebo and hypnosis, and interactions between health professionals and clients in the clinical setting, are explored.

6.4 Placebo, Hypnosis and the Relational Context

Kirsch (1994) has considered hypnosis as a non-deceptive placebo because in hypnotherapy, the expectations of the client are manipulated, without the need for an inert pill or intervention. The response expectancies are defined as anticipation of positive responses which are taking place as a consequence of particular stimuli or behaviours. Spiegel (1997) believes that the ability of the imagination is a feature which distinguishes human beings from other animals which affects their health status and plays a key role in the dynamic of the placebo/nocebo response. In his view, if the individual is placed in a situation in which some factors result in dissociation and absorption and suggestibility is maximized, then the placebo response is developed.

Considering placebo effects as meaning response also can connect this concept to hypnosis. As mentioned, hypnotic phenomena can be seen outside the hypnotic trance, and suggestibility, belief and expectancy are the key elements in their formation. In the case of placebo, the meaning of the placebo provides the suggestion needed for activation of the healing response. For example, when the doctor tells the patient that "this pill will calm your bowels", they are giving a suggestion to the patient at the time of prescribing the medicine. This suggestion is not always language based and can be non-verbal as well. For example, if the patient is going to see a well-known professor who has a long waiting list, this very fact is suggestive that anything prescribed will be effective.

Up to this point, we have tried to make it clear that the context in which a treatment or intervention is prescribed has a critical role in its ultimate outcome. Placebo researcher Kradin (2011a) believes that although elements such as the shape, color, and form of medicine and environmental factors such as space and time are important in the development of healing response, the placebo response is basically developed by mechanisms activated by the dynamic of interpersonal relationships. In other words, the interactions between the therapist and the patient are central in the development of placebo response (Rafieian 2014). The origins of these mechanisms are rooted in the early life interactions of the newborn with their caregiver. These are the soothing and relaxing mechanisms which are developed early in life so powerfully that their activation later in life has modulatory effects on the immune system, reducing anxiety and promoting healing effects (Kradin 2011b). Similar mechanisms are activated during a hypnotherapy session and are involved in the formation of its healing effects (Vandenberg 1998).

Benson (1997) believes that three components are necessary for development of placebo or nocebo effects. First, the belief and expectancy coming from the patient; second, the belief and expectancy from the therapist side; and third, the belief and expectancy which is constructed by the relationship between these two. Understanding the fact that the dynamic of interpersonal relationships is important in the formation of healing response makes it clear that the quality of relationship between therapist and patient determines success in the activation of healing mechanisms in patients. As mentioned, in the modern paradigm of biomedicine, the dominant pattern is the paternalistic relationship in which the doctor takes control of the consultation and the patient has a passive role. Many, especially in the sociology of the doctor-patient relationship, have criticised this domination and consider the doctor-patient relationship as a meeting between experts in which the doctor shares their medical knowledge and clinical skills and the patient brings their illness experience and ideas and beliefs about that particular health condition (Morgan 2008). Others such as Måseide (1991) disagree and believe that power is necessary for adequate medical practice. He claims that the doctor has the knowledge and skills needed, and they should take control of the consultation because it is the competency of the doctor that legitimizes their authority. Similarly, Schei (2006) believes that attacking the power of doctors is misguided and claims that the structural and symbolic power of doctors are necessary for actions that make the healing process possible. She argues in favour of clinical leadership and states that an imbalanced relationship is necessary during clinical consultation in order to lead the patient effectively. Kirmayer (1994) takes a balanced position and tries to describe what ideally should happen in a clinical encounter session. He argues that suffering and distress in the body is a shapeless and unformulated experience and that illness experience is dependent on interpretation and explanation. Assigning any meaning to the chaotic and unfamiliar sensations of pain and distress can give order to it, thus reducing the fear and anxiety of the patient. In fact, in his formulation, Kirmayer takes a step further from the above discussion and claims that even the process of making a diagnosis is therapeutic because it gives shape to the ambiguous and vague experience of sensations and pain and distress in the body. After listening to the patient story and looking for the signs, the doctor makes an authoritative diagnosis. Doctor should then explain the next step, which is the treatment strategy, which allows more space for improvisation. In other words, diagnosis is an authoritative basis for constructing a treatment strategy and this could be done creatively. But there is no definitive border between these two stages, and keeping balance between the two is important in clinical practice. Kirmayer (1994) explains:

> Authority is concerned with legitimation and hence with truth, while the therapeutic enterprise is fundamentally concerned with how to continue and hence with the improvisation of meaning. While authority is necessary to provide a structure (themes or modes) on which variations can be improvised, authoritative meanings inevitably restrict the possibilities for invention by clinician and patient. Seen from this perspective, the goal of the clinical negotiation between patient and physician is to create an interpretation of distress with enough closure or certainty to diminish the threat of the inchoate while preserving enough openness and ambiguity to allow for fresh improvisation. The ideal balance between ambiguity and certainty varies over time and with the characteristics of the participants in the clinical encounter as well as with the exigencies of family, work and the health care system. (pp. 183–184)

As mentioned, the main problem of health care systems in the modern world is the management of chronic health conditions. In some chronic diseases such as diabetes or heart failure, a diagnosis is easily made in the early stages of disease and the patient becomes aware of what is going on in their body. In some others, the process of making a diagnosis is not that easy. For these conditions, which are described by terms such as "functional somatic symptoms", or "medically unexplained symptoms" (MUS) (Fink et al. 2005), no definitive aetiology has been found. In fact, there is a network of different causes involved in the formation of these disorders. In any case, in a chronic health condition, after the initial diagnosis, the main topic of clinical encounters is the management and treatment of that known condition and this is actually the part in which the therapist has enough space for creativity and improvisation. This is a delicate process in which the carer should be vigilant of the patient's story, signs, and symptoms and the thoughts and emotions around them. In a sense, it could be described as improvisational acting in which the clever and experienced therapist carefully monitors the actions and behaviour of the patient and responds appropriately. But here again, similar to hypnosis, the doctor invites the patient to participate in a miniature drama and there is a chance that, for some reason, a patient may decline this invitation, fail to cooperate and refuse to share their feelings, beliefs and experiences. Even the best practitioners are not always successful and on occasions fail to develop a good rapport with some patients. Consequently, the physician-patient relationship is a reciprocal interaction which needs the active engagement of both sides. This clarifies the importance of acting in the clinical context but more needs to be said about the place of performance in medicine. It is explored briefly in the next section.

6.5 Performance and Medicine

Although the contributions of the works of American sociologist, Erving Goffman (1922–1982), such as *Asylums* (1961) and *Stigma* (1963) on the revolution in institutional care are well known, there are many inspiring ideas for medicine and health care systems in his other important work *The Presentation of Self in Everyday Life* (1992). In this book, Goffman displays the dramaturgical nature of social life and uses many examples from different social contexts to show the similarities between the elements of theatre such as front and backstage and the social interactions of everyday life. Among these, there are some interesting instances from the medical context which shed light on certain aspects of our discussion. Goffman explores the theatrical nature of the health care system in different places. Discussing different requirements of taking a social role, he states that each role should be performed in its appropriate front. In the case that a role is new for the person and for the society, sometimes the best front should be chosen from pre-existing opportunities or, if needed, a new front could be invented. He takes the example of the task of administering anaesthesia in the early stages of its development. He explains:

> In some hospitals anaesthesia is still administered by nurses behind the front that nurses are allowed to have in hospitals regardless of the tasks they perform- a front involving ceremonial subordination and a relatively low rate of pay. In order to establish anaesthesiology as a specialty for graduate medical doctors, interested practitioners have had to advocate strongly the idea that administering anaesthesia is a sufficiently complex and vital task to justifying giving to those who perform it the ceremonial and financial reward given to

> doctors. The difference between the front maintained by a nurse and the front maintained by a doctor is great; many things that are acceptable for nurses are *infra dignitatem* for doctors. Some medical people have felt that a nurse 'under- ranked' for the task of administering anesthesia and that doctors 'over-ranked; were there an established status midway between nurse and doctor, an easier solution to the problem could perhaps be found. (pp. 38–39)

Today, with the rapid progress of medical technologies, each day brings a new device or instrument to health care and this issue of emerging new roles in health care systems is even more relevant. Arguing the belief of the social actor in the part they are playing, Goffman explains that a cynical actor is a person who "has no belief in his own act and no ultimate concern with the beliefs of his audiences" (p. 28) and the actor who believes in their acts is "sincere". Then he introduces an exception: an actor who is always sincere but sometimes:

> … forced to delude their customers because their customers show such a heartfelt demand for it. Doctors who are led into giving placebos, filling-station attendants who resignedly check and recheck tire pressures for anxious women motorists, shoe clerks who sell a shoe that fits but tell the customer it is the size she wants to hear - these are cynical performers whose audiences will not allow them to be sincere. Similarly, we find that sympathetic patients in mental wards will sometimes feign bizarre symptoms so that student nurses will not be subjected to a disappointingly sane performance. (p. 29)

Consequently, Goffman categorizes the act of giving placebo to the patient with other similar behaviours seen in the society, which are seemingly insincere or even sometimes unethical but are done with the intention of helping or serving the customer. Interestingly, he also points to a situation in which a patient is doing such an insincere act for the sake of helping a health professional trainee. In a broader view, he tries to show the importance of the visible and tangible activities done by health professionals for patient satisfaction. He uses the example of nursing practice in surgical and medical wards. In surgical wards, the care of the post-operative patient includes recognizable tasks such as changing bandages and swinging orthopaedic frames. But in a medical ward, the tasks are more frequently composed of activities which are not grossly visible and understandable without explanation, such as checking the number of the breaths or the colour and tone of the skin. In the latter, it is more probable that the patient thinks the nurse is "wasting time" (Goffman 1992, p. 41).

In the clinical encounter of the doctor and patient, this element of dramatization is very important and Goffman brings some examples to show that. He points to the fact that the practitioner usually pretends to remember everything about the patient and even in the cases where the patient does not remember a point, such as the tablet prescribed in the last visit, they expect the doctor to remember it without difficulty and it is not acceptable for the patient that the doctor cannot remember either. The other instance is the process of referring a patient to a specialist by a general practitioner. Although it seems that the specialist has been chosen because they are the best option, other factors such as the ties between these two doctors may play a role in the decision (p. 58).

Using these examples, Goffman shows the essential role of dramatization and performance in interpersonal interactions in the health care system and in the context of clinical encounters between doctor and patient. As discussed, events that happen in the interpersonal space are essential for the activation of healing mechanisms and the formation of placebo response. The power of relationship in the activation of healing mechanisms is so important that the psychoanalyst and psychiatrist Michael Balint (1896–1970) speaks about the drug doctor and believes that the person of the doctor is itself a potent drug. In his research he aimed to show the function of the doctor as a drug and find out the benefits and side effects of this drug in the process of diagnosis and treatment. Exploring general medical practice revealed the problems that doctors trained in the biomedical paradigm have in recognizing the psychosocial context behind their patients' health conditions and the effort that they make to find an organic cause for any complaint, thereby failing to discover the meaning of the symptoms that emerge out of the patient's particular psychosocial situation (Balint 1957; Kirkby 2011).

In his contribution, Myers (2010) introduces placebo as performance and tries to see the act of prescribing placebo from this perspective:

> The doctor who says, 'Try this, it might help,' while prescribing a placebo, is utilizing a performance to heal the patient, for there is no healing agent other than the words issued by the doctor and the authority of those words. It is the doctor's performance, not the placebo that heals the patient. (p. 1296)

Emphasizing the concept of embodiment, he points to the fact that the biophysiological life of the body is not something detached from the social and cultural world and the effects of interactions in the social world become inscribed in the physical body. Also the subjective experience of distress and pain, which is the result of the biological processes of the body, has no meaning per se. The meaning is assigned to these experiences through the authoritative diagnosis of the doctor and the way they describe the problem according to the logic of the biomedical paradigm. The fact that placebo, in spite of its inertness, has healing effects depicts the power of rituals in medicine.

The importance of rituals in medicine has been revealed in numeroust studies. In one study, Kaptchuk and his colleague (2006) compared the effectiveness of real acupuncture with placebo treatment. In this study 270 chronic pain patients were divided into three groups and treated by acupuncture, placebo acupuncture, and placebo pill. The results of the study show that the real acupuncture was the most effective treatment but that placebo groups also had pain reduction to some extent. Interestingly, it was revealed that the pain reduction in the placebo acupuncture group was more than in the placebo pill group. To explain these results, the increased effectiveness of placebo acupuncture was attributed to the effects of the ritual of acupuncture. In another study, Ostenfeld-Rosenthal (2012) qualitatively explored the experience of the patients with medically unexplained symptoms who were treated by energy therapists. In this research, she shows that healers focus and reflect on the bodily sensations and embodied experience of patients. In their treatments, they use the imagination of the patients to induce changes in bodily experiences and ask them to describe their feelings and sensations carefully and comment and interpret this information according to the system of beliefs pertaining to the method of healing that they use. Calling the pre-reflective, pre-linguistic bodily feelings and sensations the "bodily experienced symbols", she argues that placebo response in its essence is manipulating these symbols, re-editing the body and self-image using the healing rituals. She views the primary skill of the healer is to use the power of linguistic symbols and bodily experienced symbols to help patients have the healing experience.

Apart from alternative and complementary medicine, the power of ritual has also been considered in the context of modern medicine. Wall (1996) states that the rituals of surgical procedure are important in the development of healing response. He argues that the structured process of surgical operations with all its detail is a kind of non-verbal communication which conveys the message to patients that going through this ritual helps them to move from disease to health. Similarly, Green (2006) has compared the events that happen before and after the operation with the rituals of shamanistic healing and claims that these events give suggestions to the patients that the intervention will be effective and increase their expectancy for the experience of healing. Evidence shows that in some cases the efficacy of surgeries was related to this placebo effect and not to surgical procedures. For instance, in one study, it was shown that the widely used surgery for treatment of knee osteoarthritis is actually not more effective than sham surgery (Moseley et al. 2002) and it was the placebo response resulting from the rituals of knee surgery that ultimately led to pain reduction. Moerman and Jonas (2002) claim that surgeries produce even stronger meaning response because surgical procedures have convincing rational explanations which are not present in drug treatments. In a broader view, Brody (2010) argues that rituals are present in various areas of modern medicine. He particularly speaks about two commonly seen examples: the process of physical examination and the "ward round" in teaching hospitals (p. 154). In the former the senior doctor examines the patient in search for the relevant signs and in the latter the findings are communicated to other participating doctors and students. All these events are meaningful for patients and are functional in the activation of the healing response.

Considering the above discussion, it could be claimed that although the healing mechanisms are present in the body, the patient is unable to activate them without the help of social context and interpersonal interactions. It is through the medical rituals' conscious and unconscious, verbal and non-verbal suggestions that these healing mechanisms become activated and help the patient to recover. The first step is the invitation of the patients to express their feelings, thoughts, emotions, and beliefs about the suffering they are experiencing. After that, in the process of consultation, meaning is assigned to embodied experiences. Metaphors have a key role in this context. Lakoff and Johanson (1980, 1999) have shown that the origins of the metaphors that we use in our daily life are in our embodied experiences. One example is the unconscious use of the spatial-relations concepts (Lakoff and Johnson 1999, p. 30) we use in everyday communications which are related to our spatial experience of our bodies. When we say "prices are rising", describing prices as entities which are going up, it is only a metaphor originating from our embodied experience. Metaphors can also in turn change bodily perceptions, sensations and cognition (Kirmayer 2011).To give verbal and non-verbal suggestion and manipulate the expectations and beliefs of the patient, it is important for the therapist to be a good actor and play their role in harmony with the role that the patient is taking.

Case and Brauner (2010) discuss the benefits of using acting classes and performance in medical education. They claim that using acting in medical education can improve the ability of what they define as empathic imagination which is "a cognitive skill set that helps one to imagine the experiences and responses of another person" (p. 159). This ability is necessary for understanding patients' feelings and emotions and giving an appropriate response in the process of caring and treatment. They provide different examples of applications of theatre in medical education. Acting is important for activities such as presenting a patient's history to the supervising doctor, taking a history from a person who is playing the role of the patient in an examination or participating in an educational round. Further, students consider their seniors as role models and learn the proper way of interacting with patients. Accordingly, many efforts have been made to incorporate theatre and performance into medical education. In one study, Dow and his colleagues (2007) used theatre teaching to enhance the skills of clinical empathy in medicine residents. They define clinical empathy as "the skill of recognizing a patient's emotional status and responding, in the moment, to the unique needs of the patient to promote better clinical outcomes" (p. 1114) and, in a controlled trial, showed that the skills of residents who participated in a course for teaching this skill had a statistically significant improvement compared to the control group which was assessed before and immediately after the curricular intervention. In the study conducted by Shapiro and Hunt (2003), one-person shows about AIDS and ovarian cancer were presented to a group of medical students, faculty, community doctors, staff and patients to enhance empathy and improve the understanding of illness experience in students and health professionals.

The results of their assessments were positive and showed that the participants felt they acquired better insights into the nature of illness experience. In another study (Hammer et al. 2011), theatre training was provided for medical students to improve their case presentation skills. For this purpose a group of teacher artists taught storytelling skills to medical students. All participants believed that using learning acting skills is beneficial for medical students but there was an interesting comment from one of them criticizing the course teachers saying, "Too much focus on how this relates to medicine. We will realize that later. For now, teach us the [performance] skills" (p. 20). The authors believe that this unnecessary explanation by course teachers is a sign of the tension between science and humanities and the fact that modern medicine is reluctant to accept humanities as an important part of medical practice and education. In the next section, the implications of the above discussion for research are considered briefly.

6.6 Implications for Research

Apart from the implications for medical practice and education, there are implications for research. As the placebo effect has been defined as a meaning response, to understand the mechanisms involved, it is necessary to realize the meaning for patients of elements of treatment and intervention for the patients and to focus on the qualitative aspects of these elements in the clinical setting. Also, considering the performative nature of medicine and the importance of the sociocultural context in the formation of the healing response, there is a need to develop innovative research to explore the mechanisms involved. Placebo is usually considered as a non-specific healing response but there have been attempts to specify the underlying mechanisms and some explanations have emerged based on relevant psychologicaltheories such as classical conditioning and expectation theory (Kirsch 1997). Also, research findings in the field of psychoneuroimmunology have shed some light on the complex interactions between psychological phenomena, the endocrine and immune systems and the consequent negative and positive health effects (Kirmayer 2006). Furthermore, drawing on new findings in social psychology and using various concepts and theories such as priming, client perceptions, and theory of planned behaviour, attempts have been made to clarify the ways in which placebo response could be enhanced and harnessed (Sliwinski and Elkins 2013). In spite of these efforts, there are many obscure and unexplored aspects of the processes involved and as Kirmayer (2011) states, there is still no comprehensive theory of healing. In this context, several critics argue that the research approach of the dominant model of modern medicine in which the Randomized Clinical Trial (RCT) is the gold standard, cannot adequately explore the complexity of various factors involved in the health system,because this approach ignores certain critical aspects of this complexity such as the phenomenal world of people, their health beliefs, emotions, and the important qualitative psychosocial factors which have defining effects on the health status of the population. Especially in the recent movement in biomedicine called Evidence-Based Medicine, there is an overemphasis on information coming from quantitative empirical studies in the process of clinical decision making and this further amplifies the above mentioned problems regarding the qualitative aspects of health care (Dean 2004; Rafieian 2010; Goldenberg 2006). Although using qualitative research methods is not new in health care and there is a significant repertoire of knowledge emerging from qualitative studies, there are still certain elements missing in this context. For instance, Ellingson (2006) argues that health research is rarely embodied and qualitative researchers usually do not explain their bodily experiences in research and she urges the need for embodied writing in qualitative research. To summarize, by analyzing hypnotic phenomena and placebo effects, it has been shown in this chapter that there is a complex interconnection between biophysiological, interpersonal and sociocultural phenomena in the clinical setting, and to show these complex interconnections, there is a need for research methods that include both phenomenological and first person experiences of patients together with clinical and para-clinical findings such as physical signs, test and imaging results. In other words, the researchers should seek both soft and hard data in their investigations (Adler 2000).

6.7 Conclusion

An interpersonal relationship between health care professionals and care seekers is nearly always present in the clinical context. In the biomedical paradigm, the overemphasis on the biological sciences and the neglect of psychosocial events that are inscribed into the body have resulted in the allocation of a majority of research and education tools and resources to biosciences and the emergence of new problems such as ineffective care of chronic diseases and failure to manage functional somatic symptoms. These issues have been well discussed in the sociology of medicine and health psychology. In this chapter, using the results of research in the fields of hypnosis and placebo, it was shown that performance and acting elements play a key role in the dynamic of healing response and that an effort is needed to take further steps towards the development of a theory of healing. The obvious fact is that, to understand these mechanisms, knowledge from several disciplines is required. As discussed, performance and theatrical studies are very important in this context. It was shown that teaching acting skills to medical trainees can enhance empathy in their interactions with patients and can also improve their ability to evoke the healing response. New theories of performance can help to clarify some unknown aspects of the healing process in clinical encounters. In addition, they can inform research and help to improve research methodologies in order to be better able to show the complexities of health and disease in the real world.

Finally, it was argued that placebo is a meaning response and that to understand the dynamic of healing response formation there is a need for research methods which are capable of capturing the qualitative aspects of interpersonal interactions, the embodied experience of the people involved and the phenomenal world of clients and health professionals. The current research approach of biomedicine is not equipped for this challenge and there is a need for an alternative perspective and innovation in research methodology.

References

Adler, R. H. (2000). Hard and soft data: A semiotic point of view. *Schweizerische Medizinische Wochenschrift, 130*, 1249–1251.

Balint, M. (1957). *The doctor, his patient and the illness*. London: Pitman Medical.

Benson, H. (1997). The nocebo effect: History and physiology. *Preventive Medicine, 26*(5), 612–615.

Brody, H. (2010). Ritual, medicine, and the placebo response. In W. S. Sax & J. Weinhold (Eds.), *The problem of ritual efficacy* (pp. 151–167). Oxford: Oxford University Press.

Cannon, W. B. (1942). "Voodoo" death. *American Anthropologist, 44*(2), 169–181.

Case, G. A., & Brauner, D. J. (2010). Perspective: The doctor as performer: A proposal for change based on a performance studies paradigm. *Academic Medicine, 85*, 159–163.
Coe, W. C., & Sarbin, T. R. (1991). Role theory: Hypnosis from a dramaturgical and narrational perspective. In S. J. Lynn & J. W. Rhue (Eds.), *Theories of hypnosis: Current models and perspectives* (pp. 303–323). New York: Guilford.
Cohen, S. (2014). The nocebo effect of informed consent. *Bioethics, 28*, 147–154.
Dean, K. (2004). The role of methods in maintaining orthodox beliefs in health research. *Social Science and Medicine, 58*, 675–685.
Dow, A. W., Leong, D., & Wenzel, R. P. (2007). Using theater to teach clinical empathy: A pilot study. *Journal of General Internal Medicine, 22*, 1114–1118.
Ellenberger, H. (1970). *The discovery of the unconscious: The history and evolution of dynamic psychiatry*. New York: Basic Books.
Ellingson, L. L. (2006). Embodied knowledge: Writing researchers' bodies into qualitative health research. *Qualitative Health Research, 16*, 298–310.
Fässler, M., Meissner, K., Schneider, A., & Linde, K. (2010). Frequency and circumstances of placebo use in clinical practice–A systematic review of empirical studies. *BMC Medicine, 8*, 15.
Fink, P., Rosendal, M., & Olesen, F. (2005). Classification of somatization and functional somatic symptoms in primary care. *Australian and New Zealand Journal of Psychiatry, 39*, 772–781.
Goffman, E. (1961). *Asylums: Essays on the social situation of mental patients and other inmates*. Garden City: Doubleday/Anchor.
Goffman, E. (1963). *Stigma: Notes on the management of spoiled identity*. London: Penguin.
Goffman, E. (1992). *The presentation of self in everyday life*. London: Penguin.
Goldenberg, M. J. (2006). On evidence and evidence-based medicine: Lessons from the philosophy of science. *Social Science and Medicine, 62*, 2621–2632.
Greaves, D. (2002). Reflections on a new medical cosmology. *Journal of Medical Ethics, 28*, 81–85.
Green, S. A. (2006). Surgeons and shamans: The placebo value of ritual. *Clinical Orthopaedics and Related Research, 450*, 249–254.
Hahn, R. A. (1997). The nocebo phenomenon: Concept, evidence, and implications for public health. *Preventive Medicine, 26*(5), 607–611.
Hammer, R. R., Rian, J. D., Gregory, J. K., Bostwick, J. M., Birk, C. B., Chalfant, L., & Hall-Flavin, D. K. (2011). Telling the patient's story: Using theatre training to improve case presentation skills. *Medical Humanities, 37*, 18–22.
Heap, M., & Aravind, K. K. (2002). *Hartland's medical and dental hypnosis* (4th ed.). Edinburgh: Churchill Livingstone.
Howick, J., Bishop, F. L., Heneghan, C., Wolstenholme, J., Stevens, S., Hobbs, F. D. R., & Lewith, G. (2013). Placebo use in the United Kingdom: Results from a national survey of primary care practitioners. *PloS One, 8*, e58247.
Kallio, S., & Revonsuo, A. (2003). Hypnotic phenomena and altered states of consciousness: A multilevel framework of description and explanation. *Contemporary Hypnosis, 20*, 111–164.
Kaptchuk, T. J., Stason, W. B., Davis, R. B., Legedza, A. R. T., Schnyer, R. N., Kerr, C. E., Stone, D. A., Nam, B. H., Kirsch, I., & Goldman, R. H. (2006). Sham device v inert pill: Randomised controlled trial of two placebo treatments. *BMJ, 332*, 391–397.
Kirkby, R. (2011). Who needs balint anyway? *South African Family Practice, 18*, 14–15.
Kirmayer, L. J. (1994). Improvisation and authority in illness meaning. *Culture, Medicine and Psychiatry, 18*, 183–214.
Kirmayer, L. J. (2006). Toward a medicine of the imagination. *New Literary History, 37*, 583–601.
Kirmayer, L. J. (2011). Unpacking the placebo response: Insights from ethnographic studies of healing. *The Journal of Mind–Body Regulation, 1*, 112–124.
Kirsch, I. (1985). Response expectancy as a determinant of experience and behavior. *American Psychologist, 40*, 1189.

Kirsch, I. (1994). Clinical hypnosis as a nondeceptive placebo: Empirically derived techniques. *The American Journal of Clinical Hypnosis, 37*, 95–106.

Kirsch, I. (1997). Specifying nonspecifics: Psychological mechanisms of placebo effects. In A. Harrington (Ed.), *The placebo effect: An interdisciplinary exploration* (pp. 166–186). Cambridge: Harvard University Press.

Kirsch, I. (1998). Volition as a believed-in imagining. In J. de Rivera & T. R. Sarbin (Eds.), *Believed-in imaginings: The narrative construction of reality* (pp. 156–168). Washington, DC: APA.

Kirsch, I. (1999). Hypnosis and placebos: Response expectancy as a mediator of suggestion effects. *Anales de Psicología, 15*, 99–110.

Kirsch, I. (2001). The altered states hypnosis. *Social Research, 68*, 795–807.

Kradin, R. L. (2011a). Placebo response: A consideration of its role in therapeutics. *Current Psychiatry Reports, 13*, 37–42.

Kradin, R. L. (2011b). The placebo response: An attachment strategy that counteracts the effects of stress-related dysfunction. *Perspectives in Biology and Medicine, 54*, 438–454.

Kroger, W. S. (2008). *Clinical and experimental hypnosis in medicine, dentistry, and psychology*. Philadelphia: Lippincott Williams & Wilkins.

Lakoff, G. J., & Johnson, M. (1980). *Metaphors we live by*. Chicago: University of Chicago.

Lakoff, G., & Johnson, M. (1999). *Philosophy in the flesh: The embodied mind and its challenge to western thought*. New York: Basic books.

Laurens, S. (2007). Social influence: Representation, imagination and facts. *Journal for the Theory of Social Behaviour, 37*, 401–413.

Lynn, S. J., Kirsch, I., & Hallquist, M. N. (2008). Social cognitive theories of hypnosis. In M. R. Nash & A. J. Barnier (Eds.), *Oxford handbook of hypnosis* (pp. 111–140). Oxford: Oxford University Press.

Måseide, P. (1991). Possibly abusive, often benign, and always necessary. On power and domination in medical practice. *Sociology of Health & Illness, 13*, 545–561.

Mead, N., & Bower, P. (2000). Patient-centredness: A conceptual framework and review of the empirical literature. *Social Science and Medicine, 51*, 1087–1110.

Meissner, K., Höfner, L., Fässler, M., & Linde, K. (2012). Widespread use of pure and impure placebo interventions by GPs in Germany. *Family Practice, 29*, 79–85.

Miller, D. E. (1994). The social construction of hypnosis. In N.J.Herman and L.T. Reynolds (Eds.), *Symbolic interaction: An introduction to social psychology* (p. 351). New York: General Hall Inc.

Miller, F. G., & Kaptchuk, T. J. (2008). The power of context: Reconceptualizing the placebo effect. *Journal of the Royal Society of Medicine, 101*, 222–225.

Moerman, D. E. (2011). Examining a powerful healing effect through a cultural lens, and finding Meaning. *The Journal of Mind–Body Regulation, 1*(2), 63–72.

Moerman, D. E. (2002a). *Meaning, medicine, and the 'placebo effect'*. Cambridge: Cambridge University press.

Moerman, D. E. (2002b). The meaning response and the ethics of avoiding placebos. *Evaluation & the Health Professions, 25*, 399–409.

Moerman, D. E. (2002c). Doctors and patients: The role of clinicians in the placebo effect. *Advances in Mind-Body Medicine, 19*, 14–22.

Moerman, D. E. (2006). The meaning response: Thinking about placebos. *Pain Practice, 6*(4), 233–236.

Morgan, M. (2008). The doctor–patient relationship. In G. Scambler (Ed.), *Sociology as applied to medicine* (6th ed., pp. 55–70). London: Elsevier.

Moseley, J. B., O'Malley, K., Petersen, N. J., Menke, T. J., Brody, B. A., Kuykendall, D. H., Hollinqsworth, J. C., Ashton, C. M., & Wray, N. P. (2002). A controlled trial of arthroscopic surgery for osteoarthritis of the knee. *New England Journal of Medicine, 347*, 81–88. doi:10.1056/NEJMoa013259.

Myers, W. B. (2010). The placebo as performance: Speaking across domains of healing. *Qualitative Health Research, 20*, 1295–1303.
Olness, K. (2008). Helping children and adults with hypnosis and biofeedback. *Cleveland Clinic Journal of Medicine, 75*, S39.
Ostenfeld-Rosenthal, A. M. (2012). Energy healing and the placebo effect. An anthropological perspective on the placebo effect. *Anthropology & Medicine, 19*(3), 327–338.
Pauli, H. G., White, K. L., & McWhinney, I. R. (2000). Medical education, research, and scientific thinking in the 21st century (part three of three). *Education for Health: Change in Learning & Practice, 13*, 15–25.
Rafieian, S. (2010). Cybersemiotic medicine: A framework for an interdisciplinary medicine. *Cybernetics & Human Knowing, 17*, 65–93.
Rafieian, S. (2014). Towards a sociology of placebo response: Body, emotions, and semiotics of healing. *International Journal of Body, Mind and Culture, 1*(2), 107–116.
Sarbin, T. R. (1998). Believed-in imaginings: A narrative approach. In J. de Rivera & T. R. Sarbin (Eds.), *Believed-in imagings: The narrative construction of reality* (pp. 15–30). Washington, DC: APA.
Schei, E. (2006). Doctoring as leadership: The power to heal. *Perspectives in Biology and Medicine, 49*, 393–406.
Scheiffele, E. (2001). Acting: An altered state of consciousness. *Research in Drama Education, 6*, 179–191.
Scheiffele, E. (2003). Alterations of consciousness during psychodrama and sociodrama. *British Journal of Psychodrama and Sociodrama, 18*, 3–20.
Shapiro, J., & Hunt, L. (2003). All the world's a stage: The use of theatrical performance in medical education. *Medical Education, 37*, 922–927.
Sliwinski, J., & Elkins, G. R. (2013). Enhancing placebo effects: Insights from social psychology. *The American Journal of Clinical Hypnosis, 55*, 236–248.
Spiegel, H. (1997). Nocebo: The power of suggestibility. *Preventive Medicine, 26*(5), 616–621.
Straus, R. A. (1978). Hypnosis as reality reconstruction: A sociological analysis of the hypnosis process. *Pacific Sociological Review, 21*(4), 407–422.
Upshaw, W. N. (2006). Hypnosis: Medicine's dirty word. *American Journal of Clinical Hypnosis, 49*, 113–122.
Vandenberg, B. (1998). Infant communication and the development of hypnotic responsivity. *International Journal of Clinical and Experimental Hypnosis, 46*, 334–350.
Wagner, E. H., Austin, B. T., Davis, C., Hindmarsh, M., Schaefer, J., & Bonomi, A. (2001). Improving chronic illness care: Translating evidence into action. *Health Affairs, 20*, 64–78.
Wall, L. L. (1996). Ritual meaning in surgery. *Obstetrics & Gynecology, 88*, 633–637.
Whorwell, P. J. (2005). Review article: The history of hypnotherapy and its role in the irritable bowel syndrome. *Alimentary Phannacology & Therapeutics, 22*, 1061–1067.
Zarrilli, P. B. (2011). Altered consciousness in performance: West and east. In E. Cardei\a & M. Winkelman (Eds.), *Altering consciousness: Multidisciplinary perspectives* (pp. 301–326). California: ABC-CLIO.

Chapter 7
How to Prescribe Information: Health Education Without Health Anxiety and Nocebo Effects

Farzad Goli, Alireza Monajemi, Gholam Hossein Ahmadzadeh, and Azadeh Malekian

> *Human being is more ill, less certain, more changeable, more insecure than any other animal- there's no doubt about that He is the sick animal. Where does it come from? (Nietzsche, genealogy of morals, p. 100)*

Imagine you live in a super intelligent city, in which detectors alert you via audible alarms when you approach potentially harmful stimuli, air or electromagnetic pollutants, allergens, mutagen foods, etc. You can perhaps imagine that every day would be filled with warnings and an undoubtedly large amount of beeps and rings. I think you agree with me that after a while, the alarms could be seriously harmful by themselves, not only because of the constant bombardment of information and sound pollution, but also by making you worried, preoccupied, or irritable. In addition, you may begin to adapt to the situation by adopting avoidant, pessimistic, or suspicious attitudes; or you might find yourself overwhelmed, depressed, and anxious. The overflow of warning signs turns to dangerous stimuli and overwhelming noises. We have a rather limited capacity for information processing, not to mention the possibility that signs may become distorted, misinterpreted, ignored, and/or simply perceived as ordinary noises. This is a transcendental dialectic in which the

---

F. Goli (✉)
Head of Danesh-e Tandorosti Institute, Isfahan, Iran

Energy Medicine University, Mill Valley, CA, USA
e-mail: Dr.fgoli@yahoo.com

A. Monajemi
Institute for Humanities and Cultural Studies (IHCS), Tehran, Iran

G.H. Ahmadzadeh
Behavioral Science Research Center, Isfahan University of Medical Sciences, Isfahan, Iran

A. Malekian
Psychosomatic Research Center, Isfahan University of Medical Sciences, Isfahan, Iran

F. Goli (ed.), *Biosemiotic Medicine*, Studies in Neuroscience, Consciousness and Spirituality 5, DOI 10.1007/978-3-319-35092-9_7

quality (amount, speed) of signs can induce qualitative changes in our mood, affect, attitudes, and form of life.

Now, the question is what this analogy tends to tell us. Do you find any similarities between this imaginary city and our real-life cities? Have you ever inhaled the air and had the feeling it might cause cancer? Or have you eaten fresh fruit as if it were an antioxidant or anticancer agent? It seems that the warnings are already ringing everywhere for everyone, and that a large number of warning messages comes explicitly or implicitly from the television, newspaper, and even the refrigerator! Our boundaries are surrounded by a vast variety of dangers and the tragedy is that even within these boundaries, there are still a huge number of symbolic dangers – for example, dysfunctional beliefs and traumatic memories – as well as physical ones – for example, metaplastic genes, latent disorders, and aneurysms. How can we stand to live in such conditions?

The answer is very simple: by "neglecting," As Rumi said, "negligence is the column of our world". The idea sounds deceitful and contrary to the prophecy of illumination and modern episteme. Moreover, it seems to stand in opposition to the human's will of knowledge!

According to Nietzsche, we need an informational hygiene system to protect people from information overload and mechanical use of knowledge that mislead us to no-life direction. In the *Gay Science*, he emphasized the evolutionary function of science and the fact that our knowledge must facilitate our adaptation and promote our happiness and vitality. Unlimited semiosis – the flow of signs which has constructed higher levels of organization from the lower orders – can be mentioned as the engine of life and evolution. Yet, it seems that sometimes we need to intentionally inhibit the overwhelming signs in order to help the life-drive. Hence, sometimes we need to reduce perceived significance in order to reduce excessive risk perception and worry. But we can never ignore that we live in the information era and any paternalistic control on information is regarded as censorship and a violation of human rights. Therefore, regarding health, the proper extent to which this is possible for the community education remains a dilemma. In other words, how can we balance our medical interventions in order to achieve higher levels of prevention, lower levels of health anxiety, and nocebo induction seemingly simultaneously?

While we are focused on the literal interpretations of our health education, various thought associations and consequently emotions arise and can change our mood, behavior, and even psychoneuroimunoligic responses into unhealthy ones. In other words, we are trying to figure out how to sufficiently enlighten our community without blinding it by excessive light! In this chapter, we will briefly discuss the complexities of health education and how to formulate the informational interventions on the basis of biosemiotics.

7.1 Risk and Danger

According to German sociologist and prominent thinker in systems theory, Luhmann (1993), "risk" differs from "danger", with danger being attributable to external causes (events) and "risk" referring to our decisions (actions) and a specific form of dealing with the future that has to be decided in the context of probability. Of course, this distinction is slight because "one person's risk is another person's danger", which points to the key issue of acceptance of risk decisions.

As Ulrich Beck (1992) addressed in his book *Risk Society*, the deadlock of late modernity is that as risks become more complex, because of developing reflexivity, the need for precise calculations increases, leading to growing doubts about the ability of science to control and foresee those risks. Nowadays, our control-seeking attitude has been developed far beyond our ability to control dangers. This imbalance may be due to a raised public demand for health and security as a result of hyperindividualism and hyperextension of harm-avoidant attitudes and behaviors among the public (Throop 2009, pp. 25–40).

The utopia of modern medicine is portrayed as a life without pain and sadness in which people live with optimal social functioning and utopian bodies (Foucault 1994). Believing in medical utopia, paradoxically, convinces us that we are latent/manifest ill beings and our painful life is not the real life. Therefore, releasing life from unpleasant experiences illustrates a grotesque picture of the real life of the human being. In such a worldview, "therapy" is the way toward medical salvation and its consequent terrestrial heaven. Hence, "therapy" could be distinguished as a global morality of the postmodern age. Jürgen Habermas (1987) referred to this normative therapeutic episteme with the term "therapeutocracy", indicating the process which has challenged and intervened in the autonomy of the civil society, with the consequence that it has been turned into a mode of intervention of the human state in addition to the financial outcomes and, more importantly, professional expertise.

Via therapeutocracy, even political and legal problems get reinterpreted in psychopathological and medical terms (psychiatricism and medicalism). At the moment, medicalization, healthism and therapeutocracy are very important terms in sociology and anthropology which implicate how medical norms get transformed into social and oral norms, and how this trend can distort our cognition and emotions, leading us towards maladaptive behaviors.

Any judgment on human nature implicates a sort of qualitative value, for instance, illness is interpreted not only as a hard and unpleasant condition, but also as a "bad" one. In this dualistic world, health is good and illness is bad, happiness is good and unhappiness is bad. Furthermore, the only way to be happy is to independently take the responsibility for yourself and your feelings (Throop 2009, p. 30). Egoism has resulted from, or at least been perpetuated through, the culture of therapy, more specifically of psychotherapy (ibid., p. 29). Because of the darwinistic origins of biomedicine and psychopathology, believing in the "selfishness" of organisms and even genes is the core belief in this model. Obviously, egoism has made extensive and deep impacts on the behavior of social and psychological systems.

We should also mention another factor which increases health anxiety, namely materialistic reductionism. Biomedicine translates the multi-level being of the human to a chemophysical language. This approach has been found very useful in systematizing our knowledge about the body and how to control it; yet it also has psychological implications. Biomedical instruction somehow suggests that "we are our bodies". Once one accepts such an idea, especially when the body is perceived as if reduced to a chemophysical machine, emergence of obsessive concerns about the body, like those of the ancient Egyptians, would not be surprising. Thus, darwinistic and reductionistic approaches to biomedicine and the individualistic trend of our era could be some of the main predisposing and aggravating factors for the establishment of healthism and its subsequent ever-rising health anxiety.

Now, you can imagine that a hyper-individualized person who deeply believes in healthism, just like the way his ancestors believed in their religion, is afraid of the after life and of course of the judgment of the medicine god. These days we have encountered a sort of mass harsh conscience towards health followed by illness phobias and obsessive avoidance of unhealthy behaviors, just like the epidemics of harsh religious conscience which occurred in Europe during the middle ages as a result of excessive warnings on sin and punishment.

Releasing information resources for general publication is one of the heritages of the massive enlightenment. However, a certain amount of internal and external control over the information flow through the media seems to be seriously needed in order to moderate people's risk perception and health anxiety. We need to formulate a psychologically-hygienic approach to convey health messages and instructions in order to maximize the benefits of positive perceptions (placebo effects) and to minimize the adverse results of negative perceptions (nocebo effect) among the clients (e.g., in cases of giving information about prescribed drugs or producing drug information leaflets). To achieve such an optimal health education and delivery system, we should consider different bio-psycho-social factors which mediate the response of the individuals and societies toward health information. In the next parts, we will explain social and individual aspects of nocebo responses with emphasis on health instructions and warnings.

7.2 The Social Aspects: Medicalization, Healthism, and Life Stylism

Nowadays, talking about our blood cholesterol, blood pressure, prostate-specific antigen, body mass index, etc. has become a part of our everyday lives. These issues have become so integrated, that we forget that they have only recently been medicalized. Medicalization is the process by which medical diagnostics and managements are applied to behaviors, psychological phenomena and somatic experiences not previously within the conceptual or therapeutic scope of medicine (Davis 2010; Long 2011). The concept of medicalization rests on the assumption that while some phenomena belong in the domain of medicine, some do not (Szasz 2007). Therefore, it is very different from natural sciences like physic s or chemistry that cover the entire world and do not demarcate between physical objects and non-physical ones. In other words, demarcation between physical versus non-physical objects through the lens of physics is non-sense. In this sense, everything that we do or happens to us influences the use of our body. In fact, we could treat everything that people do or that happens to them as belonging in the domain of medicine (Szazs 2007).

Careful examination of the medicalization process shows us that it rests on the basis of healthism. Healthism in extreme versions provides a justification for racism, segregation, and eugenic control, since "healthy" means patriotic and pure, while "unhealthy" equals foreign or polluted. In the weak version of healthism, frequently encountered in Western societies, the state goes beyond education and information on matters of health and uses propaganda and various forms of coercion to establish norms of a "healthy lifestyle" for all (Skrabanek 1994). Therefore, the doctrine of lifestylism, according to which most diseases are caused by unhealthy behavior, provides the required theoretical underpinning of healthism (ibid.). Consequently, human activities are divided into approved and disapproved, healthy and unhealthy, prescribed and proscribed, and responsible and irresponsible categories. Irresponsible behaviors include activities dubbed by moralists as "vices", such as "immoral" sex and the use of drugs – both legal (alcohol, tobacco) and illegal – but it can be extended to not going for regular medical check-ups, eating "unhealthy" food, or not participating in sports.

The term "healthism" was most likely used for the first time by the political economist Robert Crawford whose article, "Healthism and the Medicalization of Everyday Life" was published in 1980 (Crawford 2006). In this article, Crawford described how the new political ideology, which emerged in the US during the 1970s, "[situated] the problem of health and disease at the level of the individual". The term is also known for its use in the book *The Death of Humane Medicine and the Rise of Coercive Healthism* by Petr Skrabanek in 1994. Both authors defined healthism as a powerful ideology because – in secular societies – it fills the vacuum left by religion. The relationship between healthism and religion could be formulated in this way: everything that we do or happens to us affects the use of our body. In principle, we could extend this to belonging in the domain of medicine. Conversely, we could state that nothing that we do or happen to us belongs in the domain of religion. Such, indeed, was the case in ancient times, before people distinguished between faith healing and medical healing (Szazs 2007). Contemporary public health may be regarded as the mirror image of Christian Science.

Everything in our lives – housing, food, education, work, air, and recreation – affects our health. Therefore, everything – not only narrowly defined as health care – belongs in the domain of medicine as health care (Szazs 2007). Drawing a line between health carefulness and health carelessness is informed more by economic and political considerations rather than by medical or scientific judgment. In this sense, we must not only distinguish disease from non-disease, but also distinguish medicalization by compulsion versus by choice (Szazs 2007).

As an artificial religion, it has a wide appeal – especially among the middle classes who have recently lost their links with traditional culture and feel increasingly insecure in a rapidly changing world. Healthism is embraced enthusiastically as a path to surrogate salvation. If death is to be the final full stop, perhaps the inevitable can be indefinitely postponed. Since disease may lead to death, propitiatory rituals must prevent disease itself. The righteous will be saved and the wicked shall die (Skrabanek 1994). The narcissistic cult of youth, health and beauty, preached by health promotionists, increases the feeling of guilt and anxiety in an ageing population who would give anything for a magic mirror which would tell them that they are beautiful and needed. The pursuit of the Holy Grail of health is driven by the mistaken belief that health equals happiness. The New Age acolyte is exhorted to eat less fat, produce healthy bowel movements and buy an exercise bicycle; no more pain or love, no more suffering or despair, no more sacrifice or weeping. While gratuitous violence, terrorism and crime are on the increase, the minders of society talk about tackling the causes of this social unrest. In their beliefs, saving human lives is a noble deed. At best, they will stare at you; at worst, they will try to measure your cholesterol (Skrabanek 1994; pp. 37–41).

Health – like love, beauty, or happiness – escapes all attempts at objectification. Healthy people do not think of health, unless they are hypochondriacs, which strictly speaking is not a sign of health (Skrabanek 1994; pp. 15–53). Similarly, when our organs perform their functions perfectly, we are not aware of them. It is the absence of health that gives rise to concerns about health. The search for health is a symptom of unhealthiness. I call it health anxiety. When this pursuit is no longer an individual concern, but part of everyday life, it undoubtedly becomes a symptom o f a social sickness. Gadamer specifically pointed out this enigmatic nature of health; he correctly showed that health is not something that can simply be made or produced. He questioned the nature of health itself. Can it become an object of scientific investigation in the same way that it is for the individuals when the balance of health is disturbed? For the ultimate aim, after all, must be to regain one's health, thereby forgetting that one is healthy (Gadamer 1996). In Gadamer's view, the mystery of illness bears witness to the great miracle of health that allows us to live in the happiness of forgetting, in a state of well -being, lightness, and ease (Gadamer 1996; p. 87). Therefore, healthiness and forgetfulness belong very closely to each other; in a way that everything that makes us aware of our health makes us sick. Thus, the mystery of health remains concealed. Its concealment belongs to the preservation of good health and this consists in forgetfulness. One of the most important healing powers in our lives resides in the ability to sink into the healing sleep of forgetfulness of every evening (Gadamer 1996; p. 138).

The stated public aim of healthism is the "health of the nation", with an implicit promise of a greater happiness for all. However, there is a huge difference between attempts to "maximize healthiness" and those to "minimize suffering". As Karl Popper (1945) pointed out in *The Open Society and its Enemies*, all attempts to maximize the happiness of the people must lead to totalitarianism. Gadamer elaborated it this way: in the vast technical structure of our civilization, we are all patients. Our personal existence is clearly something which is denied everywhere and yet also always involved in the attempt to regain that balance we need for ourselves, for our lived environment, and for the feeling of being at home in the world. A very crucial point that Gadamer tried to show is that this extension is far beyond the sphere of medical responsibility and includes the integration of individuals into their family, social and professional lives. Hence, medicalization and healthism are not medicine or science, they can be categorized as a semantic-social strategy that benefits some persons and harms others (Szazs 2007). This does not seem to be an abstract task, but rather something concrete which permanently confronts us. The challenge is the continual one of sustaining our own internal balance within a larger social whole which requires both cooperation and participation. It seems that there are many situations in which we are in a position to not only identify problems which restrict us, but also to discover new possibilities for a more humane arrangement of things as they have been developed in our instrumentalized social organization. This is something we occasionally realize through an encounter with another human being (Gadamer 1996; p. 81).

The role of doctors and other health professionals should be examined and redefined carefully. Furthermore, there is a necessary need to change the role of physicians in this era. Similar to the doctor, the patient should be entrusted with a human life which must now be released from this protective care. Those who have regained health and been given back their life begin to forget the illness, but still remain bound and beholden to the doctor in a specific, if often unspoken manner (Gadamer 1996; p. 43). Jim Windolf (1997), executive editor of *The New York Observer*, wrote that the experts will not be satisfied until every last American is suffering from some kind of disease, disorder, or syndrome (Nye 2003). This pessimistic image of medical doctors in this healthism/medicalization story is very influential. The medical profession faces medicalization in a paradoxical manner. On one hand, particularly its public health branch, medicine provides the required theoretical keystone of healthism – the doctrine of lifestylism according to which most diseases are caused by unhealthy behavior. On the other hand, physicians realized that healthism stimulates the obsession with health that indubitably leads to health anxiety. The physician's role in this present-day notion of medicalization is similarly complex, as he remains an authority figure who prescribes pharmaceuticals to patients.

The role of the patient in this story is also more than a mere victim. The pioneer German neuropathologist pointed out that "the medical treatment of patients began with the infringement of their personal freedom" (Szazs 2007). This quotation rightly emphasizes the relationship between patient treatment and patients' freedom. However, it should be noted that the role of patients has also changed. Once regarded as passive victims of healthism and medicalization, patients can now play active roles as promoters, consumers or even agents of change. Healthism stimulated and aggravated the obsession towards health and, conversely, promoted illness. So how can patients break this vicious cycle? If patients succeed in taking up the same sort of dialogue as they would normally pursue when trying to reach an agreement with someone, this could help to stimulate the ongoing process of easing the relationship between pain and well-being, as well as the experience of regaining equilibrium (Gadamer, p. 137). The main issue that should be addressed here is the understanding of the role of doctors as well as patients in the techno-scientific perspective of medicalization.

The antithesis of medicalization and healthism is the process of paramedicalization, where everyday life comes to the attention of alternative medicine, traditional medicine, or any of the numerous non-medical approaches to health. The concept of paramedicalization was first presented in 1995 by a Finnish sociologist in *The*

Finnish Journal of Social Medicine. Paramedicalization refers to the trend of people placing more and more value on alternative medicine and different beliefs about wealth and health, which are not authorized by modern (Western) medical science. While for modern medicine, healthy state only means the transient absence of diseases, many alternative medicines serve a framework for understanding healthy life without any reference to diseases. Therefore, it seems that in order to manage health anxiety in this era, redefining medicalized problems in terms of alternative medicines is more reasonable. For instance, the medicalization of diet could be replaced by a more healthy type of traditional medicine.

It should be kept in mind that the process of paramedicalization runs concurrently with medicalization. On one hand, some parts of medical institutions treat alternative and complementary medicine as a pseudo-scientific enterprise, on the other hand, alternative and complementary medicine practitioners have been accepted and approved to practice beside modern medical doctors. Therefore, medicalization and paramedicalization can sometimes be contradictory and conflicting, but they also feed each other. They both ensure that questions of health and illness stay in sharp focus in defining everyday life and problems. The dialogue between modern medicine and other traditions should facilitate this process.

In conclusion, medicalization is the process by which medical diagnostics and managements are applied to behaviors, psychological phenomena and somatic experiences, which were not previously within the conceptual or therapeutic scope of medicine. Careful examination of the medicalization process shows us that it rests on the basis of healthism and lifestylism – according to which most diseases are caused by unhealthy behavior, yet, lifestylism provides the required theoretical underpinning of healthism. The medicalization–healthism–lifestylism paradigm contradicts itself by stimulating health anxiety instead of improving health society. It was proposed that paramedicalization could be an option to manage this problem.

7.3 The Individual Aspects: Medical Reality Versus Personal Realities

The perspective towards which medical science is currently headed makes doctors, patients and society unable to become distracted from the subject of disease. Health knowledge – which is distributed widely and concretely in society – has made society hyper-vigilant toward disease-related issues like diagnosis, morbidity, treatment options, medication, drug side effects, and treatment. Etiology and prevention are the two most specific subjects towards which society is hyper- vigilant approaching concretely (Manchikanti et al. 2011; Häuser et al. 2012).

People feel surrounded by pathogens in numerous ever-increasing forms like microbes (viewed as omnipresent microscopic enemies), air pollution, and other sorts of harm expected to be hidden in everything like food, water, textiles, and electronic devices among others. The medical profession, on the other side, amplifies the same attitude by giving concrete non-individualized stereotypes of advice and explanation (Houston 1938; Hahn 1995, 1997; Dunn 2005; Data-Franco and Berk 2013; Häuser et al. 2012), as well as through labeling, blaming, disputing, ignoring, guilt inducing, drug administering, ordering, recommending, alarming, medicalizing, and fear-provoking (Wells and Kaptchuk 2012).

The concrete and terrifying medical attitude towards human vulnerability makes humans of the modern age feel no more secure than their ancestors who were afraid of large animals, hunger, cold, magic, demons, dragons, swords, and oppressing kings and emperors. Medicine is now faced with a rapidly increasing number of patients who suffer from a new collection of symptoms and rule-breaking courses of illness, which do not fully comply with any definite disease category within the vast classification systems (Hellhammer and Wade 1993; Henningsen, Zimmermann and Sattel 2003). It would no longer be possible for medical science to ignore or dispute so many clients for their atypical complaint s while, at the same time, keeping its professional figure of respect, trust, and authority.

To use labels like "difficult patient" or "medically-unexplained symptoms" – among many others – would no more help the doctor to continue laying back on their old-fashioned all-powerful seat. The future perspective of the medical science has no way other than complying with human autonomy and empowering clients to make their own way towards enhanced health. Such a perspective may not be easily achievable by the medical doctor already brought-up inside the current shell of science. For all events, it may be difficult to adopt a holistic view unless we step out of the current shell. Yet, in this part, we are not going to head towards such a far-awayvague goal. Therefore, without stepping out across the current borders of medical science, later in this chapter we will track medical literature to see if there is already some evidence-based knowledge which advocates taking "realities" into account in respect to human health. For now, we will try to define the way to personal realities and how to respect individual differences in prescribing information in order to minimize the nocebo effect. To trace for nocebo as a mind reality, we can first address the "mind" itself as one of the numerous facts of which the human is made up as a "whole". Hence, we are attempting to form an idea of how far medicine has been viewing the human mind as being possibly relevant when approaching patients. Then, we will try to explore the history of medicine to see where the nocebo concept is present.

7.4 What Is "Nocebo"?

"Nocebo phenomena" referred as placebo's evil twins (Hiskey 2011), have received much less attention in medical documents thus far than their more positively perceived siblings (Enck and Häuser 2012; Tavel 2014). Indeed, it was not until the 1960s when, for the first time, the nocebo effect was documented in medical literature. Yet in terms of human history, it has much older origins. Anthropologically, the history of nocebo can be traced far back into the past, being referred to as "voodoo death" in some primitive cultures (Cannon 1942, as cited in Esther 2002), then by contributing to "mass hysteria" and "psychogenic mass illness" in the not-so-distant past (Rubel 1964).

In simple terms, nocebo phenomena can be defined as "adverse events caused by negative expectations" (Hahn and Kleinman 1983). They are mainly – but not always correctly – viewed as the opposite to placebo effects, or as their negative equivalents (Hahn and Kleinman 1983). Originally the term nocebo, Latin for "no harm", (Enck and Häuser 2012; Kennedy 1961) was used to describe clinical deterioration aroused by negative expectations towards a pill or medical intervention. It is supposed that in the absence of such negative expectations, the same drug or medical procedure could in fact be either beneficial or at least safe and neutral, but certainly non-harmful.

As you may notice, the agency of the drug or medical intervention was taken for granted in the original definitions of the nocebo effect as well as the innate safety or neutrality of the same drug or intervention (Houston 1938). However, the nocebo story goes beyond such definitions. Evidence implies that when significant pessimistic anticipations are present in one's mind about their medical condition, a clinical deterioration would be more probable even in the absence of any drug or intervention. Such a phenomenon is more precisely called a nocebo response (Kennedy 1961; Hahn and Kleinman 1983). In other words, negative expectations seem to be able to do the job "with bare hands". A typical example is the development of hyperalgesia when, in spite of all evidence to the contrary, the patient continues to perceive their medical condition as progressive and severe (Stam 1984, Stam ans Spanos 1987).

It is easy to imagine that if negative expectations create the adverse effects on their own, they would be able to do it even better if equipped by additional materials – things like the administered pills or the medical instruments applied by the therapists – which may potentially feed the patient's fears and pessimistic predictions.

Somewhat interestingly, nocebo effects have been shown to work through biological mechanisms different from those explained by the medical literature for the placebo phenomena. Therefore, one should resist the temptation to regard nocebo as a simple blocker which undoes placebo's magic spell. Rather, although being an evildoer, unlike the placebo, nocebo acts at least as much powerfully and detectably as placebo does. That is, nocebo exerts its effects independently and objectively, just like what any pill or medical intervention does (Hahn and Kleinman 1983; Benedetti et al. 2006).

The nocebo concept predictably implies that the patient's condition may clinically deteriorate or improve at a slower rate with medical practice in the event that they believe their illness is serious or progressive, or that the prescribed pill or performed intervention is ineffective or hazardous (Milton 1973). This occurs whether the pill/intervention is a real medication/procedure or an inert nocebo pill/a fake non-operating intervention, or even if, despite the patient's belief, no pill/procedure has been applied at all (Data-Franco and Berk 2013).

In a sense, it is to say that our bodies tend to behave or to suffer in the manner which we already anticipate (Häuser et al. 2012). It is particularly alarming when you come to the fact that one's expectation, as a kind of belief, can itself cause illness. We will get to this fact later in this chapter.

At this point, we will briefly demonstrate some of the controversial discussions which the medical literature has raised so far concerning the nocebo concept itself as well as the implications suggested for it in medical practice and clinical settings.

At least partly resulting from the rule-breaking and "out-of-the frame" natures of the nocebo concept and other mental realities, there have been concerns expressed in the medical literature about the emerging attention towards them. A few research studies have emphasized the important limitations which should be considered when trying to translate the nocebo effect into the medical practice (Crombez and Wiech 2011; Bromwich 2012). One of the most familiar examples of controversy is about the application of the nocebo concept in the area of communicating medical news without invading patients' basic human rights (Wells and Kaptchuk 2012; Colloca and Finniss 2012). Also, avoiding the induction of nocebo effects when talking to the patient about the drug facts in terms of side effects, while at the same time respecting their autonomy and keeping ourselves within the borders of truthfulness and trustworthiness (Häuser et al. 2012; Tavel 2014) has given rise to concerns. It should be mentioned that we do not intend to produce a practical guideline for nocebo application in the medical practice here, rather, we are attempting to discuss the concept in order to emphasize the critical need which is felt in the medical practice for taking precautions towards the patients' mental realities and trying to design practical methods to approach the patient's mind in medical practice later on. There are still other major concerns among some medical scientists in respect to the increased attention towards nocebo/placebo discussions. By a surface review of the rather small amount of literature criticizing the published nocebo studies, one can readily infer a large common concern, which has not been greatly explicitly discussed. The concern seems to be a perceived danger which may threaten medicine as a field of science. The threat they perceive is attributed to a growing new trend of ideation which can potentially make fundamental changes to the reasoning rules and thinking styles. It may put medicine – as perceived by those critiques – in danger of becoming insidiously deviated from the mainstream of science; a deviation perceived as potentially destructive, for it may destabilize at least two of the very basic fundaments on which the medical sciences have been ever standing; namely "accountability" on one hand, and the evidence-based nature on the other (Bromwich 2012; Crombez and Wiech 2011; Laarhoven and Evers 2011). Altogether, the rationale behind the critics against nocebo studies highlight some realistic concerns, as inferred below:

When entered into the research and practice area, such abstractive concepts like nocebo can settle down as a core idea around which many other abstractions would develop, most of which not potentially experimental, and hardly ever measurable with certain confidence. This may be the reason why some authors have criticized the studies which try to objectify and highlight the importance of the nocebo effect in clinical settings. Critics claim that such reports tend to describe a hallucinatory world around health events, in which health status is grounded on a non-stable foundation that is potentially responsive to mental events, just like talking about a magic world where our fears and wishes may find their way into life unpredictably and mysteriously, a world which may never get fitted into any experimental scientific framework (Crombez and Wiech 2011). The final message of such criticism is to remind the medical professionals not to overlook the valuable mainstream facts of their science. It is somehow an alarm to push the attention back to the very original version of the story. At any event, without even the slightest amount of negative expectance, one may get into trouble once they are exposed to a large enough amounts of pathogenic microorganisms, and it would be no more difficult to get afflicted by the real iathrogenesis of a hazardous medical procedure or the objective side effects of a drug. This is definitely not rationale against the nocebo consumptions in essence; rather it is a non-deniable self-evident fact, highlighted as a "take-care" suggestion by those who really care.

From another viewpoint, medical scientists try to warn their colleagues of falling into the opposite extreme of the traditional dichotomous way of thinking about mind and body. The fact focused on by these studies, however, is one in perfect accordance with the concepts of placebo and nocebo. To be clearer, we may reframe the message here to re -state it and to remind us about it before proceeding further to probe the nocebo phenomena. The following is a statement to declare our preliminary agreement: "When trying to weigh different factors which may contribute to human health, one should beware of any temptation towards raising the already neglected place of 'mental reality' from 'nothing at all' into 'everything'. Such an attitude is actually a tendency towards inclining to the opposite extreme, obviously such extreme deviation is far from adopting a holistic approach toward patients, if this is a dream we do all pursue."

7.5 Nocebo: A Response to the Supposed Enemies

Needless to say, medicine has a glorious history. By our time, it has embraced numerous brilliant stories of success. It has been significantly successful in its mission to find a reason for many of humans' physical discomforts. In other words, many enemies or faulty parts have been successfully identified in being responsible for different sorts of human illness. In the past century, micro-organisms were identified, one after another, as the pathogenic factors caused in some of the scariest diseases in human history (Bryson 2003). The big discovery of penicillin took place thereafter, followed by the discovery of other antibiotics and disinfectant agents, generations after generations. Thanks to those honorable achievements, the science of medicine proceeded to prevent and eradicate the fatal disastrous epidemics of infectious diseases.

A great amount of knowledge was also achieved about our immune system, or our body's defense army. Now, the antibiotics could be viewed as weapons that help our body's soldiers win in the battle against micro-organisms – their small yet dangerous enemies. Vaccination was the next magnificent achievement, resembling a miraculous spell which could be casted upon one's body to safeguard it against some sort of enemies forever. Now, humankind could feel itself getting closer to the so-called spring of eternal life, whose drinking water would result in immortality and everlasting life. The sweet dream of immortality seemed not so far from becoming reality. Humankind was now dreaming to be the only creature on the earth whose generation would not be faced with extinction.

Further in the course of its history, medicine extended its borders further to discover more complex concepts of health and illness beyond the concepts of battle, friend, and enemy. Genetic disorders and inborn errors of metabolism were known, and scientific strategies were developed to improve the preventive and management methods for those disorders. Medical science extended its reach further to grasp an understanding of protective internal resources of health, and about human 's allied parties who guard and improve its health. Along with the rapidly increasing knowledge about diseases and their etiologies, classification systems were developed to categorize the numerous known contagious and noncontagious diseases.

We may get the initial idea here that, by the time medicine had overcome the infectious epidemics, the question of "friend or foe" had somehow emerged as a basic assumption in the public's perception of illness and health. So, we can postulate that the friend vs. foe question was implicitly going on in people's minds at the same time as rapid discoveries about micro-organisms were made, followed by primary scientific attempts to classify them in several ways, for example according to their innate pathogenicity vs. safety for humankind; these were the human first scientific attempts to identify enemies of health.

Again, at this point, it would be worthy to pay more attention to the "friend vs. foe question", and to have examples where the answer to this question turns out to be wrong or different from what medical evidence or human common sense would predict. So, let's review some of the very simple assumptions which may become a source of misunderstanding between both popular and professional health sectors. These facts may also be a basis to explain the important contribution of personal realities in health and illness.

7.5.1 *Micro-organisms: Friends as a Rule, Enemies by Exception*

By the time the mentioned successes had been achieved by medical science, people – feeling surrounded by so many invisible enemies in the world – gradually got used to viewing microbes as enemies rather than friends. It would make sense at this time, for when you cannot readily differentiate enemies from friends, taking any stranger for an enemy would be a more prudent decision. But nowadays, human beings of the modern age still continue to have a similar attitude.

We frequently forget the fact that, as a rule, micro-organisms have always been and still are our friends and companions; they inhabit our skin, are hosted by our digestive and respiratory tracts and live friendly all over our body. Indeed, we are absolutely dependent to micro-organisms to stay healthy, or better to say, to keep living for even one more day. While our body is just one of their possible places to reside amongst almost anywhere else, we have no alternative except them to produce vitamins in our gut, to absorb our food, to provide oxygen and energy for us, and even to fight against other small foreign bodies we ordinarily swallow each day.

Hence, the "small living things" who live around us are mostly our friends – not enemies – despite the general presumption which indicates we should either hide from or kill those using antibiotics or disinfectants (Bryson 2003). From a more humble point of view, micro-organisms have been the original inhabitants of our earth who were living on it before we arrived. Indeed, once upon a time, they kindly welcomed us as friends, made rooms for us to live in their kingdom, and helped us survive happily ever after. That was one example, among several others, pointing to basic misperceptions one can have in trying to identify friends and enemies in relation to their health.

7.5.2 Sickness as a Friend's Business Rather than an Enemy's

Even when talking about a pathogen micro-organism, it is still a common mistake to attribute the illness we experience to be caused by that small enemy alone. A major part of an infectious illness does not result from what the micro -organism does to our body, rather from what our body attempts to do to the microorganism (Bryson 2003). Our immune system itself makes some inevitable harm or discomfort to our body while trying to eliminate the pathogens, as it is too hard to keep any battlefield safe from some degrees of collateral damage. So, when we feel sick, the feeling itself – the ill- experience – which is the main reason why we seek help from a doctor does not come from the small enemy itself, rather, sickness experience mostly originates from what our own immune system does to our body for the sake of our beneficence (Bryson 2003). To reiterate, the experience of illness can be more accurately imagined as being caused by a caring friend rather than by the enemy.

It is a distorted dishonest picture to attribute our symptoms to the evil harmful attacks made by small enemies inside our body. Having such a false picture in mind, the sickness state may occasionally turn into an increasingly terrifying experience for us. The false imagination may trigger a vicious cycle of interacting negative anticipations which in turn increase the severity of our suffering. On the other hand, sickness experience, even if being uncomfortably painful, should better induce in us a sense of reassurance and relief if we truly perceive our symptoms and their severity as signs, which indicate that our body is being protected through defensive endeavors of our strong immune system. In the same manner, medical treatment can be imagined as weapons to aid our defensive army. The illness would be perceived as a more tolerable, dignified and surmountable experience when the patient views suffering as a result of supportive endeavors of the body's devoted friends rather than from violent attacks of its enemies.

7.5.3 Friends Can Sometimes Get Mad and Transform into Dreadful Foes

The cascade of a disastrous illness can get fired on following behavior by a real friend rather than an enemy – by a native microbe trusted by the host's immune system rather than a foreign body. Every once in a while an unfortunate occurrence takes place in one's body. After having peacefully lived in the same body for many years, a devoted helpful friend microbe accidentally loses its way and gets into a wrong part of body where it is not welcomed from the time of its arrival. Finding itself in a forbidden area and prohibited from going further, it becomes increasingly concerned, realizing that its non-intended entrance is not going to be easily approved by the immune system. Unable to figure out a way out of the crisis, it feels increasingly frustrated and threatened. Then all of a sudden, it somehow becomes mad; in a helpless crazy attempt to save its life or to escape, it tries to grasp anything within its reach. It begins to invade the surrounding body tissue relentlessly and literally eats everything hungrily, helping the disaster develop rapidly. What it does is explicitly mad in a way; if it is not hindered by means of an extensive treatment, it almost certainly ends in the host 's death, which also leaves no chance for the microbe itself to survive afterwards.

The story above is a typical portrait of a known infectious disease of the modern age called "necrotizing fasciitis". It is either seriously fatal (as death occurs within a few hours to a few days if treatment is unavailable or proves ineffective) or terribly disfiguring (if the victim is lucky enough to survive, thanks to an extensive rigorous medical and surgical treatment).

The causative micro-organism is a bacterium that lives innocently in its own neighborhood in the human gut for years, but very rarely, it happens that it wanders about in an absent-minded state, where it then gets into the blood stream. As a result, it is brought to another part of the body, like the fibro-muscular tissues, where it becomes mad, violently invading the tissues, triggering a battle in which it keeps fighting mercilessly. Altogether, it creates a very scary disease (Bryson 2003).

This was an example of an exceptional event happening in spite of what we already expect to happen according to the medical knowledge we have acquired about the innate characteristics of a friend-labeled micro-organism. This example shows us the fact that health events are not solitary events to be defined by absolute innate traits of friendship or enmity, rather they are made of complex interactions between the potential friends, potential enemies, and their environment and the interaction itself is affected by a sort of mutually-formed perception generating reciprocal attitudes of enmity or friendship.

7.5.4 *When Insiders Are Interpreted as Strangers: Auto-immune Disorders*

As previously mentioned, at the very first step of its critical defensive role, the immune system tries to screen out all foreign cells (potential enemies) through filtering codes and causing them to be inhibited, killed, by-coated, deported or eliminated, altogether it inhibits them from further progression inside the body. From time to time, the body's army mistakenly identifies insider organisms — mostly a tissue of the body itself — for enemies or aliens.

As a result, the immune system, which is naturally designed to ensure the safety of the body's organs through fighting their enemies, imposes a war against the self-tissues. The war is called as an auto-immune disorder, that is, a disorder which occurs when the immune system fights against one's own body. Sometimes, this results in only mild to moderate auto-immune reactions and disorders, but more typically, an auto-immune disorder is an insidious, enduring, and destructive disease process. Put in the center of a shooting target, the body tissue stands helplessly with raised hands aimed at by the armed forces of its own land.

As unfair as it seems, this would inevitably be a self-defeating battle for the invader in the long run as well, for it has ended in the body's diffused destruction which would eventually also be destructive to the immune system itself. It looks like an army who devastates its own homeland by becoming preoccupied by suppressing a perceived threat from people living inside. It keeps its right eye closed to the destruction caused by itself, until the time eventually comes where it realizes that the war has left extremely little resources within the entire land – far less than what the army itself would need to survive. This is a typical example of what happens when a cascade of harmful events gets turned on crazily as a result of a falsely-perceived threat, rather than a true one. In another word, the catastrophe comes from a misperception, better to say "a perceived reality", which triggers the serial events leading to the disaster regardless of the fact that the perceived reality is not compliant with the objective reality which is often simply referred as "reality" itself – here also called the medical reality.

Emphasized in the above examples is the fact that the content of any perception is sort of a reality. The perceived reality is a real version of reality, as real as being able to go beyond other realities in making actual effects. If something is perceived as an enemy by our immune system while we perceive it as a friend in our mind, two sets of opposing mental realities are really there, one in our mind as a whole person, and the other in the virtual mind of our body. Once your body perceives an enemy – whether you and your respected medical science believe it or not – you may get into trouble by the reactions it makes.

The other arresting notion is that, very often the answer to the critical and strategic "friend or foe?" as a "perceived reality" which determines what happens next. In other words, "reality as perceived by sort of mind" defines the body's reactions. In this way, "mental reality" can be referred to as what is perceived by the whole person's mind or "mental mind", while immunity reality can be labeled on what is perceived by the virtual mind of the person's immune system or by "body's mind", and "medical reality" can be referred to what is perceived by the doctor's professional uptake and/or medical scientific evidence or by the "medical mind".

The upshot of the above discussions is to say that in order to achieve a reliable understanding of illness experience and/or to exert predictable modifications on any health outcome, we first need to figure out an overview of the interactions between different existing realities which eventually fulfill the actual answer to "friend vs. foe", "harmful vs. non- harmful", and "beneficial vs. non-efficient and/or maleficent" doubts. Hence, the final picture would be demonstrated by an overall shared answer or by the dominant reality which would provide a brighter and more predictable picture in case the answer is highly agreeable by all parties.

7.6 Paternalistic Medicine: Why Has the "Human Mind" Been Ignored in the Area of Human Health?

As an observable fact in the course of enormous scientific achievements, medical science has never been completely ignorant towards the human mind, thoughts, beliefs, conceptions, and perceptions. As discussed before, assumptions of a "perceived friend" vs. a "perceived enemy" can be found implicitly included in the documents of the young-aged conventional medicine. However, it seems surprising that the human mind, in spite of its great potentialities, has not been significantly credited by medical science as an active agent able to impact health via its perceived realities; whereas the "immune system's virtual mind" has been implicitly approved by this science in that it can exert impacts on health through perceived realities, even if those realities are in contrast with an objectively-manifested evidence-based reality.

This discriminative attitude towards the whole person's mind as compared with the virtual mind of one of their body systems may be partly due to the dichotomous approach toward mind and body in medicine. Such an approach is very common. However, when the doctor tends to examine topics in the overlapping zones of the two kingdoms, they find that the intersecting borders are extremely challenging to define.

7.7 The Nocebo Effect and the Omnipotent Doctor Picture

As mentioned before, identifying pathogen micro-organisms as major enemies of human health puts the medicine profession in a unique position. Predictably, it was increasingly viewed by the public as a scientific profession of warfare against enemies of human health. In the same way, medical doctors were viewed as commanders whose intelligent tactics in the battlefield had led humankind into the glorious victories against dreadful enemies of health. Viewing illness as a battlefield inside the body made people increasingly vigilant about their health hereafter.

One further important outcome of those significant progresses was likely the attribution of super-human potentials to medical doctors by general ideas. As an authority figure whose knowledge and expertise was likely to be sooner rather than later needed by everybody as a matter of life or death, it is not strange after this point in history to imagine the doctor perched on a throne of unchecked power, in front of which even the most merciless kings might someday bow their heads. How would you – when in desperate need of life saving help or pain relief – look upon a knowledgeable helpful figure that possesses those skills which you lack? The view would imaginably reflect a sense of respect as well as an implicit helpless obedience mixed with a hopeful reliance. This was how doctors adopted a paternalistic attitude toward their position to take care of their patients.

The paternalistic view is one which allows the doctor to take the place of a kind father feeling granted both the wisdom and the right to decide on behalf of patients as if they were his children. Such a position, aside from all the valuable power and authority from which medical experts had benefited, made them vulnerable as they were pushed to introduce themselves as omnipotent characters to feel competent in their profession.

At the same time, for people seeking help from such authority figures, no room was left to inquire about the logic behind or alternatives to the doctor's decisions. For a long time since then, doctors have covered their true human face with an omnipotence mask. However, in the following decades, the omnipotence image gradually grew to a more challenging problem, both for doctors and patients. Because while moving further, the variety of health problems referred to the medical care systems was rapidly increasing; faster than that of the new codes and terms being added to disease categories in the medical classification systems and much faster than the developed treatment protocols. Consequently, doctors were facing numerous health concerns, for the management of which they could find no magic spell in their chest; a situation which could be interpreted as a flaw in the image of omnipotence already included in the "good doctor" picture – in their own minds as well as the public's.

Approaching the present day, it seems that both medical doctors and their clients still have the image of the doctor as a kind of "omnipotent figure". For doctors, this tendency may be linked to their initial motives of their job (i.e., to save human lives and to reduce their pain). Studies have shown that when the prognosis of a disease is not promising, doctors feel reluctant to let the patient and the family see the condition as it really is unless having already been trained in specific communication skills to break bad news (Buckman 1992; Wells and Kaptchuk 2012; Schuricht and Nestoriuc 2013). In similar studies, doctors have declared they tend to keep themselves disengaged from their patients when some "less than good" medical news is supposed to be delivered and that they assume the reason to be their deep wish, as the medical doctor, to be able to manage any human suffering (Buckman 1992; Fallowfield and Jenkins 2004).

Even in the modern age, it seems that the doctors still try to fit themselves to an image they perceive as being portrayed of them in the public's mind; a portrait which reflects traits of wisdom, profound miraculous knowledge, good intentions of help, and a superior curing power; in short, an image of omnipotence. You may notice that doctors have not willingly given up their "omnipotence figure" even when faced with patients who present a very complicated illness pattern which can neither be easily assigned to any disease entity or explained by scientific medical knowledge, nor is going to get managed effectively through the evidence-based medicine. Such situations would be frustrating to doctors whose professional self-esteem is deeply rooted in the "omnipotent doctor" archetype. Bearing in mind such a professional self- image, the doctor would perceive their public creditability as being threatened whenever they find themselves uncertain or non-efficacious in diagnosis and management of a medical condition.

While any medical doctor – sooner or later during his/her career – would inevitably be faced with their own limitations (or with limitations of the medical science itself), their self-esteem would not necessarily be threatened in such situations unless they have been strongly pushed to turn their mind's distorted wishful image of the omnipotent doctor into reality. Indeed, acquiring insight toward one's personal and professional limited efficiencies would release the medical doctor from their historical discomfort towards admitting the dark sides of their own knowledge or their scientific field. It would also allow them to assume a non-conditional and non-judgmental attitude toward the patients, even towards patients whose suffering patterns do not match any known diagnostic categories or do not respond to any sort of evidence- approved treatment.

The point is that medicine is increasingly faced with patients whose physical symptoms are hardly ever compatible with the criteria of any specific known disorder or syndrome (Henningsen et al. 2003). There are also a growing number of instances where a patient's subjective complaints can be sensibly attributed to one disorder or a list of differential diagnoses; but when the patient's condition are investigated thoroughly by extensive clinical and laboratory exams, the objective findings fail to confirm any of the diagnoses or simply rules them out (Barsky et al. 2002; Colligen and Murphy 1979; Mills 2006).

Looking throughout history, one may realize that two things have so far substantially helped doctors not to get frustrated when they are pushed to their limits of professional potency: so-called technical language and scientific literature. They have helped the vulnerable medical doctor to restore at least some superficial appearance of their omnipotence figure in the public view. After all, the doctor would still remain as the man of science who knows the technical name for any mysterious illness and they are also the one who can bring wise recommendations out of his chest to be obediently followed by the patient. As a result, the public maintains the view of the doctor as the large alliance of human health who knows health enemies and diseases well enough to call them by their names and is able to classify patients according to their illness type.

As a communication aid, terms like "difficult patient" as well as descriptions such as "medically unexplained symptoms" and many others have been added to medical terminology to categorize a large group of various patients under a single name. On one hand, those labels make it possible for medical doctors and psychiatrists to communicate their shared experiences of helplessness to each other. On the other hand, it works as an excuse for the medical doctor to minimize the fact of their ignorance and despair toward these conditions in a manner of grandiosity or innocence. When frustrated, human beings tend to put a label on things, persons, or situations in order to convey a meaning of severity, strangeness, or bizarreness; such a label acts as a remedial tool to relieve their feelings of despair. In the same way for medical doctors, medical terms implying severity or complexity induce a feeling of regained mastery, as if they have somehow grasped the diagnosis or they still own the knowledge as the doctor and/or as if they were not responsible for the patient's continued suffering.

Similarly, for many systems of medical education, giving the situation a medical name has become one of the most valued aspects of the medical knowledge regardless of whether it implicates any beneficial outcome to the patient or not. Indeed, stigmatizing seems to be utilized to compensate for the times the omnipotence picture is not going to hold true. Medical doctors who care about the final mission of their job should ideally be extremely cautious when u sing terms like "difficult patient" because many professionals and patients perceive it as intended to mean something like "the non-important annoying patient". This kind of discriminative interpretation is closely linked to such labels which makes it much more difficult – not easier – for the patient to regain their health. This is one of the alarming points of which medical doctors need to be beware; causing the patient harm via stigmatizing them with medical terms.

Other than medical labeling, what most medical doctors do in such situations is to refer those patients to psychologists and/or psychiatrists, a decision which is neither readily welcomed by the patient nor perceived as an attempt intended to help; it is rather assumed as a sort of rejection or punishment as their symptoms have not been confirmed by the doctor as medically valid or as important. Moreover, even when such a patient visits a psychiatrist, it is very likely they will put the doctor in the same puzzling situation (i.e., a situation in which the patient's complaints don't indicate a clear-cut disorder and/or cannot get managed by medical interventions).

Indeed, occasions in the medical practice where the physically ill are not managed as predicted or get referred to psychologists and/or psychiatrists are unique opportunities in which medical science has to accept the "mind" as a relevant factor associated to human health. This can allow medical science to go beyond its conventional borders and acknowledge the vital presence of mind, mentality and intellectuality in the kingdom of human health. Thus far, however, medicine has not done many favors for such medical conditions except granting them a name in medical terminology. New trends, qualities, and presentations of medical conditions are now reaching such a fever pitch that we have no escape from again looking into our knowledge of human health, especially the gaps or areas where we have recurrently failed to achieve our goals. This would force us to probe medicine deeper in the typical zones where mind–body issues apparently merge into one another – where medicine is getting closer to giving up on denying the mind's contribution to human health.

There is an ancient (Islamic) quote which says "people are enemies toward what they do not understand" (Imam Ali, Nahjolbelagha). In the above historical overview of medicine, you see examples to such a wise quotation. Doctors tend to deny pain or suffering that they do not thoroughly understand. Also, they assume a rather defensive attitude towards any approach to those aspects of the whole person, to which medical science has been highly ignorant so far. And on that sobering note, it is time to tap the medical literature to see whether and how the mind's contribution to health can be acknowledged in the context of evidence-based medicine.

In summary, for medicine, the mind and realities perceived by it are among the dark sides of the human whole towards which it has been so far both ignorant and non-tolerant; while harm and benefit are the two most basic concepts in human health which are objectified through the effects of different friend or enemy factors. Those effects, as explained before, have been found not to be absolutely innate and predefined, but rather partially and reciprocally defined by (altered by/mediated through) a sort of "friendship" vs. "enmity" as perceived by a vigilant health-aware perceiver.

7.8 Contribution of Various Realities in Health, Emergence of the Nocebo Concept in Medical Literature

We are going to review the relevant medical literature in respect to the association of "harm," "health," and "mind" organizing existing knowledge – however limited – about the neglected sides of the human whole including the mind, thought and emotions. We will review the literature to figure out what else one can do/avoid to maximize beneficence and/or minimize harm in medical practice beyond what one was already trained for in the conventional diagnosis and treatment system. First, we will focus on how to avoid causing harm with a closer look at the nocebo documents, facts, and discussions. Then, we will try – if at all possible – to develop a rough preliminary evidence-derived map to exploit the human mind as one of the health alliances when approaching a patient in medical practice. Through a rather extensive review in the medical literature, we may head towards answering the following questions: (1) How can negative expectation turn into a reality? (2) Are some people more prone than others to being effected by nocebo effects? (3) How could a "nocebo-effect" become possibly induced, predicted, prevented, or stopped? In other words, we are going to examine different factors associated with the nocebo effect to obtain some strategies aimed at preventing it in the medical practice.

7.8.1 How Can a Negative Expectation Turn into Reality?

Expectance theory is so far the most inclusive theory among several theories which try to explain nocebo and placebo responses (Häuser et al. 2012). It postulates that the mechanisms, which mediate a placebo response, are activated through positive expectation linked to reward expectation traits and/or states while those of a nocebo response are induced by negative ones related to anticipatory anxiety traits and/or states.

In respect to psycho-physiological mediation, reward expectation has been linked to reward dependence traits or states, modulation of the brain reward system in the limbic area and increased dopamine and endorphin release. On the other hand, anticipatory anxiety has been associated to harm avoidance traits or states, increased brain stem and nociceptive activities, Cholecystokinin (CCK) neurohormone secretion, hyper-activation of the hypothalamus-pituitary axis, and increased blood levels of ACTH and cortisol hormones (stress hormones) (Häuser et al. 2012; Spiegel 1997; Kennedy 1961; Hahn 1997). These biological events are all in reciprocal interaction with each other and with situational precipitating and perpetuating factors as well as psychological situational factors like past experiences, emotional status, motivation, belief, therapeutic milieu, doctor-patient relationship, and received information (Brañas-Garza et al. 2010). Several documents show that placebo effects are associated with a person's tendency toward optimism and social desirability, while the nocebo effects are linked with their tendency toward pessimism (Brañas-Garza et al. 2010; Friedman and Booth-kewley 1987; Jakšić et al. 2013; Cooper and Tape 2001). The therapeutic environment can modulate both therapeutic and adverse effects of an active drug as well as the placebo/nocebo-response in antidepressant drug therapy (Henningsen et al. 2003; Bingel 2013). Those who had received placebo drugs but believed they had taken antidepressants – as they had been already told they would receive either an antidepressant or an inert drug – reported the vast majority of side effects in clinical trials. Indeed, the experienced adverse effects had been caused by the negative emotions or the psychosocial stress of anticipating receiving psychotropic drugs (Barskey et al. 2002).

In the same way, in clinical placebo-controlled trials, it has been shown that a quarter of depressive patients who did not adhere to treatment or discontinued it due to experiencing specific drug side effects were those who had received placebo. This highlights the fact that anticipatory anxiety combined with a pessimistic inclination exerts such a great impact on an individual's physiological sensations that they experience the same symptoms as if they had received the psychotropic drug itself (Cooper and tape 2001; Jakšić et al. 2012).

Depressive emotional states (or traits) are shown as the upstream mediator in nocebo effects. It has been shown that in persons who are already prone to negative expectations (e.g., in depressive patients who naturally expect that things won 't change for the better), nocebo impacts are much stronger than placebo can compensate for. Therefore, it would not be surprising that coercive therapy or patients' lack of belief in treatment, as well as a history of poor response to previous treatments, have strong adverse effects on the eventual success of antidepressant therapy (Hahn 1997; Hauser et al. 2012; Colloca 2012). Based on evidence, both anticipatory anxiety and reward expectance are modulated, activated, or precipitated by learning experiences which include information (given by clinicians) and the history of previous successful/unsuccessful therapy (Hauser et al. 2012; Hahn 1997).

7.8.2 Are Some People More Prone than Others to Be Affected by Nocebo Effects?

Individual differences in nocebo and placebo responses are a self-evident fact justifiable by that previously mentioned about mediating mechanisms as well as by several other bio-psycho-social predisposing and situational explanations; yet, valid studies have rejected the old notion that a sort of purely inherent specific vulnerability towards the nocebo response might exist. Inter-individual variation studies which identify predictors of nocebo responses are currently a major point of research attention. Early studies describing placebo-prone personality traits have been criticized by other researchers later on for methodological biases as well as flaws in their conceptual framework. In the past decades, researchers have tried to re-examine the findings of the earlier studies with more robust research frameworks.

At this point, we tend to briefly introduce the results of the third set of studies. The overall agreement over which studies have been generally congruent is that definite correlated situational factors should be specified if any individual variable is to be regarded as associated with a placebo or nocebo response (Crombez and Wiech 2011; Jakšić et al. 2013). In this way, by linking them to their specific correlated situational variable(s), old studies have been revived and the findings of even older studies re-validated. Thus, by defining the preliminary situational conditions, some predictive factors of the nocebo and placebo responses explored in those studies are now revisited and defined again (Jakšić et al. 2013).

Kennedy (1961) had emphasized that a nocebo response is a subject-centered reaction. He specifically referred to the nocebo reaction as "a quality inherent in the patient rather than in the remedy". Taking note of the nocebo definition, this is a clearly valid statement that the negative expectation – as the cornerstone of any nocebo response – is a reality in the patient's mind, not in the received pill or intervention. Yet, evidence rejects the idea that once an individual manifests a nocebo/placebo response to one treatment, they will present such a response to other treatments as well; this disconfirmation implies that in the same person, nocebo/placebo response varies according to situational and other interacting conditions. Similarly, evidence did not support the existence of a so-called "placebo-prone personality" (McGlashan et al. 1969).

Also, through a well-credited study, Lasagna, Mosteller, von Felsinger and Beecher (1954) revealed that before administering a drug or a medical procedure, one can by no means reliably pre-differentiate individuals who would manifest a placebo response from those who would not according to measurements of their individual characteristics. Moreover, hypnotic susceptibility has been disconfirmed as a predictor of a nocebo/placebo response in any individual (McGlashan et al. 1969; Stam 1984; Stam and Spanos 1987). On the other hand, there are several studies which indicate that there are major individual differences in positive vs. negative expectations towards an ongoing or upcoming event in some defined situations (Hahn 1997). As already highlighted, some specific personal characteristics have been again approved for their association to nocebo responses if mediated by specific defined situations. Specific situational variables have been also defined as mediators of specific nocebo associated traits (Stam and Spanos,1987; Drici et al. 1995; Mills 2006; Jakšić et al. 2013).

7.9 Revisited Association of Personality Traits and Nocebo Response

7.9.1 Pessimism Versus Optimism

Several studies introduced pessimism as a predictor trait for a nocebo response toward a pill. This is in line with the expectation theory (Barskey et al. 2002; Data-Franco and Berk 2013). Pessimistic individuals have been described as having more nocebo-prone personalities only if there is some sort of deceptive negative expectancy concerning the drug, but not when they are truly informed to expect positive effects along with unpleasant symptoms.

In the same way, optimistic individuals tend to benefit from placebo pills more than the pessimistic ones, provided that they are truly informed about the possibility of some positive effects, but not when they are somehow deceived by information which contains some sort of embedded negative expectation (Stam and Spanos 1987; Drici et al. 1995).

7.9.2 Neuroticism Versus Agreeableness

These are other traits found as predictors of nocebo and placebo responses respectively. Neuroticism is defined as the tendency to experience negative affect. Agreeableness, on the other hand, is a trait of accepting differences and adapting to variable degrees of the unwanted or the uncertainty in various situations.

7.9.3 Agreeableness Versus Hostility

Agreeableness is the predictive trait for a higher placebo response which is facilitated through a healthy therapeutic relationship. Conversely, angry hostility, as a facet trait of neuroticism, is highly correlated with not benefiting from a placebo analgesic treatment, even when positively introduced in general ordinary hospital settings (Mills 2006).

7.9.4 *Resiliency Versus Non-resiliency*

By definition, resilient individuals are those who are better able to recover from negative emotional experiences and flexibly adapt to the challenges of stressful situations. Resiliency and non-resiliency are known as predictive traits for higher placebo and nocebo responses respectively (Mills 2006; Jakšić et al. 2013).

7.9.5 *Suggestibility Versus Non-suggestibility*

The suggestibility trait was the most popular personality characteristic introduced by the old studies, followed by and closely related to the hypnotizability (Eysenck and Furneaux 1945; Cooper and Tape 2001). Altogether, researchers do not yet believe that there is a general factor of suggestibility yet (Mills 2006; Sedgwick 2013). However, the concept of suggestibility was an initiation for more active researches to find individual variables related to nocebo and placebo. Research studies had also proposed two different factors of suggestibility namely primary or idio-motor suggestibility and secondary suggestibility or gullibility or indirection (Eysenck and Furneaux 1945). The latter had been known as associated to a higher placebo effect; yet, there are no empirical demonstrations to support the reliability of the secondary suggestibility as well as the whole concept as a predictive factor for placebo/nocebo response (Jakšić et al. 2013; Mills 2006).

However, studies on suggestibility and hypnotizability have proposed that in stress situations, highly suggestible people tend to suspend their critical judgment and mental editing function. The mental editing function helps the person to decline the expected when it is disconfirmed through the evidence (Laarhoven et al. 2011). In extreme situations, many individuals experience a natural trance state in which they are highly suggestible (Harrington 1998; Drici et al. 1995). In other words, the error detecting mechanism in our brain is sensitive to a lack of confirmation of that which we had expected. This cognitive system may fail to do its performance in extreme stress situations; hence, the person tends to become more suggestible (Spiegel 1997; Mills 2006). This implies that inducing extreme measures of stress to human beings may make the individual suggestible enough to become enchanted into accepting any received information while their judgment system is turned off.

7.9.6 *Extroversion Versus Introversion*

In a few studies, a higher nocebo response has been found to be associated with introversion traits, while extroversion traits have been shown as associated with a higher placebo response. Again, this is in accordance with previous associations of variables like harm avoidance, neuroticism, reward dependence, and pessimism

(Mills 2006). The concepts of extroversion/introversion can be also linked to a new discussion on the internal and external locus of control.

7.9.7 Internal Versus External Locus of Control

People seem to attribute life events as happening either through their own will and behavior (internal locus of control) or to be brought about by external factors out of their control (external locus of control). The style of attribution may substantially differ for negative or positive events respectively. It is also linked to other situational factors. There might also be some individual variations in the attribution style in relation to personality traits along with situational factors. Conceptually, external locus of control can be imagined as linked to the extroversion concept; introversion traits on the other hand, seem to be conceptually close to internal locus of control, consequently, to a higher nocebo responsiveness (Hahn 1997).

When a person regards a disease to be at least partly caused by an unhealthy life style, their attribution style is one with internal locus of control. But when the person believes their illness has absolutely resulted from genetic predispositions or from air pollution, whether the belief itself is true or not, they attribute the event to an external locus of control (i.e., they perceive the locus of control as situated somewhere out of reach of self-regulation). Possible inter-relationships between internal versus external locus of control and a higher nocebo/placebo responsiveness, respectively, should be investigated more comprehensively in future.

7.9.8 Nocebo/Placebo Response and Temperamental Traits

While a reward dependence temperament has been found to contribute to placebo responsiveness, anxious emotional traits and an associated "harm avoidance" temperament have been associated to nocebo responsiveness (Hahn 1997; Colloca and Grisson 2014).

7.9.9 Nocebo/Placebo and Cultural Issues

"Ethno-medicine" as an element of culture can serve as a healing function; yet, it may be sometimes associated with higher nocebo responses in its members by giving them noxious, rigid, and rough information on the symptoms, etiology, and treatment of diseases (Hahn 1997).

7.9.10 Nocebo/Placebo and Attachment Theory

There are interesting studies which have tapped into the attachment theory in relation to placebo/nocebo responses. Studies have revealed a connection between an insecure attachment base with a low responsiveness to placebo effects (Enck and Klosterhalfen 2007, Mclean et al. 2007).

7.9.11 Nocebo/Placebo and Type A Versus Type B Personality

There are also a few studies which have proposed the "type A" personality as being prone to a nocebo response and the "type B" personality to placebo responses (Colloca 2012; Data-Franco and Berk 2013; Pietrie 1948). In another study, a lower score of time discount factor manifested as "impatience" or indirectly "time punctuality" has been found to be associated with vulnerability to "anticipation anxiety" and, therefore, a "nocebo pain response". It showed that impatient patients who value only very near events tend to take into account only negative effects of a medical procedure (costs), and a higher pain anticipation; while those patients who tend to value the far future beneficial outcomes when judging the present costs tend to have less negative pain anticipation (Brañas-Garza et al. 2012).

Concepts of impatience and low time discount seem closely related to traits of the so-called type A personality, so it seems there is an agreement between this study and the aforementioned study results. Self-scrutiny is another concept which has been linked to both the type A personality and negative anticipation leading to a higher nocebo responsiveness. From a temperamental standpoint, one can find an association between the so-called personality type A and a combination of harm avoidance and novelty seeking temperaments (Colloca and Grillon 2014). The inferable picture would be one of energizing persons who push themselves towards achieving the novel while at the same time tending to avoid any harm and hazard; you may see that such tendencies are very challenging. On the other hand, one can associate the type B personality to a temperament of lower novelty seeking and lower harm avoidance, as well as higher agreeableness, flexibility, and resiliency all in accordance with a lower nocebo responsiveness as well as a higher placebo responsiveness.

7.9.12 Other Associations to Nocebo Responsiveness

Other traits or states which have been revealed to be associated to a higher nocebo response include: higher levels of generalized distress, anxiety, depression, a heightened index of suspicion, a tendency towards somatization, symptom exaggeration, and a hyper-awareness towards bodily sensations (Friedman and Kewley 1987).

Evidently, in medical practice, patients who consider themselves as "especially sensitive to drugs" should be considered as particularly prone to manifest nocebo responses (Barskey et al. 2002).

7.10 How Situational, Temperamental and Other Psycho-biological Factors Interact When One Faces an Unwanted Event

7.10.1 Active Harm-Avoiding Individuals

Such a person can be described as one of low flexibility who attributes control to an internal locus. Such a person may tend to overestimate the uncertainty (due to negative expectance); therefore, they may become over-engaged in trying out additional prophylactic and diagnosis attempts to prevent a negative event – for example, a disease.

From time to time, the active harm avoiding person can become hostile towards related-others when facing the unwanted. It can be a kind of defensive mechanism. In other words, having a tendency towards self-blaming when faced with the unwanted, the person sometimes projects the tendency to blame themself toward others; for example, the person feels guilty for not being able to prevent the negative event. This feeling grows more and more painful and reaches a point which is unconsciously non-tolerable by their mind, so they project the self-blaming toward others causing others to feel hostile towards them, which is much less difficult than feeling hostile towards oneself. Overt hostility would get presented only if such a person has a low predisposition for becoming dependent on social rewards; otherwise, they perhaps either continue blaming themselves rather than projecting it outward or try to hide their anger and hostile feelings towards important others to avoid losing possible social rewards from them.

There is still another problem, which high harm-avoiding controlling people may face in the health area; that is, by actively detecting threat alarms and trying to control them, they may find themselves in a non-tolerable over-focused state where they become concerned about their own controlling capabilities and push themselves towards controlling the risk factors even more. If such a person has a high novelty seeking temperament at the same time, they may occasionally experience some intrusive thoughts of self-harm or recurrent impulses to behave as opposed to what they desire. This can also originate from pessimistic traits which lead them to expect negative outcomes (the anxiety aroused from anticipating the loss of control over risks and hazards). When the urge for having health hazards under control reaches an extreme, such individuals are susceptible to experience paradoxically-intentioned impulses and drives or obsessive paradoxical ruminations which severely increase their anxiety.

For example, when the individual decides to control their weight, they begin to control their calorie intake; the greater the preoccupation with controlling the calorie intake, the greater the concern of being incapable to do so. Anticipating giving up control, the opposite images become anxiously aroused in their mind, pushing them towards eating delicious and forbidden materials. The impulses and the opposing rumination may eventually make their mental challenge and the consequent anxiety feelings more than tolerable so that they may act according to the impulses; they may give up the resistance towards a mental obligation and attempt to eat big amounts of delicious high calorie foods even when not hungry. Such undesired behaviors turn into a vicious cycle in which overeating increases self-blaming thoughts and the negative affect and leads to increased anticipatory anxiety and obsessive ruminations of "beware alarms" again. The resulted anxiety and self-blame in turn amplifies the paradoxical impulses and the cycle continues. Later we will discuss the public health educational implications of such a vicious cycle.

7.10.2 *Passive Harm-Avoiding Individuals*

When a person with a high harm avoidance trait perceives the locus of control to be external, they tend to somehow passively avoid the danger, but not actively undertake the prophylactic behaviors or utilize the protective measures. Such persons feels helpless toward harm and give up trying to change the adverse situations for the better very early in the process. They tend to blame others or external factors, complain about the unfortunate events, feel hopeless, and think that there is no helpful option of behavior to improve the situation. Nevertheless, if such a person has a strong temperamental trait of reward dependence, in addition to their harm avoidance traits, they may be rather compliant, just as if attached to a supportive medical care system that may be regarded as a source of social rewards.

7.10.3 *Passivity in Low Harm-Avoiding Individuals*

As an extreme opposite to the high harm-avoiding individuals, a low harm-avoiding individual perceives an external locus of control, and due to a lower harm avoidance, may be neglectful and careless towards health, unless they have acquired deep insight and commitment through personal growth. Such a person tends to take it easy and behave carefree. They are not driven by their health care or its inherent traits; thus, they may not feel pushed towards complying with the preventive or treatment advices or drugs unless they are already justified and insightful toward them. They would also perhaps not readily adhere to the medical systems when becoming sick. Yet, if such an individual is predisposed to a high reward dependency, it is inferable that once adherent to a medical care giver (e.g., a medical doctor), they would perhaps comply with the recommendations to feel socially rewarded by creating a good relationship to their doctor.

7.10.4 *Active Controlling in Low Harm-Avoiding Individuals*

A more active non-harm-avoiding individual who attributes control to an internal locus can be a person who may realistically and reasonably analyze the cost-effectiveness issue and tailor their health-related behaviors to the standards derived from such an analysis. Indeed, healthy behaviors for such a person are majorly dependent on other inherent or acquired characteristics like the power of reward dependence traits, insight, health knowledge, and availability of a qualitative open health system.

It is time to shift our attention from the biological and psychological variables towards the social factors linked to nocebo/placebo concepts. Reviewing the literature here, we tend to figure out how all of the previously mentioned factors may be connected and re-explored to become applicable in the society and health systems. As concepts of risk, danger and cost-effectiveness are embodied in all social issues, we are going to elaborate on these concepts in relation to negative and positive expectations toward health.

7.11 Risk, Danger and Cost-Effectiveness and the Public Health Behaviors

Health protection is essentially linked to the concepts of risk and danger. As already mentioned, dangers are threats originating in the environment. Thus, they are situational factors against which we protect ourselves by regulating our distance to them. Risk, on the other hand, is another related concept rooted in our behaviors and choices. For example, sun exposure is a known risk factor for skin cancer and eye lens cataracts; here, sun exposure is the risk and cancer or cataracts are the dangers. Yet it would not be possible for us to avoid sun exposure completely. So we choose to wear sunglasses and use sunscreen, which do not provide total avoidance – these are protective factors and our choices. We tend to gather information about risk factors to apply proper protection to control the risks and, thereby, avoid the dangers.

As our knowledge about risk gradually increases, we sense more threats and expect more negative events. This will in turn increase our tendency to control the risks. As nobody can undertake all of the known prophylactic measures, the extent of prophylactic behaviors that we would apply turns to a matter of cost-effectiveness. We should assess costs and effects and figure out how and what to what extent we can control risks and to avoid dangers.

It is the exact point where the individual factors get taken into account. Like any other choice in our lives, individual 's choices and decisions for health are made by weighing costs versus effects. Needless to say, the decision about how to s et the cost versus efficacy measures is partly made according to the personality traits and characteristics. It also depends on the severity of the danger, the feasibility of risk control, and the significance of the person whose health is being considered for the one who overtakes the costs.

An instance would be as follows: A father takes his 6-year-old child to a pediatric clinic twice a year for the recommended 6-month-interval visits to the family physician to monitor the child's health status. Four years ago, the same physician had diagnosed the father as suffering from high blood pressure and recommended him to visit the clinic in 3-month intervals to monitor his blood pressure; yet, by now, he has missed all but two of the appointments for himself, while his child's regular visits have never been missed. This father seems to be readily willing to pay the needed costs for the sake of his child's health (the cost here is taking the child to the health clinic regularly and devotedly) while he tends to ignore his own health and not pay the cost of adhering to a 3-month-interval follow-up program.

When you donate your kidney to a loved one, you altruistically choose some risks which may present the danger of renal failure. However, you choose to do it for saving the life of a loved one. Even as a matter of generosity, the act of donation still depends on the situational factors; for example, if the one in need of your kidney were a 90-year-old far away relative, you might not be as generous.

7.12 Social Iatrogenesis: The Nocebo Response Induced by Social Health Systems

Could any harm be caused through the "beware of harm" messages which the media distributes amongst the public? Let us discuss the answer through an example:

Leila is a single 22-year-old female university student who lives in a big city with her parents. Leila has always sort of anxiety; she describes herself as an ever-worried girl. Temperamentally (suppose that you know), Leila has traits of high harm-avoidance combined with high novelty-seeking traits. She also feels a constant urge to put all minor and major life issues under her control as soon as possible. Regarding attribution style, Leila tends to attribute events to an internal loc us of control. She believed that she should be in control of her life; she feels responsible and tends to blame herself for all minor and major problems which arise in her own life every now and then. However, when things get out of her control, she tends to blame others as well if they have somehow contributed to the problem (albeit very slightly) through their minor mistakes and faults.

On the other hand, Leila is proud of the areas of her life which are under her control (e.g., body fitness) and dislikes people who don't care about their own body weight, fitness, and health. Yet, since a few years ago, she has been privately blaming herself for areas of personal weakness or failure (e.g., for not being able to establish a durable friendship, as well as for having lost her natural feelings of happiness a few years ago for no clear reason). Leila actively avoids exposure to health hazards as much as she can. As her mother is a breast cancer survivor, Leila found her own mind increasingly preoccupied with how to avoid the general risk factors of cancer, especially those risk factors which are almost always out of her control such as air pollution and stress. Air pollution and stress are the two risk factors, which her mother blames frequently for her breast cancer development.

Leila tries to stay at home whenever the weather reports the air pollution to be above the healthy standards. She also tries to manage her stress by participating in meditation training groups, practicing muscle relaxation techniques, and regular exercising. Yet Leila still suffers from constant distress, far beyond what is needed for keeping things under control. Leila tends to avoid stress, yet she bears substantial extra stress whenever a stressful event happens to her. When a stressful situation is over, Leila experiences even more stress than that of the situation itself; thinking over and over about why she has been faced with it and how she could have prevented it.

Leila has gathered a vast amount of information about cancer, healthy life style, stress management, air pollution hazards, and other such issues through numerous health education websites, workshops, seminars, books, magazines, T.V, and radio programs, and the hospital clinic staff who treated her mother's cancer. Recently, Leila has begun taking part in self-help groups for family members of breast cancer survivors where she receives a lot of new information in each session alongside support and empathy. She regularly visits her family physician and has shortened the interval between the visits; sometimes, she even asks for an extra visit to inquire about something she has found in her breast exam or to discuss some other bodily concerns. The family doctor is a middle-aged man who has a calm and accepting attitude and reassures Leila in each session.

In general, health hazards which one tries to avoid or to control can be classified into two groups: the first group of risk factors is tangible, avoidable or controllable; the other group of risk factors is sort of vague, general, less avoidable, an d less controllable. Not drinking non-pasteurized milk to prevent brucellosis is an example of the first group of hazards. Stress is the most typical hazard of the second group.

In Leila's case, although her mother was once advised by her doctor to avoid stress, no further information had been delivered on how much stress can be harmful or how to avoid it. Any stressful life event thereafter, including even minor daily hassles, preoccupies Leila's and perhaps her mother's minds. Moreover, from time to time, an argument happens in which Leila blames her mother for a stressful reaction towards a recent event, while her mother blames her husband as a major source of stress causing her breast cancer. Leila once talked to her family physician about the distress she experiences just while trying to avoid other types of stress, and the doctor tried to reassure her by saying that stress is unavoidable. However, Leila feels even more anxious thereafter, as she assumes she is be completely helpless towards stress as an important risk factor which, according to the doctor's comment, is always present and constantly causing harm.

Leila is an extreme example of harm-avoidant individual who tries actively to control almost all controllable health hazards she knows as far as possible; however, she is becoming more and more hyperaware of the uncontrollable ones. As a result of her pessimistic attitude toward health, Leila tends to overestimate the non-controllable risk factors and keeps trying to find a way to exert control over them. When she was an adolescent, Leila witnessed her mother's breast cancer diagnosis despite her devoted attention and care towards all hygienic advice, maintaining a healthy diet, doing regular sports, and performing other measures of prevention. At that time, Leila herself was assessed for possible genetic receptor similarities between her mother and her to determine if some preventive interventions would be needed. Leila was relieved when she was informed that the assessments had categorically ruled out such a similarity.

As for genetic vulnerability, there is no other controlling or preventative strategy, Leila tries to forget about it; yet, she feels some constant fear of the danger somewhere in the back of her mind and she compensates by increasing her prophylactic behaviors. The positive side of her awareness is that Leila is rather engaged in prophylactic and health perpetuating activities. Yet, the negative side is that Leila cannot relax even when the known controllable risk factors are already properly controlled. This is somewhat due to the result of her pessimistic and non-flexible attitude towards health, which is partly due to her general negative affect and biologic pessimistic predispositions combined with her childhood experiences as a psychological predisposing factor.

Gradually, this non-relaxing hyper arousal state becomes more complicated by disturbing signs and symptoms related to an increased autonomic response such as insomnia and palpitations. These symptoms in turn dramatically increase Leila's anxiety and concerns about stress avoidance. As an upshot, due to everything mentioned thus far, the story ends with Leila suffering from severe panic attacks and being referred to a psychiatrist for treatment of her panic disorder. This is a typical example in which negative expectation and anticipatory anxiety develops into a disturbing anxiety disorder, which in turn may be regarded by the patient as a source of risk. No matter how hard we try, we will never be able to control, prevent, or eradicate all health hazards. All we can do is to keep a reasonable distance from dangers and utilize the preventive measures.

7.13 But How Far a Distance Should We Keep from Danger? Where Can We Feel Safe?

These questions are difficult to answer because the attitude towards health is widely variable among individuals and societies. So far, there has been no rule for quantifying the reasonable amount of concerns one should have regarding a health condition. This is not a question about choosing cost versus effect, but a question of where to stand in between the two. When looking at the spectrum of cost-effectiveness, healthy behavior, preventive activities, health status, and checkup measures, among many others, behaviors are considered as costs and staying healthy is the effectiveness. At all costs, we have to accept various amounts of uncertainty in life including, to some extent, out-of-control events such as various sorts of accidents or natural disasters. We also need to be able to face various extents of the unpredicted negative events in our daily life while we continue feeling secure and confident otherwise.

The proper amount of costs to be paid for health depends on the threshold beyond which our life quality becomes impaired; in other words, seeking health is reasonable as far as it serves to maintain or improve our quality of life, and we should reconsider costs and effectiveness as soon as we find ourselves harming the quality of our life in pursuit o f better health. Flexibility as a facet of both concepts of agreeability and resiliency is a trait or a situational attitude which allows us to be comfortable with various degrees of uncertainty in various situations, or allows us to keep our basic feeling of security when facing the unpredicted. When facing the unwanted, a highly flexible person can accept, to some extent, the uncontrollable event; they are also able to and tend to face the unpredictable events even with the knowledge that may possibly fail at controlling it.

Again, we shall never know with certainty how much flexibility is enough or proper. The optimum threshold or cut -off point of flexibility would be no more specified than the point or degree beyond which our life quality would be improved or maintained while we are assuming healthy behaviors and avoiding risky ones. For example, by managing stress in pursuit of a higher health, Leila is standing in the zone of healthy health-related behaviors as far as she sticks to a healthy diet, exercises regularly, adheres to the standards of preventive measures, has desirable social contacts, continues doing leisure activities, and so on. However, she has entered the unhealthy zone as soon as she finds herself thinking over and over about the amount of stress she has experienced recently, becomes preoccupied with questions regarding the consequences of too much stress, feels unhappy and worries, tends to lose her social contacts, avoids interpersonal stressful situations, and feels dissatisfied with herself or discontent with her health behaviors.

To reiterate, among the mentioned variables associated to negative expectancy, no individual trait or emotional state can be the etiology on its own; but the interaction between such factors with different situational specificities can determine different health outcomes. Again, as an example, Leila might have developed less negative anticipation if she had not been faced with her mother's disease during her early adolescence or if she could have expressed her concerns to the doctor. In the latter case, the doctor could have educated her about the concepts of positive stress, as well as negative and optimal stress. Such insight might release her from the responsibility to abolish stress from her and her mother 's lives. There is a great deal to be discussed about the doctor–patient relationship and how to convey health-related advice. At this point, we may simply focus on a very narrow segment of the very broad issue of communicating health and medical information to people. To avoid causing harm by inducing negative expectations as health care professionals, we are going to highlight one important rule of communication, possibly among several others. The rule we intend to discuss is almost always applicable in any society or medical/health setting regardless of what the message is, who the messenger is, and with whom the message is to be shared.

7.14 How Can We Intensify Nocebo Responses?

It is time to sum up the various biopsychosocial factors which cause us to be prone to the nocebo effect. In a reverse way, we will show here how we can increase our chances of becoming sick and actualize negative health expectations. Figure 7.1 represents a schematic model of nocebo production which is based on the previous documents and reasons we discussed before.

As shown in the diagram, the main determining factors are hyperindividualism as a global trend, harm avoidant temperament, passive attitude and/or pessimism as predisposing factors, and stressful events and traumas as triggers and/or aggravating factors. The hyperindividualism draws almost all of our attention to our body and its functions as the crucial prerequisite of being a worthy individual. Acceptance of defects, losses, disorders, and death in such an egoistic worldview is very complex and challenging. Medicalization, healthism and lifestylelism and their taboos, instructions and rituals ensure the sanctity of individual life. The overwhelming health warnings induce anticipatory anxiety and subsequently avoidant trends in modern societies. To avoid potential dangers, we should keep a healthy distance by thinking and screening, along with any other reassurance seeking measures. The tragic fact is that when you mention more parameters by higher sensitivity, the uncertainty will paradoxically rise. This vicious cycle induces health anxiety and consequentially avoidant behaviors.

Avoidance of dangers is a very effective and preliminary adaptive behavior but it needs some degrees of acceptance of unpleasant events in order to organize our coping. Otherwise, avoidant behavior could become an uncontrollable maladaptive behavior which restricts access to our resources and makes our self-image more exposed and labile although it is directed towards maximum security. Therefore, excessive avoidance paradoxically makes us vulnerable and anxious. It facilitates negative expectations and, consequently, nocebo responses.

An empathic and illuminating doctor–patient communication and referring the individual to a psychotherapist in serious cases can change the client's interpretation and moderate health anxiety and its consequences. Negligent and disease- oriented approaches to these clients make them prone to nocebo effects and gradually may cause organic disorders. After a glance at nocebo semiosis and its psychophysical pathways, it is time to discuss how we can change the processing of signs towards salutogenesis and higher health.

7.15 The Way Out of Iatrogenic Nocebo Effects

In this framework, biosemiotics could be a common language which helps us follow the flow of signs through the physical, mental, and interpersonal worlds. Signs create our health conditions and forms of life in the form of electric and electronic signals, protons, molecules, and cells; concepts, emotions, beliefs, and dreams;

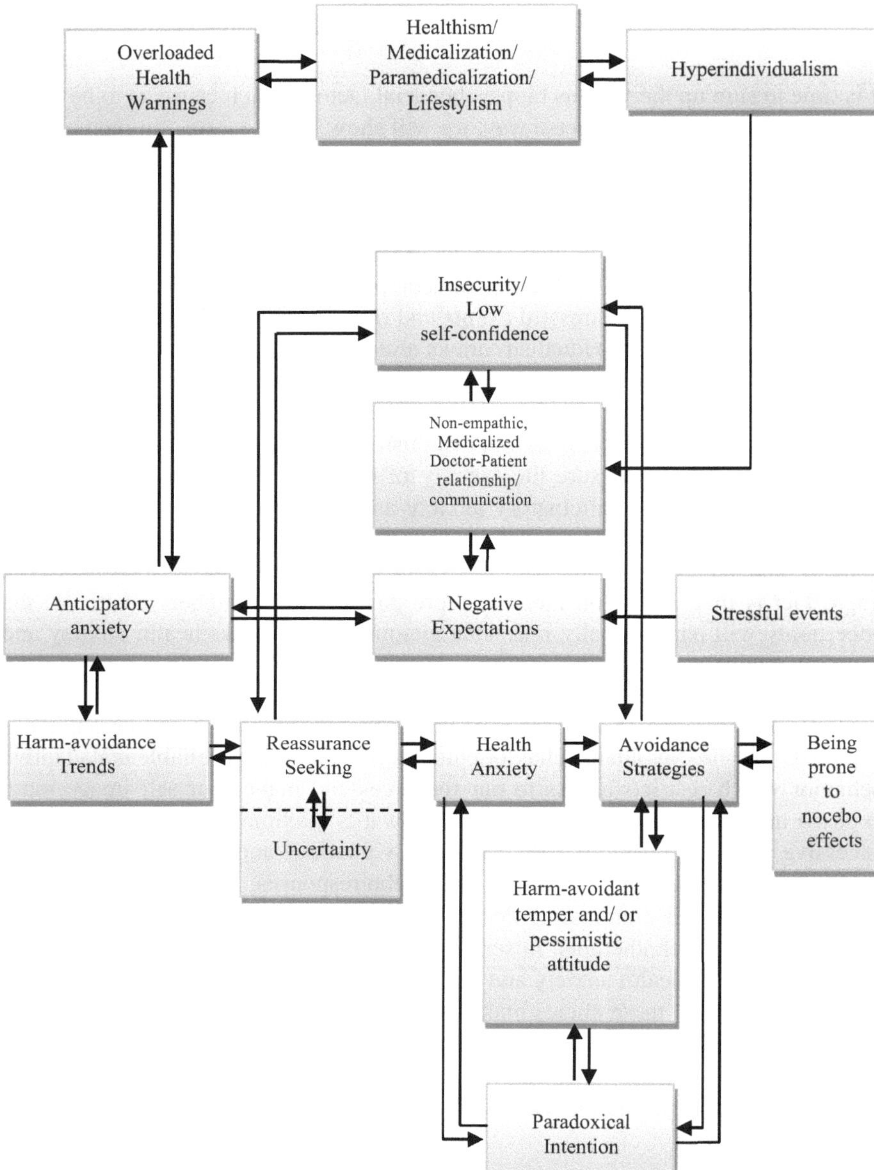

Fig. 7.1 The biopsychosocial pathways of inducing and reinforcing nocebo responses; a reverse modeling

relations, contracts, and laws. To reorganize the meaning and effect through this multilevel system, we need a more complex and integrative model for health education. Health behavior change and psychoimmune modulation as the main objectives of health education are related to cognitive-emotional reprocessing; thus, it requires a systemic model to formulate autonomic, effective, and safe informational interventions.

In the following, some of the considerations in a biopsychosocial health education to maximize the placebo responses and minimize the nocebo responses are briefly discussed, and the systemic worldview, resource-based approach, qualitative life, and health continuum are highlighted.

There is no way to neglect our experimental science and no need to abandon the objective, pathological, and disease- oriented approach to health, but it is not sufficient. Also, the interpretation of the health phenomena in this context leads us to a paranoiac worldview and brings about more insecurity and anxiety. We are seeking an integrative way to reinterpret health and illness in order to optimize health education and minimize the nocebo effect of our informational interventions. Our vast and valuable knowledge and experiences in the systemic approach to life and health – especially in the recent decades – provides an integral platform to profit by complementarity of the quantitative-qualitative knowledge, the pathologic-salutogenetic practice and the reductionistic-holistic epistemology.

7.15.1 Systemic Worldview

At first glance, it seems that these shortcomings are inevitable costs of individuality and development; however, it can be considered as a transitional condition from the traditional to the postmodern episteme. Our findings in physics and systemic biology show us that we are not alienated individuals "in" relation to the others, but we are holons; living systems which are emergently constructed from the lower levels of organization, and, at the same time, a part of higher health in the hierarchy of life (see, e.g., Wilber 2007; Luhmann 1995; Simon 1969).

Our self-actualization is related to our understanding of our biopsychosocial matrices. Beyond the local survival struggles, we can find a universal harmony, which cooperates genes (Ridley 2008; Attwater and Holliger 2012), increases significance of signs (Sørensen et al. 2012) and promotes evolution and nurtures a symbolic planet (Margulis 2008). Our hyperindividualism, alienation, selfishness, and consequently, our health anxiety could be moderated via this systemic worldview. We as holons would be more secure than lonely selves surrounded by countless dangers and threatened by numerous faults and risks.

7.15.2 Resource-Based Approach

Naming the illness experiences in the form of categorical concepts has gradually represented the disease as bad character that threatens our life, especially when we put a disease's name in the subject of a sentence. We implicate that disease is an intelligent animated entity and it is not very far from the shamanistic physiopathology.

The so-called facts like "diabetes can cause tingling and numbness" is formally very similar to this shamanistic explanation "black wind causes seizures". Our warning messages that serve to avoid dangers and dangerous behavior are more focused on the disease and its consequences. The pathological approach is appropriate for medical education and practice, especially in acute conditions; but for living with a chronic illness and community education, salutogenetic and resource-based approaches seem more effective and less harmful (see, e.g., Antonovsky 1987; Ray and Keenet 1993; Golembiewski 2010). Focusing on how to find and actualize our resources, how to change our lifestyle, and how to promote our self-efficacy would be more helpful than increasing fear of the dangers of pathogens and diseases, and the risks of maladaptive behaviors.

7.15.3 Qualitative Life

Objectivity and quantitative research are of the main values of modernity and, of course, biomedicine. The main reason for emphasizing these concepts is to avoid an illusionary world and metaphysical dogmas and to find repeatable and falsifiable facts; however, qualitative aspects of life and being values are also unfortunately neglected.

The epistemological objectification in biomedicine has gradually turned to the ontological objectification and the human being has been transformed into a naked object.

Objectification of human beings is a real threat for a qualitative life. It makes us profoundly vulnerable and prone to insecurity, meaninglessness, and alienation.

According to philosopher Martha Nussbaum (1985), a person is objectified if they are treated:

- As a tool for another's purposes (instrumentality).
- As if lacking in agency or self- determination (denial of autonomy, inertness).
- As if owned by another (ownership).
- As if interchangeable (fungibility).
- As if there is no need for concern for their feelings and experiences (denial of subjectivity).

Unfortunately, as you can infer from the cases, some degrees of all of the modalities of objectification are recognizable through biomedical research and/or practices. To de-objectify human beings, we should highlight qualitative aspects of lives and draw our attention from "doing" values (quantity of life, social function, and anthropometric indexes) to "being" values, such as wholeness, truth, playfulness, and self-sufficiency. To establish a qualitative life, we need to live our being values and develop our presence (Maslow 1968, p. 83; 1975, 44–50).

7.15.4 *Saying Yes to Life*

Acceptance of unpleasant events (dangers) and commitment to our performances (risks) is a good initiation for being present and tolerant. Acceptance and commitment therapy (ACT), as an integrative-existential therapy, emphasizes the acceptance of present-moment experiences including thoughts and feelings (Schneider 2008, p. 219). There is plenty of evidence which determines the efficiency of this approach, especially in anxiety disorders such as panic (Lopez and Salas 2009) and obsessive-compulsive disorders (Rosa-Alcazar et al. 2008).

It seems that development of self-awareness is developed only to which extent we can accept the events; otherwise, it could be turned into self-consciousness and, consequently, obsession and anxiety. Highlighting being values and empowerment of acceptance could moderate our health warnings and prevent the nocebo effect. A humanized human could be more secure and adaptive than an objectified one.

Drawing a sharp line between health and illness makes us fearful and worried about slipping into the illness world and turning into a sick person. This binary system is not only inappropriate in establishing adaptive coping strategies, but also is basically not true. A health continuum model could be more realistic and helpful – a scale which interprets our health conditions by degrees of wellness-illness. By this model, we would be more flexible and we could form a unique strategy for health and illness, health promotion – and our main objective – higher health. Even a person at the final stages can think about changing his attitude towards higher levels of consciousness and bliss. Therefore, the will to achieve higher health and consciousness evolution, as entelechy of mankind, could be the context in which we interpret all of the health and illness phenomena in it in a more active and effective manner.

7.16 Some Practical Notes

The following is a summary of the practical implications discussed in the chapter:

1. Advice for health may sometimes impair the receiver's health if it produces enough negative expectencies.
2. There are so many things in the world which may potentially harm our health. There are massive amounts of information available regarding health hazards. As innocent and beneficial such information may seem, they can burst into

harmfulness for health once they are conveyed to a health-anxious society via non-appropriate public messages or news.
3. Medical science – the profession of health – should consider delivering its scientific material with proper hygienic methods; otherwise, it may act as a source of contagious infection when spread among people.
4. The outcome of a negative health message varies significantly, at least partly depending on where it goes. If it lands on the fertile ground of pessimistic, inflexible and overanxious harm avoidance, it will almost certainly cause harm. If it find s its way to the optimistic, resilient, and agreeable ground of reward expectance, however, it may never grow into significant harm.
5. An innocent factor, which has been perceived as an enemy for health, may express some real enmity even if regarded as friend by the medical orthodoxy.
6. The key to a healthy life is not doing too much to improve health, neither is it avoiding too many things in order to prevent illness. The key is to live modestly, to keep committed toward improving our life quality at any time – whatever our health status is – and to positively look forward to a better health in the future.
7. There is no point in hiding from stress, it is always around us. Some stress is essential for completing daily tasks; all we should do is positively manage the extra amount.
8. As health professionals, we should better educate people on what to do in order to enhance their health, rather than what not to do to avoid illness.
9. As health professionals, we should educate ourselves in how to reframe medical facts into positive facts.
10. As health professionals, we should stand on the safe side of all forms of education by trying to enhance optimism and avoid creating negative expectance.
11. As health professionals, we should get trained on delivering bad news in a good yet truthful way. This skill should be regarded as one of the most important parts of educational curricula for all medical sciences students.
12. Human health control has been programmed inside humans themselves. It is neither in the doctors' hands, nor in the advanced medical schools, hospitals, books, or evidence-based knowledge. All medical professionals can do is to become familiarized with this locus of control in order to direct it towards the right outcome; this would be possible only through adequate communication. Hence, becoming skillful in the communication field is the primary footstep for anybody who is going to be a medical doctor or a health expert.

It is remarkable here that we need to figure out an evidence-based protocol to apply the concepts of mental reality (including positive or negative expectance) for enhancing human health. That is our hope and idea that we intend to convey. There was a broad area and a great deal of science to address related to the topic of this chapter – too much to thoroughly accomplish. Yet, we hope to have opened the discussion in order to stimulate the professionals' minds once again to this topic. A great deal of further communications and reflections is needed before being able to present an organized and comprehensive illustration of this topic in a not-so-far future.

References

Antonovsky, A. (1987). *Unraveling the mystery of health: How people manage stress and stay well*. San Francisco: Jossey-Bass.
Attwater, J., & Holliger, P. (2012). Origins of life: The cooperative gene. *Nature, 491*, 48–49.
Barsky, A. J., et al. (2002). Nonspecific medication side effects and the nocebo phenomenon. *Journal of the American Medical Association, 287*(5), 622–627.
Beck, U. (1992). *Risk society: Towards a new modernity* (M. Ritter, Trans.). London: Sage Publications.
Benedetti, F., et al. (2006). The biochemical and neuroendocrine bases of the hyperalgesic nocebo effect. *The Journal of Neuroscience, 26*(46), 12014–12022.
Bingel, U. (2013). The relevance of placebo and nocebo mechanisms for analgesic treatments. In L. Colloca, M. A. Flaten, & K. Meissner (Eds.), *Placebo and pain: From bench to bedside* (pp. 127–136). San Diego: Academic Press.
Brañas-Garza, P., Espinosa, M. P., & Repollés, M. (2010). Discounting future pain: Effect on self-reported pain. *Neuroscience and Medicine, 1*(1), 14–19.
Brañas-Garza, P., et al. (2012). Time discounting and pain anticipation: Experimental evidence. *Revista Internacional de Sociología (RIS), 70*(Extra 1), 73–81.
Bromwich, D. (2012). Plenty to worry about: Consent, control, and anxiety. *The American Journal of Bioethics, 12*(3), 35–36.
Bryson, B. A. (2003). *Short history of nearly everything*. New York: Broadway Books.
Buckman, R. (1992). *How to break bad news: A guide for health care professionals* (pp. 65–97). Baltimore: The John Hopkins University Press.
Cannon, W. B. (1942). Voodoo death. *American Anthropologist, 44*(2), 169–175.
Colligan, M. J., & Murphy, L. R. (1979). Mass psychogenic illness in organizations: An overview. *Journal of Occupational Psychology, 52*(2), 77–90.
Colloca, L. (2012). The influence of the nocebo effect in clinical trials. *Open Access Journal of Clinical Trials, 4*, 61–68.
Colloca, L., & Finniss, D. (2012). Nocebo effects, patient-clinician communication, and therapeutic outcomes. *Journal of the American Medical Association, 307*(6), 567–568.
Colloca, L., & Grillon, C. (2014). Understanding placebo and nocebo responses for pain management. *Current Pain and Headache Reports, 18*(6), 1–7.
Cooper, N., & Tape, D. (2001). *Nocebo phenomenon: The negative power of suggestion [A literature review]*. Missouri: Logan University.
Crawford, R. (2006). Healthism and the medicalization of everyday life. *Health, 10*(4), 401–420.
Crombez, G., & Wiech, K. (2011). You may (not always) experience what you expect: In search for the limits of the placebo and nocebo effect. *Pain, 152*(7), 1449–1450.
Data-Franco, J., & Berk, M. (2013). The nocebo effect: A clinicians guide. *Australian and New Zealand Journal of Psychiatry, 47*(7), 617–623.
Davis, J. E. (2010). Medicalization, social control, and the relief of suffering. In W. C. Cockerham (Ed.), *The new Blackwell companion to medical sociology*. Malden: Blackwell publishing.
Drici, M. D., Raybaud, F., Delunardo, C., Lacono, P., & Gustovic, P. (1995). Influence of the behaviour pattern on the nocebo response of healthy volunteers. *British Journal of Clinical Pharmacology, 39*(2), 204–206.
Dunn, K. (2005). *Vacuum virus: The nocebo effect*. Retrieved from http://harvardmagazine.com/
Enck, P., & Klosterhalfen, S. (2007). Predictors of the placebo/nocebo response in clinical trials. *Symposium on mechanisms of placebo/nocebo responses*. Tutzing, Non 28, 2007.
Enck, P., & Häuser, W. (2012). *Beware the nocebo effect: Gray matter*. Retrieved from http://www.nytimes.com/2012/08/12/opinion/sunday/beware-the-nocebo-effect.html?_r=0.
Esther, S. (2002). Walter cannon and 'voodoo' death: A perspective from 60 years on. *American Journal of Public Health, 92*(10), 1564–1566.
Eysenck, H., & Furneaux, W. D. (1945). Primary and secondary suggestibility: An experimental and statistical study. *Journal of Experimental Psychology, 35*(6), 485.

Fallowfield, L., & Jenkins, V. (2004). Communicating sad, bad, and difficult news in medicine. *Lancet, 363*(9405), 312–319.
Foucault, M. (1994). An interview with Michel Foucault. In J. D. Faubion (Ed.), *Power* (Vol. 3, pp. 239–297). New York: The New Press.
Friedman, H. S., & Booth-kewley, S. (1987). The "disease-prone personality": A meta-analytic view of the construct. *American Psychologist, 42*(6), 539.
Gadamer, H. G. (1996). *Truth and method* (2nd ed. Rev., J. Weinsheimer, & D. Marshall Trans.). New York: Continuum.
Golembiewski, J. A. (2010). Start making sense: Applying a salutogenic model to architectural design for psychiatric care. *Facilities, 28*(3/4), 100–117.
Hahn, R. A. (1995). *Sickness and healing: An anthropological perspective*. New Haven: Yale University Press.
Hahn, R. A. (1997). The nocebo phenomenon: Concept, evidence and implications for public health. *Preventive Medicine, 26*(5), 607–611.
Hahn, R. A., & Kleinman, A. (1983). Belief as pathogen, belief as medicine: "Voodoo death" and the "placebo phenomenon" in anthropological perspective. *Medical Anthropology Quarterly, 14*, 16–19.
Harrington, E. R. (1998). *The nocebo effect: A meta-analysis of the effect of suggestion on reports of physical symptoms*. Philadelphia: Doctoral dissertation, Temple University.
Häuser, W., et al. (2012). Nocebo phenomena in medicine: Their relevance in everyday clinical practice. *Deutsches Ärzteblatt International, 109*(26), 459.
Hellhammer, D. H., & Wade, S. (1993). Endocrine correlates of stress vulnerability. *Psychotherapy and Psychosomatics, 60*, 8–17.
Henningsen, P., Zimmermann, T., & Sattel, H. (2003). Medically unexplained physical symptoms, anxiety and depression: A meta-analytic review. *Psychosomatic Medicine, 65*, 528–533.
Houston, W. R. (1938). The doctor himself as a therapeutic agent. *Annals of Internal Medicine, 11*(8), 1416–1425.
Jakšić, N., Brajković, L., Ivezić, E., Topić, R., & Jakovljević, M. (2012). The role of personality traits in posttraumatic stress disorder (PTSD). *Psychiatria Danubina, 4*(3), 256–266.
Jakšić, N., Aukst-Margetić, B., & Jakovijevic, M. (2013). Does personality play a relevant role in the placebo effect? *Psychiatria Danubina, 25*, 17–40.
Kennedy, W. P. (1961). The nocebo reaction. *Medical World, 95*, 203.
Lasagna, L., Mosteller, F., von Felsinger, J. M., & Beecher, H. K. (1954). A study of the placebo response. *The American Journal of Medicine, 16*(6), 770–779 [PubMed].
Long, V. (2011). *The rise and fall of the healthy factory: The politics of industrial health in Britain*. Hampshire: Palgarve McMillan.
Lopez, F. J. C., & Salas, S. V. (2009). Acceptance and commitment therapy (ACT) in the treatment of panic disorders: Some considerations from the research on basic processes. *International Journal of Psychology and Psychological Therapy, 9*(3), 299–315.
Luhmann, N. (1993). *Risk: A sociological theory*. New York: Aldine de Gruyter.
Luhmann, N. (1995). *Social systems*. California: Stanford University Press.
Manchikanti, L., et al. (2011). Placebo and nocebo in interventional pain management: A friend or a foe – or simply foes. *Pain Physician, 14*(2), E157–E175.
Margulis, L. (2008). *Symbiotic planet: A new look at evolution*. New York: Basic Books.
Maslow, A. H. (1968). *Toward a psychology of being*. New York: Van Nostrand Reinhold.
Maslow, A. H. (1975). *The father researches of human nature*. New York: Viking.
McGlashan, T. H., Evans, F. J., & Orne, M. T. (1969). The nature of hypnotic analgesia and placebo response to experimental pain. *Psychosomatic Medicine, 31*(3), 227–246.
Mclean, L., et al. (2007). Anticipating the outcome of help: Can we predict placebo/nocebo response and side-effect sensitivity based on attachment state of mind? *Australian and New Zealand Journal of Psychiatry, 41*(1 Suppl), A56–A57.
Mills, B. S. (2006). *The nocebo effect: Identification of discrete causal personality traits in a non-clinical population*. Virginia: Regent University.
Milton, G. W. (1973). Self-willed death or the bone-pointing syndrome. *Lancet, 1*, 1435–1436.

Nussbaum, M. C. (1985). Objectification. *Philosophy & Public Affairs, 24*(4), 279–283.
Nye, R. (2003). The evolution of the concept of medicalization in the later twentieth century. *Journal of History of the Behavioral Science, 39*(2), 115–129.
Pietrie, A. (1948). Repression and suggestibility as related to temperament. *Journal of Personality, 16*(4), 445–458.
Popper, K. R. (1945). *The open society and its enemies*. London: Routledge.
Ray, W. A., & Keenet, B. (1993). *Resource focused therapy*. London: Karnac books.
Ridley, M. (2008). *The cooperative gene: How Mendel's demon explained the evolution of complex beings*. New York: Free Press.
Rosa-Alcázar, A. I., Sánchez-Meca, J., Gómez-Conesa, A., & Marin-Martinez, F. (2008). Psychological treatment of obsessive–compulsive disorder: A meta-analysis. *Clinical Psychology Review, 28*(8), 1310–1325.
Rubel, A. J. (1964). The epidemiology of a folk illness: Susto in Hispanic America. *Ethnology, 3*(3), 268–283.
Schneider, K. J. (2008). *Existential- integrative psychotherapy: Guideposts to the core of practice*. New York: Routledge.
Schuricht, F., & Nestoriuc, Y. (2013). The placebo and nocebo effects in cancer treatment. In B. I. Carr & J. Steel (Eds.), *Psychological aspects of cancer* (pp. 309–326). New York: Springer.
Sedgwick, P. (2013). The nocebo effect. *BMJ Br Med J, 347*, f6130.
Simon, H. A. (1969). *The sciences of the artificial*. Boston: MIT Press.
Skrabanek, P. (1994). *The death of humane medicine and the rise of coercive healthism*. Suffolk (UK): The Social Affairs Unit.
Sørensen, B., Thellefsen, T., & Brier, S. (2012). Mind, matter, and evolution: An outline of C.S, Pierce's evolutionary cosmogony. *Cybernetics and Human Knowing, 1*(1–2), 95–120.
Spiegel, H. (1997). Nocebo: The power of suggestibility. *Preventive Medicine, 26*(5), 616–621.
Stam, H. J. (1984). *Hypnotic analgesia and the placebo effect: Controlling ischemic pain*. Ottawa: Ph.D Dissertation, Carleton University.
Stam, H. J., & Spanos, N. (1987). Hypnotic analgesia, placebo analgesia, and ischemic pain: The effects of contextual variables. *Journal of Abnormal Psychology, 96*(4), 313–320.
Szasz, T. (2007). *The medicalization of everyday life*. New York: Syracuse university press.
Tavel, M. E. (2014). The placebo effect: The good, the bad and the ugly. *The American Journal of Medicine, 127*(6), 484–488.
Throop, E. A. (2009). *Psychotherapy, American culture, and social policy: Immoral individualism*. New York: Palgrave Macmillan.
van Laarhoven, A. I. M., & Evers, A. W. M. (2011). Response to the commentary "You may (not always) experience what you expect: In search of the limits of the placebo and nocebo effect". *Pain, 152*(8), 1931–1932.
van Laarhoven, A. I. M., Vogelaar, M. L., Wilder-Smith, O. H., Kerkho, P. C. M., Kraaimaat, F. W., & Evers, A. W. M. (2011). Induction of nocebo and placebo effects on itch and pain by verbal suggestions. *Pain, 152*(7), 1486–1494.
Wells, R. E., & Kaptchuk, T. J. (2012). To tell the truth, the whole truth, may do patients harm: The problem of the nocebo effect for informed consent. *The American Journal of Bioethics, 12*(3), 22–29.
Windolf, J. (1997, October 22). A nation of nuts. *Wall Street Journal*.p. A22.
Wilber, K. (2007). *The integral vision*. Boston: Shambhala Publication.

Chapter 8
Making Sense in the Medical System: Placebo, Biosemiotics, and the Pseudomachine

Stefan Schmidt and Harald Walach

8.1 Placebo and Biosemiotics

A placebo is, by definition, an inert substance or an inactive procedure. The placebo effect is considered as the reaction following the administration of a placebo. But, as Moerman and Jonas (2002) have pointed out, since there is no active ingredient in an inert substance, it cannot be the placebo itself that is causing the placebo effect. Rather, the placebo effect is due to the many meaningful circumstances of the placebo administration or procedure such as the information given about the likely effect of the substance, the color and branding of a placebo pill, the relationship and interaction with the person who administers the placebo, the medical context, and the background experience a person has with medical interventions in general terms, etc. It is in this sense that Moerman (2013) suggests to speak of *a meaning response* rather than of a placebo effect.

This clarification in terminology makes the underlying scientific problem of any response to placebos obvious. We see changes in the material word, for example, physiological changes as a consequence of mental activities, that is, the creation of meaning. Here we are at the heart of the mind-body problem. This is because we have a severe lack of scientific concepts and models on how these two categorically different levels of description are relating to each other or, to put it simply, how a

S. Schmidt (✉)
Institute for Transcultural Health Studies, European University Frankfurt (Oder), Frankfurt, Germany

Department for Psychosomatic Medicine, Medical Faculty, Medical Center - University of Freiburg, Freiburg, Germany
e-mail: stefan.schmidt@uniklinik-freiburg.de

H. Walach
European University Viadrina, Frankfurt (Oder), Germany
e-mail: walach@europa-uni.de

F. Goli (ed.), *Biosemiotic Medicine*, Studies in Neuroscience, Consciousness and Spirituality 5, DOI 10.1007/978-3-319-35092-9_8

change in meaning is related to a physical change. This is a problem that is usually neglected in the discussion. The problem is the following: Even if we were able to clarify the full causal physiological chain that leads from a belief to an improvement in health, it is still not clear how the very mental act of believing can indeed effect the first physiological change, just as much as it is unclear how neuronal activity creates thoughts, feelings and sensations.

One potential theoretical framework that can find appropriate descriptions bridging this mind-matter gap is *biosemiotics*. It is the application of the theory of signs and sign processes (semiotics) to biological systems. This approach was first conceptualized by Thure von Uexküll, one of the founding fathers of modern psychosomatic medicine (von Uexküll 1982; see also Goli, Raieian and Atarodi in this volume). In semiotics, as developed by Charles S. Peirce, the dyadic relationship between cause and effect in a mechanistic model is replaced by a triadic relationship consisting of a *sign*, an *object*, and *meaning*, in a more general model. According to Peirce, a causal relationship is a special case of this more general paradigm (Walach 2011) (see Fig. 8.1).

Earlier, we have placed the placebo response within this biosemiotic framework and demonstrated that this is a fruitful approach (Walach 2011). A mechanistic model can explain how a pharmacological *active* substance can result in certain physiological changes. But it cannot explain why a pharmacological *inactive* substance can result in similar or even identical physiological changes. In the biosemiotic approach, the placebo administration can be described within the respective

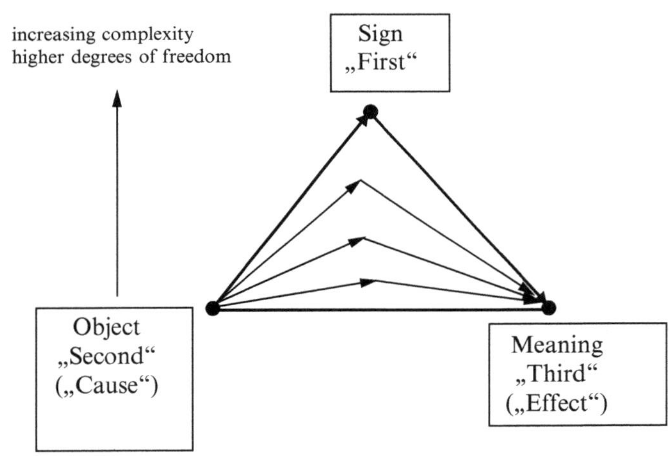

Fig. 8.1 The dual relationship of cause and effect as a special case of the tripartite semiotic relationship between Object–Sign–Meaning according to Peirce: While for simple, deterministic systems with no degrees of freedom, a mechanistic causal relationship between cause and effect is sufficient (base of triangle), systems with increasing degrees of freedom, such as simple biological systems like bacteria, or more complex systems, such as human beings, require a description in terms of a tripartite relationship (increasing angles of the triangle). Here, what may be a cause becomes a sign that produces a meaning, and hence the deterministic relationship is broken up into a relationship that allows for a variety of reactions as a consequence of this meaning-making process

medical context as a complex *sign* which creates a certain *meaning* in the recipient (and also in his or her social environment). The *object* itself is the inert pill with its inert substance (e.g., dextrose). Thus, when administrating a placebo, the object is of minor importance, but it becomes a sign for a more complex context with a certain meaning. On the other hand, when there is a pharmacologically-active substance in the pill, then the object itself is also of some importance. In the latter case we have a twofold pathway towards changes in the physiology. One is the mechanical pathway due to the active substance. Here, the pharmacological molecule becomes a sign for the system with a particular meaning, its physiological consequence. This is dependent on the genetic make-up of the organism with its metabolic specificity and capacity, counter-regulating activity, and the sensitizations history. In parallel, there is the pathway via the psychological context. This context, for instance, a medical treatment facility, becomes a complex sign with a very specific meaning that is dependent on multiple internal processes – some conscious, some unconscious – within the individual. It eventually creates a certain meaning, the complex reaction of the organism.

In a semiotic analysis, these two pathways cannot be seen as independent from each other. This is because in a semiotic model, it does not make sense to separate them, since all effects are always a complexion and a synergistic combination of material-causal and psychological meaning effects. In that sense, each intervention is a complex intervention that generates meaning in the recipient – at least if the recipient is not unconscious – and this meaning *is* the effect. That is the reason why non-active interventions can become harmful, for instance when people fall ill because of supposed and anticipated toxic effects from the environment, or why seemingly non-active interventions can be very beneficial, for instance when they are perceived as such in many cases of complementary or psychotherapeutic interventions. It makes no sense at all to ask whether there is a *real* effect for instance from psychotherapy, complementary medicine or geopathic zones. As long as there is perceived meaning, there will be an effect. This is also the reason why active interventions can lose their effectiveness completely when perceived as not important or not effective. In other words, a semiotic perspective redirects our attention from the material-causal properties of an intervention to the effects it has in the mind of the recipient. This explains why in Africa people may covet blue pills for certain types of diseases (e.g., for pain, and will find them effective), although they may be imbued with completely different meaning (e.g., as aphrodisiacs) in Western countries (Harry van der Zee, 2008, personal communication).

The important issue here is that for the triadic semiotic model *consciousness* is a necessity, yet not for the mechanical dyadic model. The latter will also work in situations where the patient is unconscious. But once the patient is conscious and able to create meaning, the result of this semiotic process cannot be predicted from the equation, since we cannot know all potential parameters entering into the meaning-making model.

One of the major implications of this approach affects our view on the generalizability of scientific statements. *Physiological processes in the human body* can be conceptualized as having a genetically-determined and thus limited variance between individuals. Of course there are differences with respect to the life cycle, genetic polymorphism, etc. but in general, the assumption is that whatever works in one human body should also work in the other, and all biomedical research is successfully relying on this assumption, at least in very general terms. *Meaning creation processes in the human mind* on the other hand show a large variation. Meaning-making is always an expansion and extrapolation of an already existing model about the world. Large parts of our world model are of course socially mediated and culturally embedded. Yet beyond this rises the individual challenge to make sense out of the world in which one lives, which is intimately tied to the individuality of each biography, its specific opportunities and individual obstacles. The consequence of this, is that in contrast to the biomedical approach, it is not so easy to generalize about individuals and to arrive at uniform statements about certain populations by quantitative research only. Especially more-refined approaches within the biosemiotic framework, as proposed here, will have to address individuality by qualitative research methods. Such an approach is less capable of adding general statements as it is usually expected when speaking about science.

8.2 The Biosemiotic Perspective on Aspirin

We would like to explain this biosemiotic perspective on the placebo response by an example from the placebo literature, Branthwaite and Cooper published a placebo study on the analgesic effects of aspirin as well as drug branding in headache patients in *The British Medical Journal* (1981). In a 2×2 design, 835 women who regularly used painkillers for headache relief received a box of tablets. These were either placebo pills in an unbranded pack (Group A), or placebo pills endorsed with the manufacturer's design in a branded pack (Group B), or 325 mg aspirin pills in an unbranded pack (Group C), or 325 mg aspirin pills in a branded pack (Group D). The resulting mean pain relief 1 h after intake can be seen in Fig. 8.2.

The pain relief of the two active groups (**C** and **D**) was significantly better than those of the two placebo groups (**A** and **B**). Furthermore, branding (groups **B** and **D**) resulted in more pain relief than no branding (**A** and **C**). The interaction was not significant.

This study shows that there are several different effects at work. The first effect is the one we can see in group A. There is a pain relief of 1.78 points 1 h after the intake of an unbranded inert pill. This effect can be due to several sources. It can reflect the *natural course* of the headache which just got better by itself after 1 h. It can reflect the action of taking a pill which would be a *placebo effect*. This effect could be explained, for example, by *expectancy* (cognitive effects), or *learning* (classical condition), in case the person is used to taking (active) pain killers for headache (Benedetti et al. 2011). Furthermore, the effect can also reflect a *change in behavior*. When the participants decided to take a pain killer for their headache, this reflects that they realized in some way that they *have* a headache which is now so strong that some action is necessary. Here the action was to take a pill, but at the

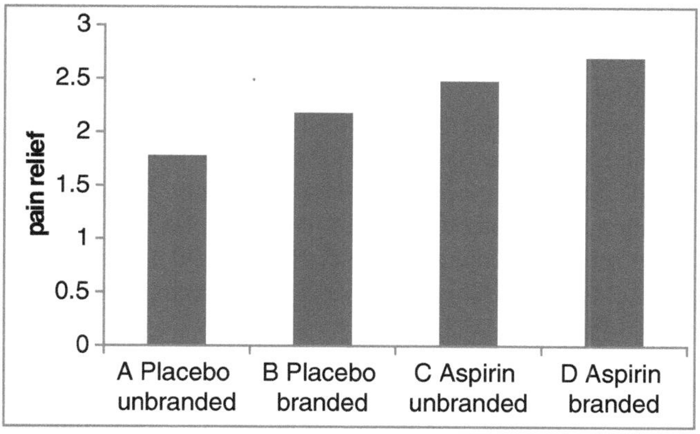

Fig. 8.2 Pain relief 1 h after intake of either aspirin or placebo being either branded or not branded. Pain relief ranges from –1 'worse' to +4 'completely better' with 0 indicating 'the same'

same time this realization may also result in other behavior changes which will help to reduce the headache, for example, drinking some water, taking a break, ventilating the room, or taking a walk in the fresh air.

This process of realization can also be conceived as a semiotic process. Here we could say that the headache is a sign which creates a certain meaning. The object would be, for instance, a dull pain in the left part of the head. The meaning of the headache will of course be quite individual, for example, "this is all too much for me, my head is already aching …", or "I have to admit that I drank too much alcohol yesterday…". Each of these processes can in turn be the starting point for the next semiotic triad. So the realization of having had too much alcohol yesterday can now be a sign to create a subsequent meaningful thought such as "maybe I should stop drinking alcohol for one week", etc.

Next to these three types of effects mentioned here, there may be even more effects at work, with some of them also related to the fact that the data was obtained within a scientific study (e.g., Hawthorne effect). However, we cannot disentangle these different effects (natural course, placebo effect, behavioral change, others) from each other with the study design applied here. One way to assess them would be to have a fifth group in which participants, for example, instead of taking a pain killer, wait for another hour, and then note their pain before taking the pill.

If we now look at group B, we see an improvement of 0.4 points to a mean of 2.18. This is obviously caused by branding the inert pill. This effect cannot be explained within a pharmacological causal framework because no pharmacological agent was present. It is solely due to the semiotic process, which requires a conscious person able to create meaning. Here the sign is the branding, and the object may be the inert pill. The meaning created by the sign will also be individually different, but in this case may go in the direction that this will be a powerful pain killer because the branding is well – known, there is a lot of advertising for the pain relieving effect of this brand and also many people in the social environment will have given positive accounts after using this brand.

If we jump from group A to group C, we see the effect of the pharmacological drug which raises the pain relief by 0.7 points to 2.48. This effect cannot be explained within a semiotic triadic framework since the difference between groups A and C (both unbranded) cannot be detected by the participants. Thus, this effect needs to be explained within a pharmacological causal framework. In the case of aspirin, the pharmacological active ingredient is acetylsalicyclic acid which suppresses the production of prostaglandins and thromboxanes by inactivating the enzyme COX-1 and modifying the activity of COX-2.

Finally, group D should show all three sources combined, that is, initial effects (group A, placebo effect, natural course, etc), branding effect, and pharmacological effect. Under the assumption that these effects are not interacting, they could be added up. Then D should be $1.78+0.4+0.7=2.88$. The value measured is a little smaller with 2.7, thus reflecting some minor interaction.

The interesting point here is that only approximately one quarter of the pain relief measured in this study can be attributed to pharmacological processes. If we assume that more or less the same mechanisms are at work if aspirin is taken in daily life and not within the framework of a study, then this is an astonishing fact. It reveals that three quarters of the pain relief of aspirin are due to biosemiotic processes beyond the pure pharmacological action of acetylsalicyclic acid. On the other hand, the lay user of aspirin will likely attribute the whole effect to some pharmacological mechanisms which is also of course a semiotic process. This, by the way, tallies nicely with the result of a meta-analysis of all kinds of long-term pharmacological interventions. This resulted in a correlation between improvements under placebo and treatment of $r=0.78$ which means that across different treatments and diseases approximately 60 % of the variance in treatment effects is explained by all sorts of effects, including the meaning effect, and only 40 % of the variance is attributable to a causal effect of the pharmacological intervention (Walach et al. 2005).

8.3 Biosemiotic Pharmacology

If we generalize these conclusions to pharmaceutical therapy in general, then we can assume that a large portion of the therapeutic effects seen in general are misattributed to pharmacological mechanisms only. Or in other words, large parts of the effects are only working in conscious and meaning creating drug consumers, because the consumption of drugs has a certain culturally, historically and scientifically-produced meaning. This means that in order to describe pharmacological effects adequately, the standard dyadic causal models are insufficient. Biosemiotic descriptions are more appropriate since they are able to add meaning to the framework. Let's take a look at some more examples to illustrate this perspective.

Both of the two following examples employ the so-called open/hidden design. The standard design of the randomized controlled trial cannot determine the size of the placebo effect. Even the above presented 2×2 design of the aspirin study is not able to do so; this is why we have suggested a waiting condition. Another option to assess the size of the placebo effect is the open/hidden design (Amanzio et al. 2001; Bingel 2013; Levine et al. 1981). Here the same pharmacologically active ingredient is given either openly in full view to the participant, or in disguise. The difference between these conditions represents the placebo effect; the pre-post difference in the hidden condition represents the pure drug effect.

Benedetti et al. (2006) investigated placebo effects in patients suffering from dementia due to Alzheimer's disease in an experimental pain paradigm (venous puncture). If large parts of drug effects are due to semiotic processes, then they should decline with ongoing dementia. Thus, Bendetti et al. correlated the size of the placebo effect to pain application with cognitive status as measured with the Frontal Assessment Battery. The placebo effect was measured by applying a local anesthetic to the skin either openly in full view of the patient, or covered with a tape. Thus, in both conditions patients received the same analgesic treatment (and the same pain stimulus), but only in one condition were they aware of this fact. The results of their replication testing 1 year after a first test when Alzheimer patients showed further cognitive impairment can be seen in Fig. 8.3.

The cognitively not-impaired controls showed a pain reduction of 66 % in the open condition. The hidden condition reveals that only 16 % of this reduction is due to the pharmacological substance, and the remaining 50 % is due to the placebo effect. Like in the Aspirin study, approximately three quarters of the overall pain reduction cannot be accounted for by the causal effects of the drug. In Alzheimer patients, the reduction due to the drug in the hidden condition is 23 %, but the placebo effect is obviously reduced, with only 41 % pain reduction in the open condition. Furthermore, there was a significant correlation between cognitive status and

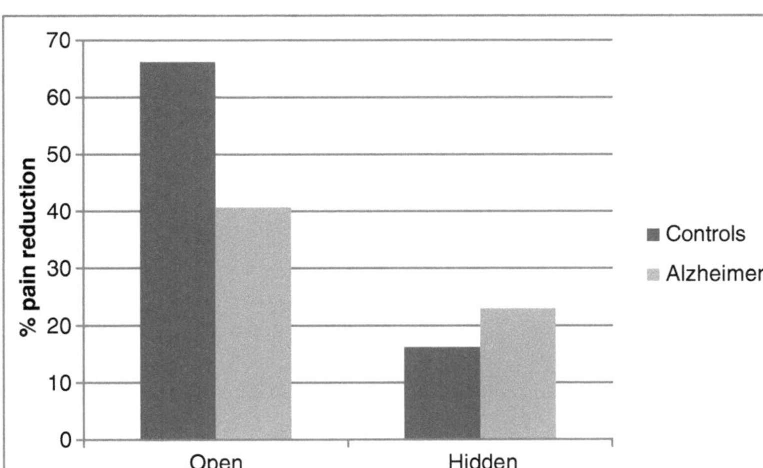

Fig. 8.3 Self-reported pain reduction in venous puncture by an analgesic that was either openly administered or covertly applied. Patients were either suffering from dementia due to Alzheimer Disease or healthy controls with no dementia

pain reduction in the open condition of $r=-45$, indicating that with a more impaired cognitive status, the pain reduction declines. These findings clearly underline the assumptions that large parts of pain relief after pharmacological therapy is due to meaning creating processes in conscious patients.

In a second study, Colloca and Benedetti (2005) also applied the open/hidden design, this time for the assessment of the placebo effects in analgesic drugs on postoperative pain. Patients after a thyroidectomy were randomized in two groups, and both groups received the same analgesic treatment. In the open condition, a doctor injected it in full view of the patient. In the hidden condition, the same dose of the same drug was administered by a computer controlled infusion pump at a preset time unknown to patients and careers. In addition, two analgesic drugs were compared regarding their ability to reduce postoperative pain after surgery – Metamizol, also known as Novalgin, and Buprenorphine, an opiate. The results can be seen in Fig. 8.4.

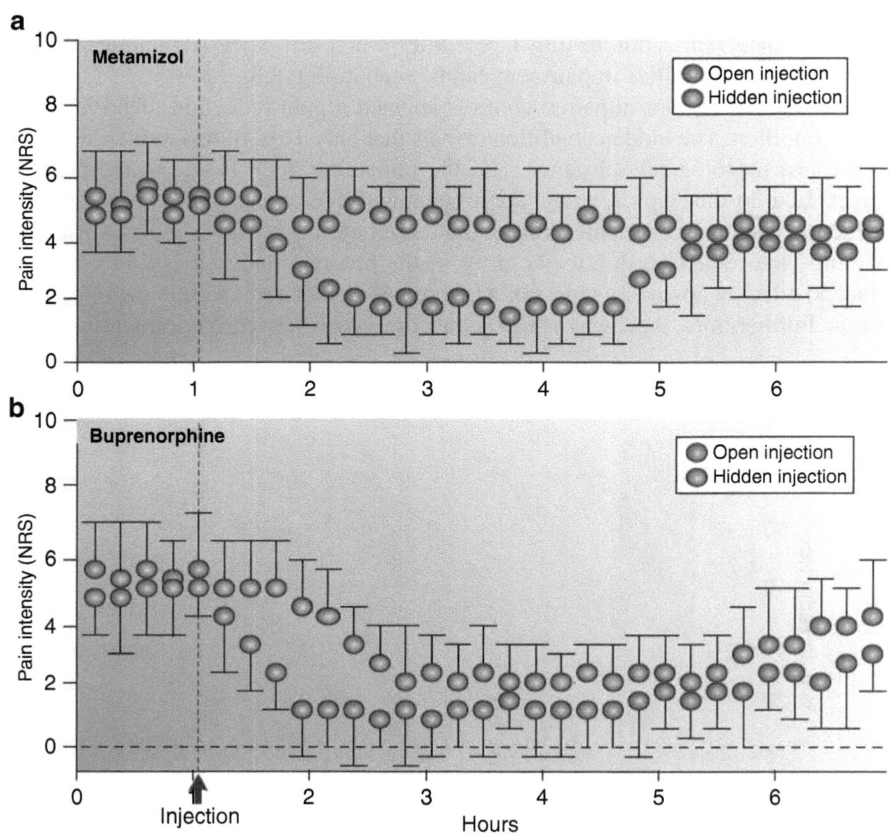

Fig. 8.4 Pain intensity rating on a numerical rating scale (NRS) from 0 to 10 in patients suffering from post-operative pain. The time course of the pure pharmacological effect of either Metamizol (**a**) or Buprenorphine (**b**) is reflected by the hidden injection. The placebo effect can be inferred as the difference between the open and the hidden injection (From Colloca and Benedetti 2005, p. 550)

On the lower panel, Buprenophine shows an overall drug-induced pain relief of approximately 2 points on the numeric rating scale (NRS) in 12 patients, 2–4 h after the injection in the hidden condition. If the same drug is given in full view of the patient (open injection) the analgesic effect starts much earlier and causes a placebo effect of approximately 3 points 1 h after the injection. The pharmacological drug effect is obviously much slower in onset than expected. If Burprenophine results in instant relief, this is mainly due to a placebo effect. On the upper panel, one can see that there is hardly any pharmacological effect in ten patients receiving Metamizol in the hidden condition. Pain reduction takes place only in the open condition when the patients are aware that they are indeed receiving a painkiller. Based on this data, one could conclude that Metamizol, which is a frequently-used analgesic medication, has no specific, that is, pharmacological effect at all, at least in patients with post-operative pain. This is a rather unexpected finding since Metamizol is a well-studied standard analgesic.

For a proper interpretation, it is, first of all, important to realize that these data have to be seen as preliminary in some respect. There are 22 patients in two groups, the study has not yet been replicated so far, and the publication lacks most of the methodological details. In another study applying the open/hidden paradigm, Metamizol showed a small drug -related effect in post-operative pain after 1 h (Amanzio et al. 2001). However, there are no data reported on pain relief beyond the first hour.

But let us assume that the data of Colloca and Benedetti (2005) are reliable. How can this complete lack of a pharmacological effect be explained? Since Metamizol is a licensed analgesic drug, we can infer that it has demonstrated a significantly stronger effect than a placebo in some randomized controlled trials (RCTs). So how can it be more effective than placebo in an RCT, but on the other hand, show no specific effect in the open-hidden-design? The authors suggest as an interpretation that the drug itself has no analgesic effect but enhances the release of placebo induced endogenous opioids. In other words, the idea is that this substance improves the placebo effect and thus, can only work when the patients are consciously aware that they receive a drug. Again, a pure pharmacological model cannot explain this finding. It can, however, be described within a biosemiotic framework. The other important point that can be drawn from this example is that drug and placebo effects are not necessarily independent, but may interact with each other. On the other hand, the RCT, which is the standard design to demonstrate specific effects of pharmacological substances, relies exactly on the assumption that placebo and verum do not interact, but are simply additive in effect. But from the data presented here, and many others (Kleijnen et al. 1994), we have to conclude that this assumption of additivity and a lack of interaction is wrong.

Finally, a third example investigates the interaction between pharmacological and placebo induced effects more formally. This is the field of *active placebo*, which is still underrepresented in the currently fast-growing placebo literature. An active placebo is defined as a pharmacologically-active substance used as a control condition in an RCT. This can be best explained by an example. Antidepressant medications have very clear and well-known side effects. Thus, in RCTs of such agents, the participants are often unblinded during the course of the trial, because they can infer from the presence or absence of the side effects whether they have been randomized to the verum or placebo condition. To avoid this unblinding, researchers apply active placebos in the placebo condition, which are able to produce similar side-effects but lack the specific pharmacological substance (Enck et al. 2013).

But this logic of the active placebo can also be applied for investigating the placebo effect itself. In such an active placebo design, a pharmacologically-active substance is given which results in some physiological effects that can be noticed by the participants of these experimental studies (Flaten 2013). In addition, this drug administration is also combined with different types of information regarding the effect of the drug. The idea here is that the process of sensing the physiological effect of the drug in the body will interact with the information given.

Flaten et al. (1999) administered Carisoprodol, a centrally-acting muscle relaxant which induces drowsiness. They combined the administration of the drug with the information that this is (1) a relaxant, (2) a stimulant, and (3) no information on the drug was given. Three more groups also received the capsules of the same form and color with the same information, but in this case, the capsule contained only lactose and acted as an inactive placebo. The resulting changes in difference between the self-reported relaxation and tension, measured on a visual analog scale (VAS) for all six groups can be seen in Fig. 8.5.

One can see that participants of the "no information" and the "relaxant information" group showed some relaxation in the course of 2 h following the administration of either placebo or carisolprodol. But the most interesting finding is the group which received the active placebo and the information that it was a stimulant drug. They showed a compatible strong increase in tension while receiving at the same time apharmacological substance acting in the opposite direction. This third example

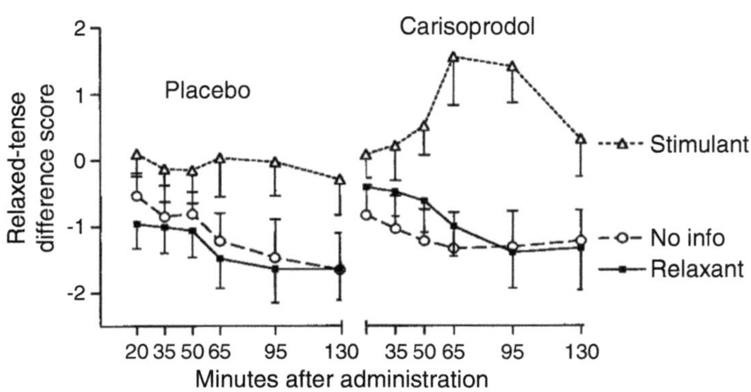

Fig. 8.5 Self-reported change regarding relaxation and tension after the administration of either a placebo (*left panel*) or a muscle relaxant, Carisoprodol (*right panel*). Data of six different groups are displayed. They received in a 2×3 design either placebo or carisoprodol with the additional information that this would be either a stimulant, or a relaxant drug, or they received no information at all (From Flaten et al. 1999, p. 253)

shows, once more, that the standard model of simple pharmacological effects which are not mediated by any context factors can no longer be maintained. Furthermore, the difference between the two groups getting the information that they have received a stimulant demonstrates that placebo effects are larger if any kind of physiological effect of the drug is noticed at the same time. This show, like in the other examples, that pharmacological and placebo effects interact with each other. Thus, they cannot be conceived as being independent from each other.

8.4 The Pseudomachine

We have seen that the biosemiotic process of assigning meaning is a powerful factor in inducing physiological changes in relation to either a pharmacological substance or to the administration of an inert pill, and that this process is governed b y a somewhat complex dynamic. We have also seen that a simple dyadic cause-effect model is not able to explain these processes. Interestingly, the patients and participants in these studies are very often making simple causal assumptions regarding their effects, which is a biosemiotic process in itself. In the aspirin study, the women taking the drug most likely may have assumed that aspirin works because there is a pharmacologically active agent, which in some physiological cause-effect mechanism unknown to them, eliminates the physical cause of the headache. It is less likely that they assumed that the simple effect of taking a pill, may it be inert or not, will substantially reduce their headache anyway. So the causal assumption of the consumer is only partially correct if an active substance is taken; it is completely incorrect if an inert pill, that is, a placebo is taken. However, it is exactly this causal attribution towards the pill which is responsible for the resulting effect to a large extent.

What is happening here is that people make incorrect assumptions about the causal mechanism of some treatments. This attribution process in turn results in effects confirming their causal model. The headache disappears because of the belief in a causal pharmacological process, although there was no such process in the case of a placebo.

Walter von Lucadou (2002), a German physicist and psychologist, calls this process a *pseudomachine.* He differentiates between machines and pseudomachines. A machine – by his definition – is a technical device or a causal process having a well-defined goal, and mostly amplifying or transforming properties, for example, a snatch pulley. In some machines like in hair dryers, bikes, or cars, their mechanisms are obvious to the user. In other machines, for example, in computers or microwaves, the mechanisms are more complex, and many users will not understand how the effect comes about in detail. It is especially this latter aspect that allows for the attribution of some causal effect to a device although we do not understand its precise mechanism. Such complex machines seem to be magic in some respect. They heat our dishes although they do not get hot themselves or they fly through the sky although they are very heavy. According to von Lucadou, it is mainly this latter experience which results in the *attribution* of causal effects to machines that are not able to causally affect them in reality. Von Lucadou's real world example of a pseudomachine is a magnet attached to the fuselage of a car sold for the purpose of fuel saving when driving a car. According to the "scientific" description of the manufacturer, the magnet "aligns the molecules of the fuel" so that hidden energy potentials can be used once the magnet is fixed to the gas tank. However, this assumed mechanism is impossible from a physical point of view, and the underlying theory is flawed. Nevertheless, people buying the magnet and attaching it to the gas tank, report needing less fuel when driving. The mechanism behind this effect is most likely that the car-drivers buying such a magnet change their driving behavior with respect to fuel consumption. Sometimes such a magnet is even sold together with a CD providing information about fuel saving by changing one's driving style. So the magnet does indeed do what the drivers expect. It helps in saving gas, but the attribution of the effect is incorrect. The effect is actually caused by psychological processes, not by physical mechanisms. This is what von Lucadou calls a *pseudomachine*. Important for the function of a pseudomachine is that the effect is attributed externally to the machine and not internally to the user. Furthermore, it is important that the attributed mechanism is confirmed or at least partially confirmed in reality to maintain the attribution pattern.

From the perspective of learning theory, this would be a kind of operant conditioning. We know from learning theory that next to regular enforcement, an intermediate enforcement works best to maintain the attribution. Operant conditioning is often made responsible for magical thinking. An ill person gets an amulet, which has, according to the person handing it over, magical powers. The person recovers and attributes the healing to the power of the amulet, and, lo and behold, another pseudomachine is born.

Von Lucadou further separates between *classical* and *non-classical pseudomachines*. In classical pseudomachines, the physical and psychological effects can be clearly separated from each other. The magnet and the amulet are such examples, but also the intake of an inert placebo, especially when there is a learning history as per the aspirin example. A non-classical pseudomachine, on the other hand, is a procedure or apparatus where the physical and psychological effects may interact with each other or are entangled, and where it is not so easy to describe the effects of the "machine" in solely-physical or psychological terms. An example here is the active placebo where the physical sensing of the carisoprodol was related to larger change according to the (incorrect) information that the drug is a stimulant.

As we have seen above, each drug intake by a conscious and meaning-making consumer will result in some effect due to the interaction between pharmacological and psychological mechanisms, and thus the very process of taking a pill can be described as a non-classical pseudomachine. This concept of the pseudomachine is a fruitful one for describing all kinds of activities within the medical system, not only for oral medication, which served as a blueprint here. One may justly ask whether some kinds of operations, such as arthroscopic knee surgery for osteoarthritis (Moseley et al. 2002), or some instances of stent operations for stable angina (Stergiopoulos and Brown 2012), should not be conceived as pseudomachines.

Within complementary and alternative medicine, there are many "magic" devices sold, for example, machines measuring the "energy of meridians" by applying electrodes to various acupuncture points or machines on computerized biocommunication, or for measuring the "energy field" of the body. While the users report very good effects and are convinced by the practical results of applying the machine, double-blind testing reveals often that no physical effects are involved in this process, so they can be considered classical pseudomachines.

On the other hand, a recent meta-analysis revealed that the analgesic effect of acupuncture seems to be a non-classical pseudomachine. Until recently, the literature showed contradictory evidence for acupuncture having specific physiological effects. Specific effects were, for example, demonstrated in a model on blocking adenonsin receptors in mice (Cressey 2010; Goldman et al. 2010). On the other hand, there are several clinical studies in which sham acupuncture proved to have the same analgesic effects as real acupuncture (Cherkin et al. 2009; Haake et al. 2007). Sham acupuncture is a treatment that tries to mimic acupuncture (McManus et al. 2007), for example, in the German Acupuncture Trials (GERAC) where acupuncture needles were placed in purported "inactive" points rather than in specified acupuncture points (Diener et al. 2006; Haake et al. 2007; Witt et al. 2005). In a study on back pain, Cherkin et al. (2009) applied toothpicks on the back which do not penetrate the skin. The individual patient data meta-analysis by Vickers et al. (2012) was large enough to demonstrate that both mechanisms contributed to the overall analgesic effect of acupuncture. Since real acupuncture resulted in significant analgesic effects compared to sham acupuncture, a specific effect can be assumed which is beyond placebo effects. On the other hand, comparing acupuncture to standard treatment or usual care showed larger effects than in comparison to sham control; this points towards the interaction of placebo and "real" effects of acupuncture.

8.5 External Causal Attribution as a Special Biosemiotic Process

What we can learn from these examples and the fruitful concept of the pseudomachine is that human beings are clearly looking for (external) causal mechanisms and explanations regarding inner states in general and their health status in particular. The causal pattern is one of the most basic cognitive patterns in order to create meaning. Furthermore, ascribing a certain causal mechanism to a certain procedure may result – similar to a self-fulfilling prophecy – in the expected effect although we can show from a mechanistic point of view that the attributed causality is wrong. The strongest changes seem to occur in cases where the procedure or machine to which the causality is attributed does indeed show some small causal effect, especially with respect to bodily sensations. Such sensations seem to function as a kind of proof for the assumed causal mechanism and can thus act as a powerful amplifier. In this case, it is even possible that the causal effect can be overridden by the semiotic attribution processes in the opposite direction, as has been shown in the Carisoprodol example.

In this sense, making causal external attributions can be considered as a special type of a semiotic process, that is, a special type of creating meaning in the world. In the case of a pseudomachine, classical or non-classical, the "machine" or procedure can be seen as the sign, which will result in this special causal and external attribution type of meaning.

At this point, we have to take care that we do not fall into the trap of conceiving this meaning pattern as individually invariant. We can assume that many of our causal attributions are deeply rooted in our culture, for example, that taking a pill will result in a physical change, and that it will be almost impossible to escape from this pattern. We know from cultural studies that the only way to do so is to become aware of one's own inculturation, which is not an easy process. On the other hand, we have to see that on a more refined level of attribution, people will be different from each other with respect to which attributions are meaningful for them. Within some native cultures from the Amazon, there will be hardly any possibility for an individual to escape the idea that an amulet will have powerful forces. Within a more Western industrialized culture, the opinions will be split. Many will consider attributions towards amulets as superstitious, but others will stick to them although they will not always disclose this. As with any semiotic process, one's individual internal model of the world will be the starting point, and the individual will only construct attributions that fit this mod el. On the other hand, we have to acknowledge that large (if not all) parts of an individual's world-model are due to his or her cultural embedding.

The culturally most independent part in this process is the causal pattern itself. Inferring linear causal connections is one of the earliest cognitive patterns resulting from sensorimotor integration in the newborn. Indeed, it is likely to be rooted in the evolutionary success of our whole heritage as mammals and primates. It was William Ockham, and later on Hume, who argued convincingly that the cause is not necessarily a mechanical event taking place in the outer world, but a cognitive inference of the human mind connecting regular and contingent observations.

According to Ockham, causality is not a property of things, but a result of observations of regularity and hence a property of our mind. We only observe correlations: "Where smoke is, there is fire", we infer and attribute a causal property to the fire itself, although all we observe is the correlation (Goddu 1984). Hence, Ockham (1957) defined a cause as something given that it is taken away, the supposed effect does also not happen, and given it is, the effect happens (p. 629). Hume (1977), later in the eighteenth century, took up the same line of argument postulating that the idea of a cause is an abstraction of our mind. It is formed once we observe that (1) causes precede their effects, (2) in close proximity, and (3) and regularly. Thus, the concept "cause" is formed in our mind. But it is important to realize, according to Hume, that there is no cause in the outside world, but only in the model we construct of it. Kant, disturbed by this analysis, considered causality a condition of our mind and a precondition for understanding. With the advent of evolutionary theory, we can assume that the concept of causality is something which is an evolutionary a-priori of our existence. It helped us to understand contingencies, avoid dangerous ones and exploit propitious ones, and hence made us what we are. But we should not forget that the causality we attribute to the world is in fact one constructed by us.

If children or members of aboriginal cultures show what is usually called superstitious beliefs or magical thinking by ascribing causality to processes which have no causal connection from the perspective of modern science, we smile at this because we assume we have a superior and more refined understanding. If within our societies, some people believe in the healing powers of certain machines and healing rituals which we consider to be inert and of no mechanical causal relevance, we also react with depreciation since we assume that – from a scientifically informed world view – any effects due to these procedures are "nothing but" mere suggestion and self-deception and cannot compete with a "real" healing process employing physiological causal processes explained in scientific terms.

But let's step back for one moment from this line of reasoning. If the dominating scientific model in the medical science is looking for the "real" causal pattern, then what is the difference from the lay person making their own causal attributions of the world? Isn't it the case that this type of science replicates the same intrinsic nature of us humans on a larger scale to find meaning by creating causal descriptions of the world? Or in other words, by explaining effects in the field of medical science with simple mechanic linear cause and effect descriptions similar to the ones of physics? One would argue that the difference here is that the causal mechanisms can be proven by experiments. For instance, it can be demonstrated by a double blind randomized control trial that an effect is taking place which cannot be related to the mindset of the patient if the blinding was appropriate. But let us come back to our first example, the aspirin study. We have seen that only approximately, one quarter of the pain relief can be unequivocally attributed to the pharmacological process. What about the other three-quarters? In the case of pain relief, these three-quarters are the crucial part. The mechanistic biological model usually attributes all effects related to drug intake to the pharmacological process and neglects or ignores any placebo or biosemiotic effects. But isn't this also a crude misattribution? Isn't this just a replication of the laymen's behavior of looking for simple mechanistic cause-effect models and to ignore the more complex relationships? Obviously, the difference between the native attributing pain relief to the amulet and the scientist attributing it only to the COX-1 inhibition is not as big as expected after all.

From the point of view of semiotics, we could reframe science, like many have done before (Foucault 1991; Latour and Bastide 1986; Latour 1999; Shadish and Fuller 1974), also as a social meaning creating process rather than a procedure in order to find the truth about the world. It is obvious that science – as it is conducted today – is a social practice governed by certain rules and basic assumptions. It is also clear that many of these rules are due to social agreement rather than due to objective proof (whatever that could be). Take for example the rule that values smaller than or equal to .05 ($p \leq 0.05$) are considered as significant while larger p-values are indicative of no significant finding. We all know the importance of this fine line. But of course this is not a given fact, but simply a social agreement. Or as Rosnow and Rosenthal put it: "God loves the 0.06 nearly as much as the 0.05" (Rosnow and Rosenthal 1989, p. 1277). Of even larger importance is that at the very heart of our modern science, there are many basic assumptions which are unproven (Walach and Schmidt 2005). And some of them are relating directly to our topic, for example, the assumption that all effects in the world are of a mechanical nature. This implies that all changes are brought about by the local impact of material parts which is efficient causation. But this presupposition that the world is mechanical and causal in its core is unproven. If we take this for granted and make causal explanations to a criterion for scientific proof, we are creating a dogma (Sheldrake 2013). In this case, an unproven presupposition turns into a belief and science shifts towards scientism.

From such a philosophy of science point of view, there is space to complement the mechanistic causal explanation pattern with a semiotic one. The first one may be the more dominant one when dealing with unconscious items such as in physics, but the latter may likely give us the better explanations in all instances when consciousness kicks in. The field of medicine surely belongs to the second group.

8.6 Healing Due to Semiotic Processes: Is It Allowed?

The dominant mechanistic model in the medical sciences, combined with the assumption that all change is due to direct causation, results in another strange misconception. Healing processes that cannot be explained within such a causal model are not taken seriously. If somebody underwent some medical procedure unable to be explained currently within such a framework, any resulting healing process is depreciated, if not negated. If somebody benefits, for instance, from a treatment in homeopathy, or from a visit to a spiritual healer they will often hear comments like: "Well, of course you might feel better, but that is *only* a placebo effect". What happens here is that the ability of finding a scientific explanation is rated higher than the benefit of the patient, or the experience of an individual. If we do not understand how a placebo effect works, then benefiting from a placebo is considered as being not real. Or, in other words, the dominating scientific model discriminates between accepted and unaccepted healing processes, which is rather strange from the patients' perspective and may also in part explain the longstanding debate about the role of complementary and alternative medicine in our society.

The blueprint for this line of reasoning is that the placebo-controlled RCT is the standard to evaluate the efficacy of any drug or procedure. Here, the idea is to control against the placebo and this means that only the difference between the placebo effect and the verum effect is taken seriously while everything else is ignored. You may benefit 90 points from a placebo and additionally 10 points from the verum. Then only the 10 points are considered to be a 'true' improvement while the other 90 points are neglected. We have shown elsewhere that this line of reasoning may even result in the strange case that a more efficient procedure is neglected in favor of a less efficient one (efficacy paradox, see Walach 2001, 2011).

But meanwhile, the climate is changing ever so slightly. Since the since the mid-1990s, researchers have started to recognize the power of the placebo. The placebo concept is now shifting from being a control condition that needs to be ruled out in order to find a 'true' effect, towards a valuable treatment in itself. This also has to do with the discovery that placebo effects are mediated by neurobiological processes, for example, by neurotransmitters binding to the same receptors as pharmacological substances (Benedetti et al. 2011; Colloca and Benedetti 2005; Price et al. 2007). Hereby, large parts of the placebo response can be incorporated into the mechanistic model. The distinction between so call ed 'specific' and 'unspecific' effects slowly melts away (Linde 2006). If after the intake of an inert pill, the idea of having received an aspirin results in the release of endorphins and thus analgesia, this may be considered a very specific process.

8.7 Is the 'Open Placebo' the Future?

If we extrapolate this development for another 20 years, we may imagine a medical system that makes heavy use of the large potential of semiotic processes, for example, by designing hospitals and procedures which optimally support healing processes (see, e.g., Jonas and Chez 2004; Ulrich 1984), or by developing communication and treatment strategies which are known to maximize placebo effects by initiating positive semiotic processes and meaning constructing. This sounds promising, but there is one major problem associated with many of these ideas. Once we have understood that a positive semiotic process might be due to certain causal assumptions about the world which cannot be maintained from a physical point of view, it will not be ethical to communicate them any longer. Or, in other words, if we start to understand that some classical pseudomachine is at work, then we have the ethical obligation to inform the patient about this fact, while this may at the same time result in the placebo effect to disappear. Going back to von Lucadou's example of the magnet, it will be fine to sell these magnets if you are personally convinced that they have a causal effect. But once you are aware that the description of the magnet aligning the molecules of the fuel is wrong, you should no longer tell this to your customers since this is deception. In the same line, psychiatrists should no longer tell their patients that selective serotonin reuptake inhibitors (SSRI) will improve depression once they have understood that almost all of their effects cannot be attributed to the process of serotonin reuptake inhibition (Kirsch et al. 2008). But, on the other side, being honest here is to the disadvantage of the customer or the patient. It looks like two ethical principles are in contradiction here, that is, being honest and acting in the patient's best interest (Kaptchuk 2002).

The solution of this dilemma may be a surprising one: "the so-called open placebo". In 1965, Park and Covi published a paper entitled "Nonblind Placebo Trial" (1965). In this study, 14 patients attending a psychiatric outpatient department were offered to take a sugar pill which is as they were told "a pill with no medicine in it at all" (p. 337). This offer was combined with the statement "…Many people of your kind of condition have also been helped by what are sometimes called 'sugar pills', and we feel that a so called sugar pill may help you, too" (p. 337). After 1 week the patients showed reasonable improvements on a symptom check list and a generic self-report scale. This is a surprising result, since it is usually assumed that a placebo will not work anymore once it is known to be a placebo. But this assumption may be wrong. It took 45 years until this finding was replicated in a more stringent study. In 2010, a publication by Kaptchuk et al. (2010) reported about a randomized open placebo trial in 80 patients suffering from Irritable Bowel Syndrome (IBS) . They were either randomized to a no treatment control condition or to an open placebo condition receiving "placebo pills made of an inert substance, like sugar pills, that have been shown in clinical studies to produce significant improvement in IBS symptoms through mind–body self-healing processes" (Kaptchuk et al. 2010, p. 1). Patients in the open placebo arm showed significant improvement compared to controls in the main outcome criteria (symptom severity, global improvement).

How can these results be explained? Until now, there is no conclusive model that can account for these findings. So far we thought that placebo effects were elicited by the expectancy that one would receive a powerful drug. But in the case of the open placebo, the deception that is usually employed to convey the expectancy was disclosed. Obviously the authors of these two studies were able to maintain positive expectations despite the lack of deception. When reading the two statements, one can see that both are relying on prior positive experiences with treatment or placebo by telling the patients that these sugar pills have helped many other patients before. So it looks like the expectation this time is not tied anymore to some assumed pharmacological process, but to the placebo itself. What is conveyed is the message "this placebo will help you because it is a placebo and we know that placebos are very powerful". This is the pseudomachine reloaded by itself and back onto itself. One can furthermore assume that the powerful ritual of taking a drug which is very well established in our society also assisted the process through unconscious learning processes (Jensen et al. 2012). Or to put it the other way around, the idea that taking a pill will result in no change at all seems to be nearly impossible.

If we try to interpret this finding from a semiotic point of view, we can see that certain types of expectations, once they are out in the world, cannot be just switched off like in a causal model. This is, in fact, a situation that is seen very often. There are some ideas about certain mechanisms in circulation but one tends not to believe them, for example, amulets protecting from evil. Nevertheless, it proves difficult to eliminate these ideas completely once they are known to be there or shared by others. This is often reflected in statements of the type "Actually I don't believe in x but why not give it a try?" Obviously, meaning making processes do not follow linear models but integrate many different perspectives and they can also take up ambiguities and contradictions and still come up with a coherent view. Here we are only beginning to understand how humans, based on their prior experience and their world models, create meaning and how this meaning making then interacts with physiological indicators. Based on other research (Jensen et al. 2012), we also can assume that this is not an entirely conscious activity but will also tightly interact with many non-conscious processes.

At any rate, this perspective and the empirical data to support it have shown that a semiotic model is more useful for understanding therapeutic effects in humans than a causal-mechanistic model. Thus, the placebo effect teaches us, and medicine at large, that humans are not machines, that therapy is not a reparation process, and that it is clever to understand and appeal to meaning-making processes also in the treatment, if not in the understanding of disease.

References

Amanzio, M., Pollo, A., Maggi, G., & Benedetti, F. (2001). Response variability to analgesics: A role for non-specific activation of endogenous opioids. *Pain, 90*, 205–215.

Benedetti, F., Amanzio, M., Vighetti, S., & Asteggiano, G. (2006). The biochemical and neuroendocrine bases of the hyperalgesic nocebo effect. *The Journal of Neuroscience, 26*, 12014–12022.

Benedetti, F., Carlino, E., & Pollo, A. (2011). How placebos change the patient's brain. *Neuropsychopharmacology, 36*, 339–354.

Bingel, U. (2013). The relevance of placebo and nocebo mechanisms for analgesic treatments. In L. Colloca, M. A. Flaten, & K. Meissner (Eds.), *Placebo and pain: From bench to bedside* (pp. 127–136). San Diego: Academic.

Branthwaite, A., & Cooper, P. (1981). Analgesic effects of branding in treatment of headaches. *British Medical Journal, 282*, 1576–1578.

Cherkin, D. C., Sherman, K. J., Avins, A. L., Erro, J. H., Ichikawa, L., Barlow, W. E., & Deyo, R. A. (2009). A randomized trial comparing acupuncture, simulated acupuncture, and usual care for chronic low back Pain. *Archives of Internal Medicine, 169*, 858–866.

Colloca, L., & Benedetti, F. (2005). Placebos and painkillers: Is mind as real as matter? *Nature Reviews Neuroscience, 6*, 545–552.

Cressey, D. (2010). Acupuncture for mice. *Nature, 465*, 538–538.

Diener, H. C., Kronfeld, K., Boewing, G., Lungenhausen, M., Maier, C., Molsberger, A., & Meinert, R. (2006). Efficacy of acupuncture for the prophylaxis of migraine: A multicentre randomised controlled clinical trial. *Lancet Neurology, 5*, 310–316.

Enck, P., Weimer, K., & Klosterhalfen, S. (2013). Balanced placebo design, active placebos, and other design features for identifying, minimizing and characterizing the placebo response. In L. Colloca, M. A. Flaten, & K. Meissner (Eds.), *Placebo and pain: From bench to bedside* (pp. 159–173). San Diego: Academic.

Flaten, M. A. (2013). Placebo responses, antagonistic responses, and homeostasis. In L. Colloca, M. A. Flaten, & K. Meissner (Eds.), *Placebo and pain: From bench to bedside* (pp. 103–113). San Diego: Academic.

Flaten, M. A., Simonsen, T., & Olsen, H. (1999). Drug-related information generates placebo and nocebo responses that modify the drug response. *Psychosomatic Medicine, 61*, 250–255.

Foucault, M. (1991). *Die Ordnung des Diskurses. Inauguralvorlesung am Collège de France, 2.Dezember 1970*. Frankfurt a. M.: Fischer.

Goddu, A. (1984). William of Ockham's arguments for action at a distance. *Franciscan Studies, 44*, 227–244.

Goldman, N., Chen, M., Fujita, T., Xu, Q. W., Peng, W. G., Liu, W., & Nedergaard, M. (2010). Adenosine A1 receptors mediate local anti-nociceptive effects of acupuncture. *Nature Neuroscience, 13*, 883–U130.

Haake, M., Muller, H. H., Schade-Brittinger, C., Basler, H. D., Schafer, H., Maier, C., & Molsberger, A. (2007). German acupuncture trials (GERAC) for chronic low back pain: Randomized, multicenter, blinded, parallel-group trial with 3 groups. *Archives of Internal Medicine, 167*, 1892–1898.

Hume, D. (1977). *A treatise of human nature*. London: Dent.

Jensen, K. B., Kaptchuk, T. J., Kirsch, I., Raicek, J., Lindstrom, K. M., Berna, C., Gollub, R. L., Ingvar, M., & Kong, J. (2012). Nonconscious activation of placebo and nocebo pain responses. *Proceedings of the National academy of Sciences of the United States of America, 109*(39), 15959–15964.

Jonas, W. B., & Chez, R. A. (2004). Toward optimal healing environments in health care. *The Journal of Alternative and Complementary Medicine, 10*, S–1–S–6. doi:10.1089/acm.2004.10.S-1.

Kaptchuk, T. J. (2002). The placebo effect in alternative medicine: Can the performance of a healing ritual have clinical significance? *Annals of Internal Medicine, 136*, 817–825.

Kaptchuk, T. J., Friedlander, E., Kelley, J. M., Sanchez, M. N., Kokkotou, E., Singer, J. P., & Lembo, A. J. (2010). Placebos without deception: A randomized controlled trial in irritable bowel syndrome. *PLoS ONE, 5*, e15591.

Kirsch, I., Deacon, B. J., Huedo-Medina, T. B., Scoboria, A., Moore, T. J., & Johnson, B. T. (2008). Initial severity and antidepressant benefits: A meta-analysis of data submitted to the food and drug administration. *PLoS Medicine, 5*, e45.

Kleijnen, J., de Craen, A. J. M., van Everdingen, J., & Krol, L. (1994). Placebo effect in double-blind clinical trials: A review of interactions with medications. *Lancet, 344*, 1347–1349.

Latour, B. (1999). *Pandora's hope: Essays on the reality of science studies*. Cambridge, MA: Harvard University Press.

Latour, B., & Bastide, F. (1986). Writing science – fact and fiction: The analysis of the process of reality construction through the application of socio-semiotic methods to scientific texts. In M. Callon, J. Law, & A. Rip (Eds.), *Mapping the dynamics of science and technology: Sociology of science in the real world* (pp. 51–66). Houndmill/London: Macmillan.

Levine, J. D., Gordon, N. C., Smith, R., & Fields, H. L. (1981). Analgesic responses to morphine and placebo in individuals with postoperative pain. *Pain, 10*, 379–389.

Linde, K. (2006). Der spezifische Placeboeffekt. *Bundesgesundheitsblatt–Gesundheitsforschung–Gesundheitsschutz, 49*, 729–735.

McManus, C. A., Schnyer, R. N., Kong, J., Nguyen, L. T., Nam, B. H., Goldman, R., & Kaptchuk, T. J. (2007). Sham acupuncture devices–practical advice for researchers. *Acupuncture in Medicine, 25*, 36–40.

Moerman, D. E. (2013). Against "Placebo": The case for changing our language, and for the meaning response. In L. Colloca, M. A. Flaten, & K. Meissner (Eds.), *Placebo and pain* (pp. 183–188). San Diego: Academic.

Moerman, D. E., & Jonas, W. B. (2002). Deconstructing the placebo effect and finding the meaning response. *Annals of Internal Medicine, 136*, 471–476.

Moseley, J. B., O'Malley, K., Petersen, N. J., Menke, T. J., Brody, B. A., Kuykendall, D. H., & Wray, N. P. (2002). A controlled trial of arthroscopic surgery for osteoarthritis of the knee. *The New England Journal of Medicine, 347*, 81–88.

Ockham, W. V. (1957). Expositio in libros Physicorum Aristotelis. In G. I. Etzkorn (Ed.), *Opera philosophica* (Vol. 5, pp. 616–639). Franciscan Institute, St. Bonaventure, NY.

Park, L. C., & Covi, L. (1965). Nonblind placebo trial. *Archives of General Psychiatry, 12*, 336–345.

Price, C. J., Finniss, D. G., & Benedetti, F. (2007). A comprehensive review of the placebo effect: Recent advances and current thought. *Annual Review of Psychology, 59*, 565–590.

Rosnow, R. L., & Rosenthal, R. (1989). Statistical procedures and the justification of knowledge in psychological science. *The American Psychologist, 44*, 1276–1284.

Shadish, W. R., & Fuller, S. (1974). *The social psychology of science*. New York: The Guilford Press.

Sheldrake, R. (2013). *The science delusion*. London: Hodder & Stoughton.

Stergiopoulos, K., & Brown, D. L. (2012). Initial coronary stent implantation with medical therapy vs medical therapy alone for stable coronary artery disease: Meta-analysis of randomized controlled trials. *Archives of Internal Medicine, 172*, 312–319. doi:10.1001/archinternmed.2011.1484.

Ulrich, R. S. (1984). View through a window may influence recovery from surgery. *Science, 224*, 420–421.

Vickers, A. J., Cronin, A. M., Maschino, A. C., Lewith, G., MacPherson, H., Foster, N. E., & Linde, K. (2012). Acupuncture for chronic pain: Individual patient data meta-analysis. *Archives of Internal Medicine, 172*, 1444–1453.

von Lucadou, W. (2002). Die Magie der Pseudomaschine. In W. Belschner, J. Galuska, H. Walach, & E. Zundel (Eds.), *Transpersonale Forschung im Kontext* (Jahresband 2 Des DKTP, pp. 77–100). Oldenburg: Bibliotheks-und Informationssystem der Universität Oldenburg.

von Uexküll, T. (1982). Semiotics and medicine. *Semiotica, 38*, 205–216.

Walach, H. (2001). Das Wirksamkeitsparadox in der Komplementärmedizin. *Forschende Komplementärmedizin und Klassische Naturheilkunde, 8*, 193–195.

Walach, H. (2011). Placebo controls: Historical, methodological and general aspects. *Philosophical Transactions of the Royal Society, B: Biological Sciences, 366*, 1870–1878.

Walach, H., & Schmidt, S. (2005). Repairing Plato's life boat with Ockham's razor: The important function of research in anomalies for mainstream science. *Journal of Consciousness Studies, 12*, 52–70.

Walach, H., Sadaghiani, C., Dehm, C., & Bierman, D. (2005). The therapeutic effect of clinical trials: Understanding placebo response rates in clinical trials – A secondary analysis. *BMC Medical Research Methodology, 5*, 26. doi:10.1186/1471-2288-5-26.

Witt, C., Brinkhaus, B., Jena, S., Linde, K., Streng, A., Wagenpfeil, S., & Willich, S. N. (2005). Acupuncture in patients with osteoarthritis of the knee: A randomised trial. *The Lancet, 366*, 136–143.

Chapter 9
Medical Practice in/with the Semiosphere

Farzad Goli

Fantasy is non-being-like in the world
Observe! The world is floating on the Fantasy (Rumi)

After a brief contemplation on the biosemiotic approach to medicine, it seems that we live "in" a world of signs. And medicine, like one of the guides of Alice in Wonderland, helps us to interpret the signs of body and environment, not just assisting in finding a way "out" of deviational pathways of pains and illness. However, it should be noted that we as open, multilevel, self-organizing, and self-narrating sign systems live "in/with" the semiosphere. The semiosphere is the totality of semiotic processes occurring in our planet (Hoffmeyer 1998, p. 470). From this viewpoint, illnesses and conflicts do not only appear inside our life scenario as intruders and threatening characters, but can also be developmental forces and essential signs for evolution. Medicine in this perspective is an intentional, intersubjective system, which extends our natural adaptation to the levels of personal and social consciousness. Medical discourse is a sign system which differentiates signs to healthy, unhealthy, pathogenic, and salutogenic.

Placebos and nocebos are pure symbolic signs which show the no-matter aspect of healing and meaning response but as we pointed out in Chap. 5, that even the responses of organisms to material and energic stimuli are semiotic (Kull 1998). Physicochemical agents are interpreted by psychoneuroimmunologic systems and form the bodily, emotional, and cognitive responses. Thus, all of the medical phenomena could be explained in a biosemiotic frame of reference. We can therefore say that medical discourse is a health-related semiotics. The origin of semiotics may go back to the early stages of history of medicine and may refer to the crucial role of semiotics as a fundamental adaptive capability of human organisms (see Uexküll 1982). According to Hippocrates in the last aphorism of prognostic, symptoms are global signs which we need to follow and interpret properly:

F. Goli (✉)
Head of Danesh-e Tandorosti Institute, Isfahan, Iran

Energy Medicine University, Mill Valley, CA, USA
e-mail: Dr.fgoli@yahoo.com

F. Goli (ed.), *Biosemiotic Medicine*, Studies in Neuroscience, Consciousness and Spirituality 5, DOI 10.1007/978-3-319-35092-9_9

> However, one must clearly realize about sure signs, and about symptoms generally, that in every year and in every land bad signs indicates something bad, and good signs something favorable, since the symptoms described above prove to have the same significance in Lybia, in Delos, and in Scythia. So one must clearly realize that in the same districts it is not strange that one should be right in vast majority of instances, if one learns them well and knows how to estimate and appreciate them properly. (Hippocrates 1962, p. 55)

It is, of course, Hippocrates who remains the emblematic ancestral figure of semiotics – that is, of semiology, in the narrow sense of symptomatology– although he took the notion of clue from the physicians who came before him (Eco 1980, p. 277, as cited in Sebeok 1994).

Interpreting the symptom signs in order to diagnosis and recognize harmful signs in the ever-changing environment, as well as healing signs which are able to recognize human systems are the main semiotic tasks of medicine and in some respects, the basic vital drives of human beings. It seems that medicine and semiotics are historically and even ontologically interconnected.

9.1 Life-Oriented Medicine

Corresponding to Dyson's (1985) saying that "life resides in organization, not in substance" (cited in Harold 2003, p. 12), medicine as an organization manifests and develops self-preserving and self-referential properties of human life. Nonetheless, many great philosophers and thinkers like Canguilhem (1991), Focault (1973) and Illich believed that medicine is a social system which interferes the natural will-to-live in order to control human life in an ideological manner. For instance, Ivan Illich (1982), in his influential book *Medical Nemesis* criticized the professional monopoly and the scientism in health care. He coined the term *medicalization* to explore how medical systems address hypernormalization by excluding pain, illness, and death from the normal life; how overwhelming cautions and fears of personified diseases and the preference of the economic interests of medical systems to the health care necessities are also addressed. From this viewpoint, society created the systems and institutions such as medicine to serve society, but these systems have their own needs and demands and sometimes they may prefer to serve their needs instead of the society. This is why the medical system becomes counterproductive and produces many social, cultural, and clinical iatrogenic problems (Illich 1982).

According to Mishler (1981, 1984), the communicative side effects of the biomedical model are stated as ignored or blocked lifeworlds. He used Habermas's Communicative Action theory in medicine. Mishler explained how goal-directed rationality of the medical system has surrounded the value rationality of the lifeworld and why we need to give voice to it in the medical discourse (Mishler 1984; Barry et al. 2001).

It seems rather confusing, that medicine is either the epiphany of the will to live in human societies, or a social system which misleads the life drive by its rational and emotional biases. The ambiguity of these two roles medicine plays in life is because of the paradoxical nature of life. Life is not a linear story; it creates its antitheses in order to synthesize new orders and levels of organization. In order to establish a post-industrial medicine, we should consider both the paradoxical roles of medicine as an intentional extension of life and its role as an artificial system which resists against natural events such as illness and death.

This life-oriented approach to medicine requires a common language for the life system, the medical system, patient – doctor communication (as subsystems of social system), and also the lifeworlds of the consciousness system. In line with Luhmann's (1982) ideas, these three systems have their own autopoietic worlds and each one is also an environment (or context) for the others. For example, life and social systems are formed by the consciousness system and, reciprocally, consciousness systems are embodied through the life and social systems.

Thus, life for medicine is something more than an object; medicine is from/for/to life and would be a sustainable living system. For this purpose, medical discourse should follow the signs beyond the categorical and positivistic framework of disease to the lifeworld and phenomenal contexts. Actualization of the evolutionary role of medicine is correlated to reframing it in the transdisciplinary framework of the semiosphere; the worldwide web of signs which forms beings, relations, possibilities, and knowledge.

9.2 A Clinical Story

In order to show how we can follow the signs beyond the context of diseases and how to change psychosomatic dynamism by reframing the story, let us start with a clinical story. First, I will present a descriptive report of the anamnesis, which is adjusted by distinction of the contexts.

Nazanin R., who had suffered from a severe chronic ulcerative colitis for 9 years, was a 35-year-old married teacher. She had non-stop rectorrhage even after total colectomy. Her hemoglobin rate was critically low (6) and she was a candidate for colostomy, which would have meant carrying the burden of wearing a colostomy bag for the rest of her life. Her mother and sister were also affected by ulcerative colitis and her father died of cancer. Her symptoms became exacerbated by stress and irritant foods, especially spices and milk products. She was extremely anergic, pale, sad, and shy and immensely fearful and transient about eye contact. After following the signs from "disease context" to the "illness context", I realized that she felt deeply guilty because of her sexual obsession after having eye contact with men. She believed that her disease was a punishment from God because of her sin. Sometimes she thought suffering from this sickness was a way to salvation. She was depressed and could not stand such a dirty and hard condition; she was not contemplating suicide but greatly preferred death to such a "dark life".

Her sexual obsessions had become aggravated after the unexpected deaths of her father and her 5-year-old daughter in 1 year, about 4 years before. She believed her symptoms had become worse after this period, but her illness flared up in a typical manner after the last colonoscopy a few months before the colectomy. In one of her visits to her physician's office, her physician showed her the ulcers in the monitor and told her that "these sores are due to your raging anger and the more you eat yourself up over it, the greater the number of the ulcers". It was to be predicted that due to her control issues, this visual and audible warning was a new source of worry and, of course, anger because she could not control her emotions. She believed she exacerbated her illness further because of her poor emotion regulation. She had the obsessive imagination of ulcer formation in her gut, especially when she was tense. She attempted to control her stress and paradoxical increasing agitation, but as a result had pushed herself into a vicious cycle of negative self-suggestions, worry, control, failure, anger, and more negative self-suggestions. It quickly made her illness worse and the amount and frequency of her bleeding increased almost twofold after a few weeks.

In her "personal context", she is a religious Muslim Iranian woman who covered her hair tightly and dressed elegantly. She was ambivalent between her novelty seeking and harm avoidance temperaments. Her forgotten favorite pastimes were music, dancing, novel reading, and travelling. She became even more passive especially after the loss of her child in an accident and the exacerbation of her illness signs and symptoms. Her schedule was categorized by very limited activities: teaching at school, cooking, housework, watching TV with her husband, visiting her mother on the weekends, and sometimes walking from school to home. Just before visiting me, she had a novel experience. One day, while she was looking out of the window, it started to rain. In one miraculous moment, she got to observe the hard and dangerous way she had chosen to live her life and she decided that it must be changed – it is important to point out that rain in Esfahan (her residing city) is rare and implicates blessing and hopefulness.

In the "family context", she felt that they were "living in limbo"; they had lost their young daughter after an accident, for which nobody could be held accountable for, and because of Nazanin's health condition, it seemed that there was no chance to have another child. Her unreasonable anger towards her husband had caused her husband to treat her gently but to keep his distance to protect himself from her anger. Most of the time, they watched TV, drove or walked together in silence. Their sexual relations, especially after the child loss, were extremely low (once in 2–3 months). It was often without orgasm because of her husband's premature ejaculation. Her sexual desire was more or less normal, but because of her despairing and vanity holds, too intimate for her. In her parental family, she was the second child of a five-person family. She had restricted communications with her sisters and mother. She described her family as dogmatically religious and intolerable. She had very close relations with her late father, but he also sometimes blamed her – milder than the mother and sisters – for her reformist religious beliefs and behaviors. Having fallen ill, she found her mother's and sisters' behavior became more gentle and tolerable with her; however, she still felt estranged most of the time. On the other hand, the younger sister, who was only 1 year younger than her, always received kind support from her family, as opposed to her.

The patient had nice but shallow relationships with her colleagues and neighbors as well. In the first years of teaching, she forgot all of her problems in the classroom, but in the last 2 years, she found the class to be a heavy burden. The headmaster and her colleagues had facilitated her program after her operation, but she felt their pitiful gaze and support making her weaker and more pathetic. She believed that the colleagues, like her family, are supportive of her but that nobody cares about her achievements. Her only close relationship was with one of her colleagues. They saw each other during breaks or frequently on the way to school. In the beginning she seemed more or less insecure but motivated. Sometimes she was detached and spoke with a low monotone voice, as if she were speaking to herself without any hope to be heard and noticed. At the end of the first session, she was more open and more determined to continue her psychosomatic care.

9.3 Reading the Patient as a Multilayer Text

Before reading the content of this story, let us review the account formally. We can follow the signs among the contexts of disease and illness – in addition to the personal, domestic, social, and ecological contexts. The disease context is formed in the medical discourse. A set of signs and symptoms can be interpreted from a global category named as a disease. Etiology, pathology and/or clinical features can make the differentiation between the diseases. At this stage, the signs are interpreted in the objective framework of physical levels of organization. Nosology and the related diagnostic criteria are very useful in establishing coordinated and generalized management programs but the problem is that we encounter very different, complex, and chaotic features of illnesses in clinics compared with the medical school setting (Boet et al. 2012).

Each individual responds in a unique manner to pathological states, or even forms signs and symptoms in their own incomparable way based on a unique biopsychosocial matrix. Therefore, diagnosing the disease category is not the end of the clinical semiosis. For an expert clinician, the disease is a sign, which signifies an illness experience, or the subjective aspect of a disorder; the attitudes, beliefs, emotions, and behaviors of the patient in an unstable health condition (see, e.g., Hurwitz et al. 2004; Pescosolido et al. 2011). A patient has their own explanations, attributions (Brickman et al. 1982), coping strategies (Martz and Livneh 2007), and management plans. The patient's lifeworld is something more extensive than their illness experiences, and a patient is more than a disease (Chan et al. 2010). It seems like an axiom but it is, most of the time, evidently ignored in the disease-oriented approach of biomedicine.

Personal context includes traits, temperaments, life skills, self-concept, body image, interests, internal recourses, life plans, worldview, and spirituality. The basic trends in interpretation and the behavior in environment are shaped in our personal context. According to Buber (1958), "in the beginning was relation". We are formed, and form our works in a communicative network. Our beliefs, behaviors and even our genomes are unique combinations from the others. The family context represents the family games (Palazzoli et al. 1989), sexual relationship, narratives (Gouldrup 1987), attachments (Bifulco and Thomas 2013), economy (Serido et al.

2010), and alliances (Nicholas and Schwartz 2004). Family is the most common and immediate relational framework. We can follow the signs of the lower levels of organization in this level and indicate the genetic, epigenetic, and symbolic themes which affect or determine the psychophysical responses (Wisching and Stierrlin 1979).

The sociocultural context is the next level which constitutes doctor–patient relationships, health-delivery systems, friendships, workplace relations, cultural schemas, political systems, and our memberships in formal and informal social groups and networks. Social function and fulfillment have an essential role in our well-being (Larkin 2011). Finding the social resources, restraints and conflicts, cultural schemas (Nishsid 1999), and political state (Brown 2010) are very important in health behavior change and even the overall health condition.

These signs can be followed to the ecological context, and to the point that the climatological factors including natural and artificial components can affect or even determine our health (Jirtle and Tyson 2013). Briefly, the signs make up and flow among the above-mentioned contexts and form our health and illness.

Signs emerge in the form of reflective, symbolic, and physical variations. In other words, a history could be interpreted as the consciousness–information–energy–matter interactions and transformations. Information is the organizational power of creating structure and system in nature (Stonier 1997). As Wiener (1961) addressed, "information is information, not matter, nor energy" (p. 132). The negentropic power of life is also manifested in the symbolic world of language, society, and culture. From the pan-information worldview, there is no need to build a slippery bridge between the mental and the physical because both sides are built upon information.

At this point, we will follow signs from the abstract world of disease to the phenomenal worlds of the illness, person, family, and society. From one perspective, we will examine how psychosocial factors influence the symptom formation, and from the other, how psychoneuroimmunologic response affects the pathologic factors and therapeutic agents (Vedhara and Irwin 2005). We are dealing with two narratives here; the patient's view (see Porter 1985) and the psychosomatic specialist's view. In the patient's narrative, we determine how signs are interrelated and how habitual relationships not only control the patient's feelings, thoughts, and behavior, but also the immunologic function of the body and the placebo response to specific situations. The patient's narrative demonstrates how the physical and symbolic signs are interwoven to create their present biopsychosocial situation. It should be noted that the patient's narrative might seem very paradoxical and illogical but it does not devalue its importance because it is the most important narrative and frame of reference available for the occurring events. Hence, we carefully consider every wording, metaphor, attribution, and explanation that the patient uses in the narration–just like a person who is interpreting the Bible. Now let us examine the patient layer by layer.

In the biomedical narrative (disease context), Nazanin's case can be more or less displayed as following:

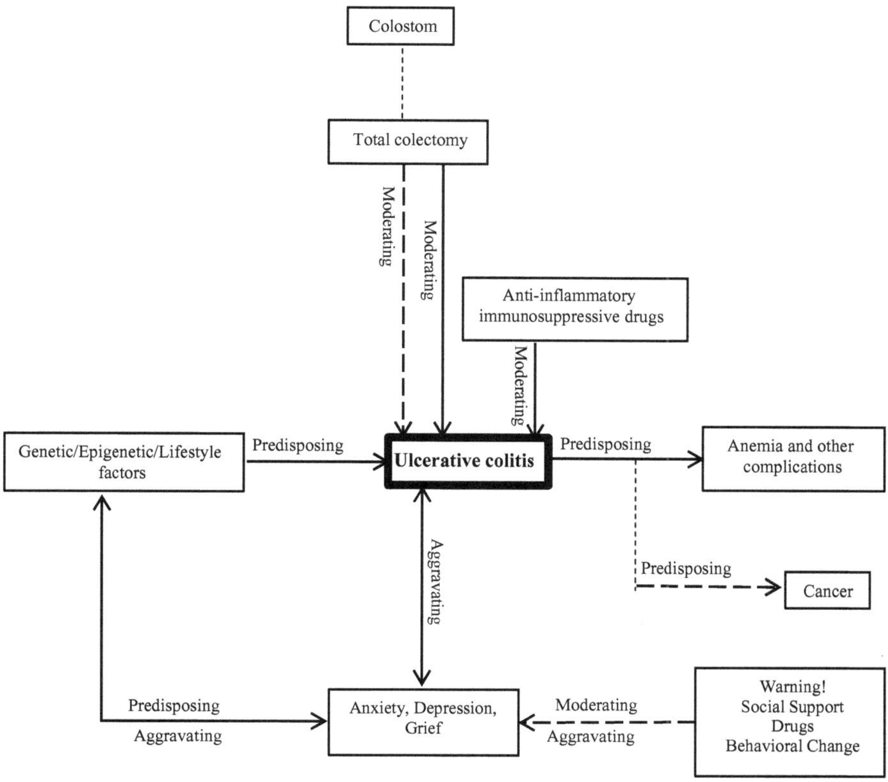

Fig. 9.1 A disease-oriented feature of Nazanin's story

> A 35-year-old woman with an active chronic ulcerative colitis from 9 years ago, with a history of a total colectomy (3 month ago) whose rectorrhage has not been under control even after the operation and receiving intensive medication (Sulfasalazine 3 g/day, Cyclosporine 50 mg/day, Prednisone 40 mg), and under a strict diet regimen. She has fatigue, abdominal pain, severe anemia (Hb; 6 g/dl), also anxiety and depression. She is a candidate for colostomy.

In a more sophisticated biomedical approach, we also attend to the predisposing, aggravating, moderating, and triggering factors of the illness. We know that Nazanin's anxiety and depression could be among the aggravating factors of the illness. From the objective perspective, even in the best case, the disease would be treated as a biochemical disorder and one or more psychoactive interventions would be prescribed; from a more comprehensive perspective, she would be trained in stress-reduction exercises. Although, this had not been performed on Nazanin and she had merely been warned to avoid anxiety.

Figure 9.1 briefly illustrates how Nazanin's story is interpreted in a disease-oriented approach. The mechanical problem of rectum's ulcers disorganizes the homeostasis and some symbolic factors disturb it. The disease-oriented solution is a biochemical intervention to moderate pathologic procedures and eliminate the irreversible pathologic tissue. This algorithm is to some extent what we expect to read from the disease context.

In the phenomenal contexts (illness, person, family, society), we deal with two main parallel narrative; the patient's narrative and the psychosomatic physician's narrative, which is actually the narrative of the patient's narrative of the illness and of course the objective signs. The physician goes beyond the categorical context of disease and follows the physical and symbolic signs that function in the formation and aggravation of the symptoms, in addition to symptoms that have an active and critical role in the psychoneuroimmunologic functions. Consequently, the specialist can restate and interpret the patient's narrative. Nazanin's narrative of her lifeworld could be summarized this way:

> … I suffer from this disease because of my shortcomings in life. This disease is the torment of my sins, especially my sinful sexual thoughts and dirty looks. I am a bad person and not worthy of living. After losing my daughter and father everything went from bad to worse. Since the time my doctor showed me how I am creating more ulcers by eating up myself over it, my disease has flared up and I came to know better what a worthless and weak person I am to make myself sick this way. My parental family is supportive but distant and strange. My husband is also supportive but especially after our late daughter, we are becoming sexually and emotionally disinterested. Indeed, we live in limbo. I do not even have the power to imagine carrying a bag of stool for the rest of my life. This will be the end of my story. A few days ago, I had a miraculous moment, I was mindful of my manner of living and I became determined to change my life.

This is an abstract of patient's report, which can be seen as a descriptive interpretation. In practice, we usually review in our mind what we perceive from a patient's account before analyzing its interpretation. Evidently, such a review or description would be selective, transformed, transposed and also biased by the healer's countertransferences and conditioned by previous experiences. But real practice is something entirely different from clinical trials (Yalom 1980, pp. 21–28) and some of what we label as biases in research are, in fact, pattern recognitions and expert intuitions (Quirk 2006) within the practice framework. At any rate, a psychosomatic practice would be a circular reinterpretation of the procedure where the healer and the individual being healed, interpret each other's and their own interpretations towards an approximation to reality (Gadmer 2004); or better said, towards the construction of a new reality. Even irrelevant interpretations in rapport groundwork can be inspiring and motivational for the circular hermeneutic procedure of psychotherapy. After all, psychotherapy is the unlimited reading of readings.

On the other hand, the psychosomatic specialist's narrative (analyzing interpretation) was different in attitude, relation, and attributions. Signs in his professional mind implied something else and were interpreted differently:

> Nazanin is a 35-year-old teacher with active ulcerative colitis and severe anemia who has been advised to seek a doctor due to her fear of colostomy surgery, which is the result of her persistent ulcerative colitis despite medicinal intervention and a total colectomy 3 months ago. She is pale and shy. In the beginning, she did not make eye contact and her voice was weak and trembling. She is elegant but has covered her hair tightly. She has had many conflicts with her family because of her reformist trends. It seems that she is severely ambivalent towards novelty seeking and harm avoidance temperaments. She has had the difficult

experience of losing her father and child within the past 3–4 years and has not been able to adjust. Her libido has increased in confrontation with death as a compensatory sexual desire to oppose death, especially after her child's death; nonetheless, her sexual drive is not fulfilled in her marriage. Her sexual drive has shifted into a visual drive and emerged in the form of sexual fantasies.

Nazanin tried hard to control her looks and thoughts because her religious background and perfectionism, which lead her to feel as if she were committing a sin. But her sexual obsessions are exacerbated due to paradoxical intentions and, consequently, her anxiety and sinful feelings are worsened. She assumed her disease is God's punishment for her sins and at times did not consider herself worthy of living. She has had ulcerative colitis for 9 years and has kept it in check with medicine, but the disease got out of control and became aggravated due to the stress resulting from grieving over deaths alongside her severely-increasing anxiety resulting from obsessive thoughts and psychoneuroimmunologic instability. During her last endoscopic visit, the gastroenterologist, while showing her ulcers on the monitor, told her "the more you beat yourself up, the more ulcers you will form". The gastroenterologist, having not known her obsessive background, caused her increased health anxiety and decreased self-esteem and self-efficiency. The obsessive thoughts of causing the ulcers to increase were added to her sexual obsessions. This made her anxiety ever-present, and the feelings of impotence, self-hate, worthlessness, and frustration were also gradually added to her sinfulness and made her psychoimmune condition severely unstable. Bleeding and other symptoms were immensely growing, which led to a colectomy surgery 3 months prior.

At this point, because of her critical health situation, she was a candidate for a colostomy surgery; but surgery at this point, where her life-drive was drastically decreasing, would have meant the end of her life. She oftentimes described her life as being hard, dark and dirty. For her, carrying a bag of stool would be a confirming metaphor for uncovering her dirty thoughts and life. Based on her general conditions, temperament improvement and obsession control might provide the opportunity for healing through psychoimmune modulation. She had a reserved relationship with me, severely- weakened self-esteem, and feeble physical and psychological conditions; nonetheless, she was optimistic of her healing through this psychosomatic treatment because of a presence experience a few days before and her friend's positive suggestions.

This psychosomatic narration implies that she has been introduced to the phenomenal context of illness to some extent and has recognized which illness-prone mechanisms are actively at work reproducing and aggravating the disorder. Symbolic factors such as obsessive thoughts, pictures of God's punishment, the picture of a feeble and worthless self; in addition to unpleasant feelings of anger, anxiety and depression had disrupted the psychosomatic balance and led to a lower quality of life, wasting of life force energy, limited access to resources, and a psychoneuroimmunologic disorder which ultimately resulted in the disease aggravation.

The importance of following the signs from the disease context to the phenomenal contexts may have become more evident thus far. Obviously, treating Nazanin's symptoms – abdominal pain, rectorrhage, and/or anemia – without diagnosing her disease would be a symptomatic and incomplete treatment, in the same way, the symptoms had to be stated in the illness context to reach a more meaningful pattern. The same relationship existed between the disease and illness. If the illness behaviors and placebo/nocebo responses of the patient are not assigned to accidental or previously conditioned patterns, then the disease needs to be interpreted in the illness context.

Illness experience is part of patient's sign-world; our phenomenal world is placed in sign-worlds and includes resources, cognitive maps, and emotional regulation models. Illness experience is the cognitive, emotional, bodily, and behavioral response to perceived disorders; hence, the genetic and epigenetic factors in the personal context clearly determine the illness, but the meaning of personal context is not complete in itself. Rather, a more complex context, called family, determines the attachment pattern, personal and cognitive paradigms, basic patterns of self-care, alliances, support resources, and primary conflicts of the relationships that are formed in it.

Social context is in a similar mutual relationship with the previously stated contexts. On one hand, personal and familial relational patterns determine our social skills and patterns, on the other hand, social institutions' structure, discourses and cultural schemas determine forms of life (Barker and Galasinkski 2001) and meaning-making frameworks in other contexts. By redefining and readjusting the story of the patient in different contexts, the psychosomatic specialist not only supports the understanding and explanation of the individual's health, but also reveals the biological, psychological and social resources. This way, the healer can facilitate more adaptive and salutogenic narratives for the patient and reprocess the patient's cognitive, emotional, bodily, and behavioral responses.

Furthermore, through the communicative interventions in any context, the doctor could change the meaning -making procedure in that context, as well as in the other contexts, as a way to connect with the internal and external resources which in turn effectively actualizing the salutogenesis stream. Consequently, the healer, the individual being healed, and other related persons in our story create a different reality in a circular hermeneutic approach, a story that might be much more dynamic and creative.

In order to avoid becoming pedantic by explaining every single context (detail) of this story, I have briefly reported the patterns that emerged in the context through the following signs and summarized the events that happened during Nazanin's treatment. Figure 9.2 is an example of a clinical storyboard. I specifically utilize this method for complex psychosomatic patients in order to simply and logically follow their story and identify the resources and obstacles. In this illustrative figure, the higher levels of organization encircle the lower levels. It could be permissive in some senses, but we should again mention that the higher levels (like society) are reflected in the lower levels (such as person), and the lower ones are also an environment for the higher. Our findings and achievements during therapy can be added to the storyboard, and we can follow how the contexts change.

Social Context	Family Context	Personal Context	Illness Context	Disease Context
○ low tension relationships ○ Shallow social relations ○ Feeling her job as a heavy burden ○ Teaching was a powerful resource for her ○ Despite of her critical and psychological problems, she is still working ○ An intimate but restricted friendship ○ Positive attitude in D-P relationship. ○ + Supportive Coleagues	○ Missing her dead father ○ Feeling herself far and strange from her family of origin ○ Cold relationship with husband ○ Unreasonable anger toward husband ○ Low & unfulfilled sexual relationship ○ Supportive- distant husband ○ Getting more care from her family	○ Religious ○ Ignoring favorites ○ Stereotypic life ○ Reformist ○ Elegant dress up ○ Wish to refound her favorites ○ Trying to establish her own values ○ Becoming aware of her coping/ attribution ○ Transient experience of hopefulness	○ Guilty feeling ○ Sexual obsessions ○ Avoid eye contact with men in public ○ Anger ○ Negative automatic thoughts ○ Try to control thoughts and emotions ○ Abdominal discomfort ○ Anergy ○ Death desire	○ Rectorrhage ○ Bowel active ulcers ○ History of total collectomy due to ulcerative collits ○ Non-response to drugs ○ Obsessive thoughts ○ Lack of eye contact ○ Pale skin ○ Hb= 6 g/dl ○ Depressive mimic/ posture ○ Sighing ○ Grief
Patient's interpretation: ■ "My job is a heavy burden." ■ I'm a pathetic person. ■ Nobody cares about my achievements" Doctor's interpretation: ■ Social anxiety/ avoidant trait. **R:** ■ supportive colleagues **R:** ■ Good social function **R:** ■ Ability to establish intimate relation. ■ Cultural beliefs: "Sick person is pitiful, help her to bring heavenly reward."	Patient's interpretation: ■ "We live in a limbo" ■ My parental family are supportive but far and stranger. Doctor's interpretation: **R:** ■ ↑Social support ↓Emotional support Blockage of orgonic energy **R:** ■ ↓Relatedness ■ Mutual avoidant strategy with her husband: Fear of anger [and death] has suppressed trust and love.	Patient's interpretation: **K:** ■ "I'm bad." ■ "I'm not worthy of living." ■ "I must change my life." Doctor's interpretation: ■ Being ambivalent between novelty seeking and harm avoidance temperaments **R:** ■ A mindful change in the expectations & self efficacy. **R:** ■ She is on the edge of individuality and mindful being. **R:** ■ Normal libido. **R:** ■ Good self-care.	Patient's interpretation: ■ " Dark life" **K:** ■ " Punished by God" **K:** ■ " I make sick myself" Doctor's interpretation: ■ PNI instability due to adjustment disorder, health anxiety, self- defeating and low self efficacy	Doctor's interpretation: ○ Active Ulcerative colitis ○ Depression/ adjustment disorder ○ OCD ○ Anemia **Management plan** **Strategies**: Acceptance of automatic thoughts as meaningless signals/ stress reduction/psychoimmune modulation / self expression in family/ promotion/ Conciousness evolution and individuation. **Techniques**: ACT, Hypnotherapy, Narrative Tx, CBT, SportTx, Massage, DrugTx.

Fig. 9.2 A clinical story board which shows the patient's and doctor's interpretation of signs through the disease contexts. "***K***" marks the key signs which may lead us to the core of patient's narrative, and "***R***" indicates resources which by spotting and reinterpreting them, we can make the story more generative and lively

9.4 Deconstructing the Story: Is This the Sickness unto Death?

This is a shocking question that Kierkegaard, quoting from the Bible, posed at the beginning of a book of the same name (Kierkegaard 1983). The word "death" does not denote physical death here, but rather the downfall of spirit; a case in which the human sees themselves separated and alienated from the world and whatever exists in it; individuals see themselves isolated from God and from the cosmos. Nazanin's worries were about the kind of death that swiftly occurs before her physical death. Considering her obsessive background, perfectionism and the religious meaning of cleansing and purity, the revealing of her dirt and carrying of her filth was the last episode in the symbolic painful torment that her God had inflicted upon her because of her sins. What made this tragedy more painful were her paradoxical intentions, and her strong control-seeking tendencies had caused her to ascend deeper into it the more she tried to distance herself from it. This epigenetic psychosomatic pathway had neutralized the placebo effects of her treatment – highly increasing the nocebo effects. Having considered the limited timeframe for attempting psychosomatic treatments, we had to find an effective shortcut for the deconstruction of the narrative; an escape from the cannibalistic labyrinth.

Because of her religious background, we had to figure out a way to cleanse her and set her free from the cruelty of her superego. Therefore, in our first session, after listening to her account, I told her the story of Nasavi, an Iranian Wiseman. Following is our conversation:

"I've heard your story. Now, I would like to tell you a mystic story. This story might provide us with some clues to how you have been destined to this torment." I was guessing the meaning of "destined to torment" is that brick of the prison that will destroy the wall if removed. Anyway, we should focus on the determinant signs and the proper gates through which we can enter into the patient's world sign.

She looked at me for a moment and asked me to start by nodding her head.

"'Nasavi stated, "once, while I was meditating, my animal soul dropped out of me in the form of a small fox. I grabbed my cane and beat it. The more I beat it, the bigger it became until the fox turned into a monstrous creature. Having felt frustrated and exhausted, I stopped beating it. I asked the fox to reveal its mystery. It said, 'opposition is my food'."'" (Hojviri 1982, p. 259). I became silent. After a few seconds, she looked at me confused.

"So what?" she asked.

"Do you know this feeling of frustration and exhaustion?" I continued.

"It is exactly how I feel." she said as if she were talking to herself. Then, she looked at me and said, "But what should I have done?"

"Let the fox be as small as it is." I continued, "You mistook automatic thoughts with intentional thoughts and interpreted them as immoral; by fighting with these thoughts you made them stronger and stronger and yourself weaker and weaker day by day. Automatic thoughts are like muscle tonus. Thought is a natural flow, just like water or air, and does not have anything to do with ethics; it could be sexual thoughts that pass through everybody's mind or thoughts about your ulcers. You have not committed any sins. Your problem is not due to your weakness. You are decent. You have only been scared by shadows."

From Nazanin's face, I could say with certainty she had accepted and believed this new narrative. I said, "first, you are allowed to eat yourself up." (In the Persian language the expression "*hers-khordan*" is used in this context which would be translated "eat annoyance" word by word. The metaphoric use of the word "eat" in Persian might have worsened the relationship between anger and her ulcers). "It could be even energizing too. Throughout the day, whenever these thoughts enter your mind, let them come and go, just like the flow of a river that allows anything to float on it. The point is that you know it does not mean anything and should not be interpreted in anyway. If you ever become engaged with your thoughts, it is not important. Distance yourself gently. Observe and shift your attention gently to your work or things around you again. If you remember it later, it does not matter either."

I asked her to include gentle and light swimming and pampering of her body at least twice a week. At the end of the session, I pointed out to her the promising experience of rain she had a few days back.

"Did you have a better feeling in the past few days?" I said.

"Yes. I think I am more hopeful."

"You had made your decision to recover before visiting me. Your organismic sense, your unconscious mind had promised you recovery."

By focusing on her, I could emphasize that "you should change yourself", as you are inspired, and "all you need is in your own hands". It would help her to bring her focus of control towards the inside. Because of her critical health conditions and negative and passive attitudes towards her family, I preferred not to engage her family in the acute phase. The next appointment made me certain that the black sky had burst and that she no longer had a sickness unto death. The light she had in the deep gaze of her eyes was getting close to me; making eye contact was eminently easier and she had less frequent and less amounts of ulcer bleeding. Her surgeon agreed with her psychosomatic treatment and she decided to continue it. In the following 7 weeks, her symptoms were improving and in the ninth appointment, her bleeding, pain, and other symptoms ceased altogether. Her anemia had clearly ameliorated (Hb 10 g/d) and, in the later months without any medication, she did not experience any symptoms even when she consumed irritant foods.

Throughout her psychosomatic treatment, relaxation and imagery techniques were utilized to reinforce her health-generating imaginations, calmness and self-esteem. She made better use of her relationship resources; after several years, she could freely talk to her husband about the feelings, thoughts, and expectations they had after the death of their daughter. She talked about her sexual expectations with her husbands as well. They reached a rather satisfying sexual relationship with no clinical intervention.

In the final sessions, I could hear her voice had a pleasant tone; blood was flowing under her white matte skin making her cheeks rosy; you could see that sun had risen in her world and made all the shadows and silhouettes disappear. She felt connected to herself, others, her God, and being. Her story was interwoven with story of life.

9.5 Towards a Semiotic Art of Healing

According to Davidson's (1970) theory of anomalous monism, there is no psychophysical law. It means that several symbolic agents can be associated with a certain physical agent and vice versa. These irregular conditions of psychological interaction can be found in our story, as well. Through the pathogenesis phase, "a domestic patterns of expressing stress intestinally", "a grief,", "sexual obsessions", and "a warning" were various irrelevant symbolic signs which resulted in ulcer formation. On the contrary, a vast variety of signs such as the threat of colostomy as a terminal condition, an unexpected rain, and the interpretation of fantasies as a natural stream led the psychoneuroimmunogic approach to heal the ulcers. By changing the narrative, the consciousness–information–energy–mater flow was modified and the meaning-making procedure changed physically and mentally. The mechanical explanations are evidently useless in these clinical conditions because certainty, predictability, and objectivity are suspended in this view. The psychophysical rules are formed through a unique communicative hermeneutic procedure and we should also seek help from semiotic explanations in order to understand and change health conditions properly. It seems that we need a heuristic and creative approach which synergizes two worlds toward healing. Lifeworlds of doctors and patients are significantly different in clinical medicine. The doctor's world is primarily one of disease, while the patient's world is one of lived illness (Toombs 1992, Carman 1952). It is a phenomenological principle that Husserl, Heidegger, and Gadamer are striving to articulate; human life is embedded in a meaning-structure; a horizon of meaning that surrounds every act, action, articulation, or reading (p. 154).

Two beings in the world come together and after a while create a two-person system, the unhomeliness experience of the patient is experienced and reinterpreted in the doctor's world; interpretations and feelings, and healing expectations are reinterpreted in the patient's world. The fusion of these horizons creates new horizons, lifeworlds and new psychophysical pathways – a new reality (see Fig. 9.3).

Not only the response to a pill, but also the pharmacodynamic verum effect of it can be explained as meaning responses. The symbolic and/or chemophysical signs can make meanings such as a certain motivating salutogenesis and/or blockage in the pathogenesis. Therefore, various signs in the form of matter (*Msign*), energy (*Esign*), symbol (*Ssign*), and consciousness (*Csign*) alternations construct our psychophysical narratives: Msigns such as coticosteroids, immunosuppressive molecules; Esigns such as music, an electromagnetic pulsation; Ssigns such as a metaphor, an attribution; and Csigns such as a mindful perception of unexpected rain, a pause in conceptualization or a suspension in conditioned stimulus-response chain.

For instance, Nazanin's obsessive thoughts which obviously aggravate her colitis (disease context) are resonated in her attribution to God's punishment and intensive attempts to control them (illness context). Considering her obsessive traits and ambivalence towards her values in the personal context, these uncover her excessive control seeking and the violence of her superego. Feeling abandoned from her parental family explores her ambivalence and torment and why she believed she was not worthy to have a child. Her avoidant social relations make her more dependent on the family. We can also follow the implications of the disease changing her self-concept, family, and social relations. Perhaps several sign processes stream among the contexts simultaneously, and each chemophysical or symbolic intervention changes the meaning-making processes and changes the whole narrative. Figure 9.4 shows the reciprocal flow of signs through/by the different levels of organization, and how each interpretant could be a sign for another interpretant.

It seems that from such a semiotic view, there is no supposition of a subject-object or mind-body dichotomy. Merleau-Ponty (1986) indicated this non-dual panorama, as "my body is neither internal to my consciousness nor external to me in the environment". Therefore, body is the embodied mind, and mind is the self-organizing order of the body. In other words, lived body is nothing but a multilevel autopoietic mind by billions of replaceable elements; a unique, dynamic "point of view on the world" (p. 70).

We can imagine ourselves as proactive/reactive waves of an omnipresent ocean of the semiosphere with a myriad of transformative signs which form feelings, organisms, habits, symbols, and images. What would healing practice be like if the real world were so subtle, chaotic, and creative?

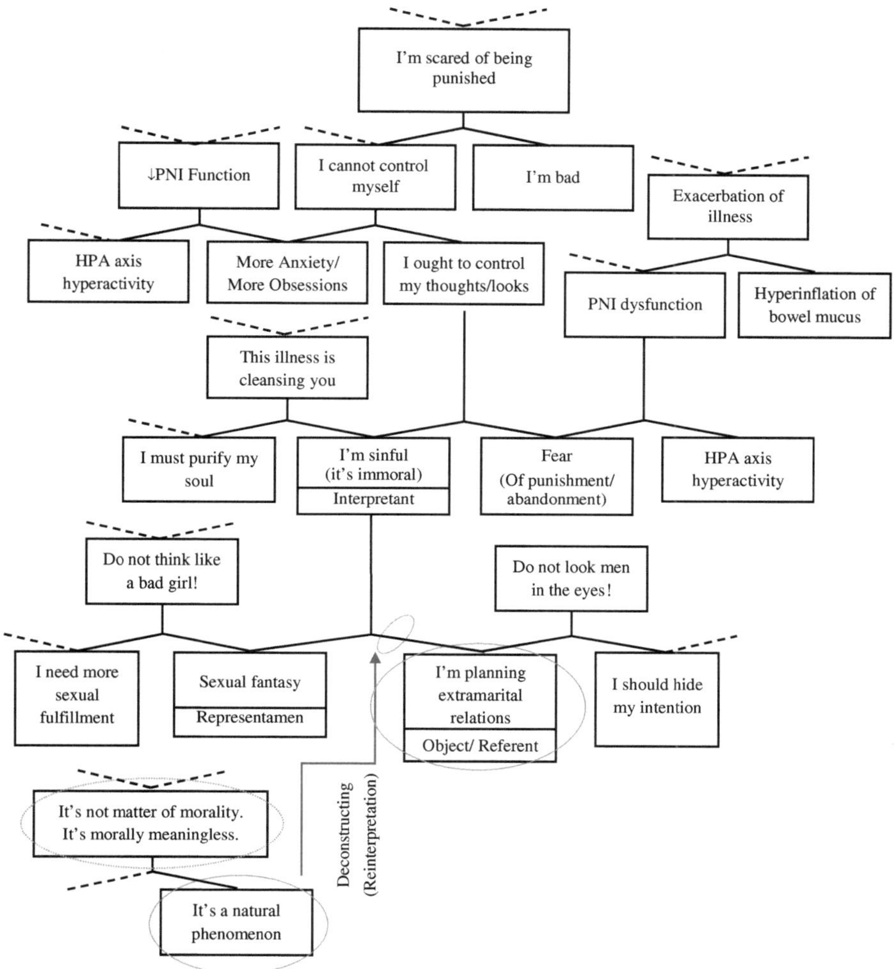

Fig. 9.3 A schematic illustration of a window to Nazanin's sign processes. This two-dimensional diagram shows how we can deconstruct a sign world and block a meaning-making system by sub-situating, and that a new belief (a symbolic sign) such as a semiogram is not so precisely abstract and generalizable like cognitive maps. It could be a clinical tool which can bring into view some aspects of the client's phenomenal world and reveal the proper points for deconstruction. To simplify the diagram, some of the characters are mentioned in fact as a set of representamens, objects (referents), and interpretations

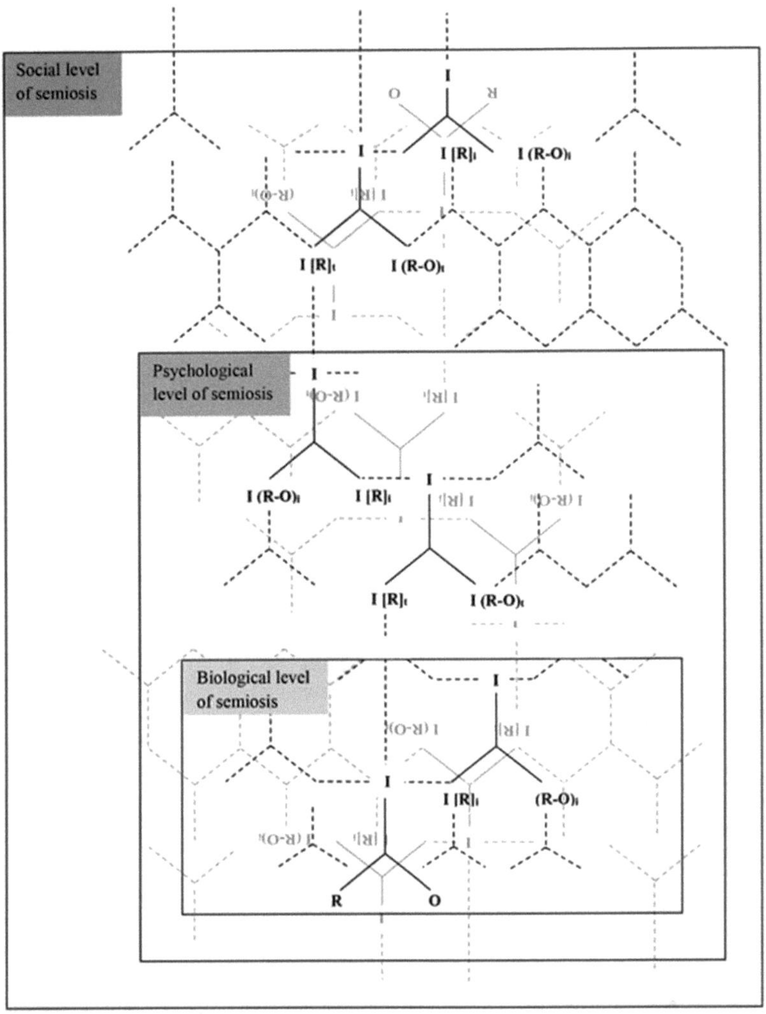

Fig. 9.4 The semiosis through and beyond the levels of organization. *R* representation, *O* object, *I* interpretant. The interpretant (I) of the first sign turns to be the representamen ($I[R]_i$) of the further sign as an object $(R\text{-}O)_i$. An interpretant could be interpreted simultaneously by several interpretants. Some signs escape from one level to another level $I[R]_t$ like bodily sensations, which are conceptualized as emotions in the psychological level. The semiosis forms the systems, and the whole dynamic sign systems (holons) mutually coordinate the meaning formation through the systems

9.6 Living in/with the Semiosphere

In a very fundamental sense, health is about communication. From this viewpoint, illness is the misinterpretation of an internal sign, like what we see in cancers and autoimmune disease (Hoffmeyer 2010, p. 22) or a series of attempts toward autopoietic reinterpretation of external signs such as traumas and microbes. Of course, pathogenic signals in the form of mechanical, biological, chemical, physical, or symbolic agents block or break down the functional cycle (see Uexkül and Pauli) of the organism and create a disorder in the self- referential order and induce some misinterpretations in the system. The psychoneuroimmunologic system usually interprets them as disturbing factors and tries to find its way back to its conditioned/structured order, but sometimes it heuristically finds a new way of autopoesis even by setting some capabilities aside; of course sometimes the only way is to death.

This procedure may be associated with the darwinistic struggle of living systems; the war of membranes which tries to exclude or absorb the others in order to find a greater chance for survival. We can find some evidence that there is such a struggle even between selfish genes of a certain cell (see Richards 2007, pp. 155–165).

Is selfishness the motivating engine of evolution? Is chance the fuel of this machine? Is this the unique scientific explanation of life? By acceptance of these assumptions, medicine would be ambivalent between a narcissistic cathexis to attain more time and more pleasure and an altruistic cathexis to help sick and weak people. Thus, medicine from a Darwinian view is an evolutionary-antievolutionary system; it seems crazy! Now a question arises "could this paradoxicality be a matter of perspective?"

In the recent decades, plenty of new systemic theories such as autopoietic systems (Maturana and Varela 1987), synergetic (Haken 1978), chaotic structures (Prigogine and Strngers 1984), the Gaia theory (Lovelock 2000), biosemiotics (see Barbieri 2007), cooperative genes and naturalgenetic engineering (Shapiro 2005) uncover the cooperative and agapistic features of life based on myriad of reliable facts. The scene of life theatre has been changed by shifting our focus from membranes to signs, from the living beings to the living systems. Signs stream freely in/between systems and construct more coherent and complex systems beyond the selfish wars of the membrane systems. What seems to be a war story in a surrendered gaze appears to be a love story of the accelerating meaningfulness of signs in the form of new synergetic living systems. Death seems to be the terminal of living systems but is also the desire – the will to live. In Pierce's (1923) philosophy, we can find such a complementary approach to evolution: evolution by chance (tychism), necessity (anancism), and love (agapism).

The mechanical necessities of reactions are formed in a ground of fortuitous variations and tend to form more synergetic systems and be under the restraint of the higher levels of order and ultimately the whole universe – the nurturing love. At first sight, the concept of love in medical discourse reminds us that, in addition to the instrumental rationality of technomedicine, we need to imperiously communicate rationality in medical discourse in order to approach what Habermars (1987) named ideal speech interaction – communication without any distortion or blockage. In my opinion, it would be a communicative definition of health in its vast meaning, from intracellular to global communication. Thus, agapism seems to fit the state of being synergistic, open and responsive to the other systems of the same level of organization and the others.

According to physicist and theoretician Bohm (1980), each part of the universe is a four-dimensional picture of an undivided wholeness, and each part is formed and behaves in relation with the other parts as a whole. He says:

> The relationships constituting the fundamental law are between the enfolded structure that interweave and inter-penetrate each other, through the whole space, rather than between the abstracted and separated forms that are manifest to the senses (and to our instruments). (p. 235)

Bohm's holonomy seems to be unanimous with Pierce's agapism; both of them emphasize a cosmic harmony despite local disharmonies and highlight the top-down organization. We should borrow this convincing worldview for medicine instead of struggling with the current model, that is, the omnipresent wars between life and health, disease, and body, medicine and disease, and doctor and illness. The war metaphor of biomedicine, as we mentioned before, brings rise to a culture of horror and risk especially in the modern hyper-individualized societies and systematically induces sociocultural iatrogenesis and empowers nocebo responses.

The love metaphor and a systemic worldview can help us diagnose disorders as communicative problems and manage it not only in the disease context, but also in the higher levels of organization. Therefore, a healing system would be what reprocesses the unlimited semiosis by physical, chemical, energic, symbolic, and/or mindful signs; towards an ideal intra/inter/transpersonal communication.

From this standpoint, love is neither a decorative element for human life nor exclusively the erotic drive for reproduction and producing pleasure but it is the big picture of life. Thus, we can remodel medical discourse based on love. Health, illness, and healing should be redefined. Thereby, health is being attuned interpersonally (mind–body coordination), interpersonally (self–other integration), and transpersonally (wholeness experience). Wholeness and integrity would be key concepts of this definition of health, where healing as the will-to-be- wholly-healed and whole have the same ethnological roots. When a part comes out of the holomovement, as Bohm says, it creates a disorder and the best way to regain health is by facilitating the system to behave and organize itself and as a whole – usually a new one. Healing is therefore something more than control of disordered functions or any other local approach to the pathologic organ processes. The bottom-up control of pathologic-based therapies should be complemented and elaborated by the top-down organization of salutogenic-based healings.

A systemic approach to health would actualize health not only in the person, family, and society levels, but also in the whole life system along with the evolution of consciousness. Illness and death could be interpreted as evolutionary signs in the medical discourse as they function within the life system. Moreover, healing is a creative intra/inter/transpersonal sign system which reinterprets illness and death to retrieval and redirect significations towards reorganizing whole organization. This worldview is not so individualistic but individuals as holons can be actualized properly.

Therefore, health is not the silence of organs, as the great physician Bichat described, but the harmonious flow of signs through the body's organon; the organon made of signs. Health would be functioning, working and being throughout the whole body – our inter/intra/transpersonal body. Thus, being healthy is being in love.

9.7 Practice in/with the Semiosphere

Following Hoffmeyer (1995), the semiosphere is a sphere comparable to the atmosphere, which penetrates through the other spheres and consists of communication: sounds, odors, movements, colors, electric fields, waves of any kind, chemical signals, touch, etc. (p. 35). Love can be defined as the trend of these miosphere moving towards increases in meaningfulness and creating more complex forms and networks. It would be a common teleological criterion for life, health and morality.

Signs flow along these miosphere by their nature and/or intentions; Stuart (2010) named this intentional layer the ethiosphere. She writes:

> Our senses open us to the reception of these forms of communication, but it is not a passive reception; it is a reciprocally effective, intentional, co-agential, concernful, enkinaesthetic communication in which we are able to affect others and be affected by them, to move and be moved within the sphere of ethical engagement, that is, within the ethiosphere. (Stuart 2010, p. 321)

The patients' and the therapists' experiences include both actions and events. Patients could have an active role in their pathogenesis and salutogenesis and doctors would normally experience various affections and non- intentional interpretations (see Balint 1957). But therapists' training is focused on how to intentionally interpret the intentional and non-intentional signs of the patient's world and their own life world as well. Because medical practice is formed within the ethiosphere, ethics are therefore not an additional aspect. Conversely, medical practice is intentional and ethical by the nature.

Although fear and greed are the main resources of the preconventional and conventional stages of moral development, the higher stages of morality (autogeneous levels) arise from consciousness (See Kohlberg 1982). In the highest level – the transpersonal level – love is the resource of morality. Love could be mentioned as the highest level of the ethiosphere, the autogenic openness to the whole harmonizing our intention with the accelerating meaningfulness of signs.

From this non-dual view, intra/inter/transpersonal egoism is what had been named altruism from the dualistic view of identity. We can still be selfish, but by following the path to actualization of ourselves in/with/of others, we can still say "I", but as a holon, it is a part/version of the whole.

Being in love, full of life, actualized, healthy, and moral seems to be the same as looking through biosemiotic glasses. Morality for human beings is not exclusively a social contract, but it is self-actualization in an intercorporeal, co-experiencing and co-consciousness context. Thus, morality and spirituality could be mentioned as the interpersonal and transpersonal extensions of our vitality and corporeality.

According to Stuart (2010), the boundaries which seem to separate us from our worlds open us to those worlds and reveal how inseparable we are from them. She emphasizes the biosemiotic and kinaesthetic aspects of ethics. Kinaesthetic activity, with its temporal–spatial–energic qualities, is always affectively laden. Through the formation of intercorporeal resources, activity necessitates enkinaesthetic entwining with the agents and objects in our community with which we are in perpetual relation. Kinaesthetic structures and melodies represent the dialogical nature of the feeling of being as the feeling of being-with or being-among (pp. 305–8).

Indeed, medical practice is an intercorporal and being-with experience towards healing. In order to form a powerful response, the signs – from architecture, decoration, rituals, bodily and verbal messages to the chemophysical interventions – should be consistent and harmonized. A biosemiotic pharmacy that formulates harmonious meaning responses allows the organism to behave as a whole and then heals it as a whole by its autopoietic power. Thus, the will-to-heal can be embodied in our body, behavior, language, discourse, and institutions. The coordination of these sign systems can provide a culture of love, in which each lived system tries to be open to the others in order to expand its territory by being with others, not just by being as others.

Noë (2009, p. xiii) describes this holistic view as follows:

> Human experience is a dance that unfolds in the world and with others. You are not your brain. We are not locked up in a prison of our own ideas and sensations. The phenomenon of consciousness, like that of life itself, is a world-involving dynamic process.

Autonomy, benevolence and beneficence – the axioms of medical ethics – can be translated into this systematic language as the principles of meaning response formulation. Each clinical intervention should respect the client's agency and help the person to develop their consciousness as the highest level of human functionality. Benevolence indicates the interconnectedness of the self-other interests and implicates the concept of beneficence maintaining the direction of our intentions and actions toward eros – the desire to live. Thus, setting our intentionality by life orientation, self-other integration and consciousness evaluation helps us harmonize our biosemiotic prescriptions.

The biosemiotic model integrates biomedicine, psychology, ethics, sociology, anthropology, and spirituality in a common groundwork as meaning systems, which intentionally forms our narratives of our life and induces meaning responses in human systems. Consciousness and signs (symbolic, energic, and material) are reprocessed by our knowledge and create the forms of life, followed by our feelings, conceptions, illnesses, healings, narratives, and behaviors. We create our collective and individual realities intentionally or non- intentionally, and our cellular, systemic, personal, domestic, social, and ecologic meaning responses are formed within these narrations.

Practicing in/with the semiosphere, which is made of the soft reality, would be similar to composing a symphony or a poem; Listening to the agapistic song of signs harmonizes our manner of healing or being healed, doctor-patient communication, and prescriptions and manipulations, with this universal a melody. This is the way to the art of healing. Being a healer, as Michael Balint (1957) emphasized, is sometimes different from practicing medicine; the doctor is a drug when they care for their being and allow the supportive, informative, cathartic, and salutogenic dynamisms be performed by the *dasein* (The quality of being there). Therefore, healing is a "motor intentionality", a kind of embodied poise or readiness, which Merleau-Ponty (1986, p. 110) calls "habit". Thus, over everything, healing is a whole body experience which can be expressed intra/inter/transpersonally.

The semiotic approach to medicine can provide a salutogenic wisdom for the medical discourse by integrating our subpersonal science into the personal, suprapersonal, and transpersonal knowledge. Such a paradigmatic shift can change our worlds, education, technology, and practice to a more holistic, humanistic, hermeneutic, and – of course – consciousness-based health system. Recovery which means to regain one's health, is meaningless jargon in this meaning-nature creative world. The problem is how to create more sustainable, healthy systems, and compose more beautiful stories.

References

Balint, M. (1957). *The doctor, his patient and the illness*. London: Tavistock Publications.

Barbieri, M. (2007). *Introduction to biosemiotics: The new biological synthesis*. Dordrecht: Springer.

Barker, C., & Galasinkski, D. (2001). *Cultural studies and discourse analysis: A dialogue on language and identity*. London: Sage publications.

Barry, C. A., Stivenson, F. A., Britten, N., & Bradley, C. P. (2001). Giving voice to the life: world more humane, more effective medical care? A qualitative study of doctor–patient communication in general practice. *Social Science and Medicine, 53*, 487–505.

Bifulco, A., & Thomas, G. (2013). *Understanding adult attachment in family relationships: Research, assessment and intervention*. London: Routledge.

Boet, S., Sharma, S., Goldman, J., & Reeves, S. (2012). Medical education research: An overview of methods. *Canadian Journal of Anaesthesia, 59*(2), 159–70.

Bohm, D. (1980). *Wholeness and the implicate order*. New York: Routledge.

Brickman, P., Rabinowitz, V. C., & Karuza, J. (1982). Models of helping and coping. *American Psychologist, 37*, 368–384.

Brown, L. D. (2010). The political face of pubic health. *Public Health Review, 32*, 155–173.

Buber, M. (1958). *I and Thou*. New York: Scribner's Sons.

Canguilhem, G. (1991). *The normal and the pathological* (C. R. Fawcett & R. S. Cohen, Trans.). New York: Zone Books.

Carman, T. (1999). The body in Husserl and Merleu-Ponty. *Philosophical Topics, 27*(2), 205–226.

Chan, G. K., Brykczynski, K. A., Maline, R. E., & Benne, P. (2010). *Interpretive phenomenology in health care*. Indianapolis: Sigma Theta Tau International.
Davidson, D. (1970). Mental events. In L. Foster & J. Swanson (Eds.), *Experience and theory*. London: Duckworth.
Dyson, F. (1985). *The origins of life*. New York: Cambridge University Press.
Eco, V. (1980). The sign revisited. *Philosophy and Social Criticism, 7*, 261–97.
Facault, M. (1973). *The birth of the clinic: An archaeology of medical perception* (A. Sheridon, Trans.). London: Collins Press.
Gadmer, H. G. (2004). *Truth and method* (2nd ed., J. Weinsheimer & D. G. Marshall, Trans.). New York: Crossroad.
Gouldrup, L. P. (1987). *Writing the family narrative*. USA: Ancestry.
Habermas, J. (1987). The theory of communicative actionlk. In *Lifeworld and system: A critique of functionalist reason* (Vol 2, T. A. McCarthy, Trans.). Boston: Bacon press.
Haken, H. (1978). *Synergetics: An introduction to nonequilibrium phase transitions and self-organization in physics, chemistry and biology*. Berlin: Springer.
Harold, F. M. (2003). *The way of the cell: Molecules, organism, and the order of life*. New York: Cambridge University Press.
Hippocrates, O. (1962). *Hippocrates collected works* (Vol. 2, W. H. S. Jones, Trans.). London, Massachusetts: Cambridge Harvard university press.
Hoffmeyer, T. (1998). Semiosis and biohistory: A reply. *Semiotica, 120*(3/4), 455–482.
Hoffmeyer, J. (2010). A biosemiotic approach to health. In S. J. Cowley, J. C. Major, S. V. Steffeusen, & A. Dinis (Eds.), *The signifying bodies: Biosemiosis, interaction, and health*. Braga: The Faculty of Philosophy of Braga/Portuguese Catholic University.
Hojviri, A. (1982). Kashf-ol-mahjub. In V. Zhukovski (Ed.), *Revelation of the veiled* (p. 259). Tehran: Amirkabir Publications.
Hurwitz, B., Greenhalgh, T., & Skultns, V. (2004). *Narrative research in health and illness*. Massachusetts: Blackwell Publishing.
Illich, I. (1982). *Medical nemesis: The expropriation of health*. New York: Pantheon.
Jirtle, R. L., & Tyson, F. L. (2013). *Environmental epigenomics in health and disease: Epigenetics and disease origins (Epigenetics and human health)*. New York: Springer.
Kierkegaard, S. (1983). *The sickness unto death: A Christian psychological exposition for upbuilding and awakening* (H. V. Hong & E. H. Hong, Trans.). Princeton: Princeton University Press.
Kohlberg, L. (1982). Moral development. In J. M. Broughton & D. J. Freeman-Moir (Eds.), *The cognitive developmental psychology of James Mark Baldwin: Current theory and research in genetic epistenology*. Norwood: Ablex Publishing Corp.
Kull, K. (1998). On semiosis, umwelt, and semiosphere. *Semiotica, 127*(1/4), 115–131.
Larkin, M. (2011). *Social aspects of health, illness and health care*. New York: Open University Press.
Lovelock, J. (2000). *Gaia: a new look at life on earth*. New York: Oxford university press.
Luhmann, N. (1982). The world society as a social system. *International Journal of General Systems, 8*(3), 131–138.
Martz, B., & Livneh, H. (Eds.). (2007). *Coping with chronic illness and disability: Theoretical empirical and clinical aspects*. New York: Springer.
Maturana, H., & Varela, F. (1987). *The tree of knowledge: A new look at biological roots of human understanding*. Boston: Shambhala/Newscience Library.
Meleau-Ponty, M. (1986). *Phenomenology of perception* (S. Colin, Trans.). London: Routledge & Kegan Paul.
Mishler, E. G. (1981). Viewpoint: Critical perspectives on the biomedical model. In E. G. Misheler, L. R. Amarasingham, S. T. Hauser, S. D. Liam, & R. O. Ewaxler (Eds.), *Social contexts of health, illness and patient care*. Cambridge/New York: Cambridge University Press.
Mishler, E. G. (1984). *The discourse of medicine: The dialectics of medical interviews*. Norwood: Ablex.

Nicholas, M. P., & Schwartz, R. C. (2004). Structural family therapy. In *Family therapy: Concepts and methods* (10th ed.). Boston: Allyn and Bacon.
Nishsid, H. (1999). Cultural schema Theory. In W. B. Gudykunst (Ed.), *Theorizing about intercultural communication* (pp. 401–418). Thousand oaks: Sage publications.
Noë, A. (2009). *Out of our heads: Why you are not your brain, and other lessons from the biology of consciousness*. New York: Hill and Wang.
Palazzoli, M. S., Cirillo, S., Selvini, M., & Sorrentino, A. M. (1989). *Family games: General models of psychotic processes in the Family*. New York: Norton and Company.
Pescosolido, B. A., Martin, J. K., McLeod, J. D., & Rogers, A. (2011). *Handbook of the sociology of health, illness, and healing: A blueprint for the 21st century*. New York: Springer.
Piercem, C. S. (1923). *Chance, love and logic: Philosophical essays edited and introduced by Morris Rapheal Cohen*. New York: Harcourt/Brace and company.
Porter, R. (1985). The potential's view: Doing medical history from below. *Theory and Society, 14*(2), 175–198.
Prigogine, I., & Strngers, I. (1984). *Order out of chaos: Man's new dialogue with nature (foreword by Alvin Toffler)*. New York: Bantam.
Quirk, M. E. (2006). *Intuition and metacognition in medical education: Keys to developing expertise*. New York: Springer.
Richards, J. R. (2007). *Human nature after Darwin: A philosophical introduction*. London: Routledge.
Sebeok, T. A. (1994). *An Introduction to Semiotics*. London: Pinter Publishers.
Serido, J., Shim, S., Mirshra, A., & Tang, C. (2010). Financial parenting, financial coping behaviors, and well-being of emerging adult. *Family Relations, 59*(4), 453–464.
Shapiro, J. A. (2005). A 21st century view of evolution: genome system architecture, repetitive DNA, and natural genetic engineering. *Gene, 34*, 91.
Stonier, T. (1997). *Information and meaning: An evolutionary Perspective*. Berlin: Springer.
Stuart, S. (2010). Enkinaesthesia, biosemiotics, and the ethiosphere. In S. J. Cowley, J. C. Major, S. V. Steffeusen, & A. Dinis (Eds.), *The dignifying Bodies: Biosemiosis, Interaction, and health*. Braga: The Faculty of Philosophy of Braga, Portuguese Catholic University.
Toombs, S. K. (1992). *The meaning of illness: A phenomenological account of the different perspectives of physician and patient*. Dordrecht: Kluwer Academic Publishers.
van Uexküll, T. (1982). Semiotics and medicine. *Semiotica, 38*(3/4), 205–215.
Vedhara, K., & Irwin, M. (Eds.). (2005). *Human psychoimmunology*. New York: Oxford University Press.
Wieher, N. (1961). *Cybernetics or control and communication in the animal and the machine*. Cambridge: MIT Press.
Wisching, M., & Stierrlin, H. (1979). Family dynamics and family psychotherapy of psychosomatic. *Psychotherapy and Psychosomatic, 32*(1–4), 128–33.
Yalom, I. D. (1980). *Existential psychology*. New York: Basic Books.

Index

A
Acceptance and commitment therapy (ACT), 189
Active agent, 4, 167, 205
Active interventions, 197
Active placebo, 203, 204, 206
Actor-network theory (ANT), 34, 35
Affective states, 7, 137
The affect theory, 7
Agapism, 233, 234
Agency, 4, 67, 76, 119, 120, 160, 188, 236
Ainsworth, M.D.S., 87, 100
Altered states of consciousness, 136
Analgesic drug, 92, 202, 203
Anisotropic agent, 12
Anticipatory anxiety, 172, 179, 183, 185
Asymmetry of the relationship, 88
Attachment theory, 86, 88, 99, 177
Attribution meaningfulness, 208
Authoritative diagnosis, 140, 143
Autogeneous level, 235
Autogenic role, 130
Autoimmune disorders, 233
Automaticity, 135
Autonomy, 2, 6–9, 137, 138, 153, 159, 161, 188, 236
Autopoietic model, 52
Autopoietic systems, 33, 52, 64, 68

B
Balint, M., 5, 14, 88, 100, 143, 235, 237
Barbieri, M., 55–58, 67, 233
Behavioral-PNI modulation, 109
Behavioral response, 97, 100, 109, 226
Belief system, 86, 92–93, 95, 97–99, 106, 109, 129
Benedetti, F., 108, 160, 198, 201–203, 211
Beneficence principle, 6
Benevolence, 236
Bichat, 235
Biochemical intervention, 223
Biocommunication, 207
Biological response pattern, 88
Biomedical approach, 15, 133, 198, 223
Bio-medical discourse, 92
Biomedical interventions, 9
Biomedical model, 10, 218
Biomedical narrative, 222
Biomedical paradigm, 4, 23, 143, 147
Biophysical model, 58
Biopsychological response, 120
Biopsychosocial dynamism, 120
Biopsychosocial health, 187
Biopsychosocial matrix, 187, 221
Biopsychosocial model, 10, 11, 13–15
Biopsychosocial pathways, 186
Biopsychosocial response, 99, 118
Biopsychosocial therapies, 121
Biosemiotics, 2–4, 12–18, 23–78, 97, 152, 185, 195–213, 233, 236
Biosemiotics approach, 4, 63, 70, 196, 217
Bodily experiences, 89, 128, 143, 146
Bohm, D., 234
Bohr, 43, 74
Bottom-up control, 234
Boudewijnse, B., 118, 119, 121, 130
Bowlby, J., 87, 99
Brier, S., 15, 17, 23–78, 187
Buber, M., 221

F. Goli (ed.), *Biosemiotic Medicine*, Studies in Neuroscience, Consciousness and Spirituality 5, DOI 10.1007/978-3-319-35092-9

C
Canguilhem, G., 218
Cartesian dualism, 38, 62
Cathartic effect, 121
Causal model, 3, 4, 32, 62, 200, 205, 210, 212
Causal-mechanistic framework, 209, 210, 213
Chemophysical language, 153
Classical conditioning, 146
Clinical context, 2, 9, 12, 14, 122, 123, 141, 147
Co-construction, 91
Code-emergence philosophy, 57
Code model, 56
Cognitive factor, 101
Cognitive patterns, 207, 208
Cognitive responses, 217
Colloca, L., 4, 6, 8, 161, 172, 176, 177, 202, 203, 211
Commitment to growth, 121
Common language, 12, 185, 219
Communications systems, 52
Communication-theoretical approach, 53
Communicative action theory, 218
Communicative interventions, 226
Complementary alternative medicine (CAM), 18
Complementary interventions, 197
Conditioning, 7, 9, 88, 106–109, 206
Consciousness system, 54, 219
Consilience theory, 48
Construction of meaning, 85, 90
Constructivism, 30, 32, 40, 46, 72
Context effects, 107
Conventional medicine, 1, 17, 167
Coping strategies, 86, 95, 101, 128, 189, 221
Counter-transferences, 224
Creative function, 118
Cultural context, 125–130
Cultural schemas, 222, 226
Cultural system, 104
Culture-bounded rituals, 123
Culture-specific disorders, 17
Cybernetics, 16, 43, 50, 51, 54, 58, 66, 76
Cybersemiotics, 15–17, 23–78
  model, 15
  star, 24, 34–43, 47, 64, 69, 71, 76, 77

D
Damasio, A.R., 57, 60, 65
Danesi, M., 56
Darwinistic approach, 154
Darwin's theory, 47
Davidson, D., 12, 229
Defensive mechanisms, 178
Deterministic systems, 196
Developmental neuropsychology, 99
Developmental trajectory, 88, 91
Difficult patient, 159, 169, 170
Disease context, 219, 221, 222, 224, 225, 227, 230, 234
Disease-oriented approach, 221, 223
Doctor/patient communication, 2, 5, 8, 90, 100, 185, 237
Doctor-patient interaction, 85, 90, 122, 133
Doctor-patient relationship, 7, 15, 85, 88–89, 92, 140, 172, 184, 222
Dramatization, 142, 143

E
Efficacy, 6, 8, 18, 119, 120, 122, 127, 144, 180, 210
Egoism, 153, 235
Einstein, 43
Ellis, A., 97
Embodied experience, 143–145, 147
Embodied language, 120
Embodiment theory, 12
Empathic imagination, 145
Empathy, 145, 147, 182
Empiricism, 28, 61
Engel, G.L., 10, 11, 15
Ethiosphere, 235
Ethnomedicine, 123
Evidence Based Medicine (EBM), 17
Evolution, 31, 36, 42, 44, 47–50, 52, 53, 55, 64, 65, 68, 69, 72, 76–78, 152, 187, 189, 217, 233, 234
Evolutionary theory, 47–49, 51, 59, 74, 208
Expectance theory, 7, 171
Expectancy approach, 98
Expectancy effects, 102, 107, 108
Expectancy framework, 98
Expectation theory, 146, 174
Experiential consciousness, 39, 59–64, 70
Experiential world, 25, 27, 48, 65
Expressive function, 118

F
Fallibilism, 28, 67
Family context, 220, 221
Firstness, 28, 31, 37, 60, 61, 64, 67, 68, 70–72, 74
First person experience, 15, 26, 49, 53, 146
Flaten, M.A., 8, 101, 204

Fleck, L., 15
Focault, M., 123, 153, 218
Formality, 119
Friend *vs.* foe experience, 163
Functionalist approach, 56
Functional somatic symptoms, 141, 147

G
Gadamer, H.G., 34, 36, 38, 46, 156, 157, 224, 230
Gaia theory, 233
General systems theory (GST), 10
Goffman, E., 141–143

H
Habermas, J., 53, 153
Haecceity, 36, 37
Hard data, 17, 147
Harm avoidant temperament, 185
Harm avoiding person, 178
Hawthorne effect, 107, 199
Healing belief system, 95, 96, 103, 127
Healing effects, 5, 86, 88, 90, 134, 136, 139, 143
Healing expectation, 123, 230
Healing mechanisms, 139, 143, 144
Healing metaphors, 125–130
Healing narratives, 130
Healing response, 7, 9, 11, 12, 96, 99, 117–130, 134, 137, 139, 144, 146, 147
Healing rituals, 103, 106, 109, 117, 118, 120, 121, 123, 126, 128, 130, 144, 209
Healing symbols, 122
Health anxiety, 151–190, 225
Health care system, 18, 103, 133, 140–143
Health condition, 6, 96, 133, 134, 137, 140, 141, 143, 183, 185, 189, 220–222, 229–230
Health continuum model, 189
Healthism, 153–158, 185
Health psychology, 11, 147
Health system, 130, 146, 180–183, 237
Heidegger, 130, 230
Helman, C.G., 118, 121
Hermeneutics, 29, 34, 36, 44, 46
- approach, 85, 226
- liberty, 130
- paradigms, 60
- response, 2

Higher health, 184, 185, 187, 189
Hippocrates, O., 218
Hoffmeyer, J., 13, 40, 41, 55, 66, 71, 233
Holon, 187, 232, 235, 236
Holonomy, 234
Hume, D., 208
Hunt, L., 145
Husserl, E., 27, 28, 31, 44, 45, 68, 230
Hyperindividualism, 153, 185, 187
Hypnotherapeutic intervention, 128

I
Iathrogenesis, 162
Iatrogenic problems, 218
Illness behavior, 95–97, 106, 127, 225
Illness context, 219, 225, 230
Illness experience, 99, 128, 140, 145, 167, 188, 221, 226
Illness narratives, 91
Illness-prone mechanisms, 225
Illness theory, 89
Inculturation, 208
Individuality, 119, 187, 198
Inert agent, 1
Instrumental-pragmatic view, 58
Integrative medicine, 17–18
Integrative model, 12, 187
Intentionality, 4, 27, 31, 39, 40, 42, 43, 45, 52, 57, 58, 236, 237
Interpersonality, 137
Interpersonal relationship, 139, 147
Interpretant, 3, 32, 33, 40, 57, 69, 74, 96, 230, 232
Intersubjective context, 91
Intersubjective systems, 217
Intra/intercorporal fields, 230
Intra/inter/transpersonal fields, 119, 120
Irritable bowel syndrome (IBS), 8, 102, 134, 212

J
Johanson, 144

K
Kant, E., 61, 77
Kirmayer, L.J., 103–105, 118, 123, 124, 128, 140, 145, 146
Kreinath, J., 119
Kuhn, T.S., 15, 46, 60

L
Lakoff, G. J., 123, 144
Language games, 58, 59, 72
Languaging, 36, 59, 64
Latour, B., 34, 35, 209
Levels of consciousness, 189
Lévi-Strauss, C., 118
Life experiences, 86, 99
Life-oriented medicine, 218, 219
Life scripts, 85
Life stylism, 154–158
Linguistic communication, 58, 67
Linguistic relativity, 62
Linguistic-symbolic behavior, 58
Linguistic symbols, 144
Lived illness, 230
Locus of control, 95, 176, 179, 190
Luhmann, N., 29, 52, 53, 152, 187

M
Materialistic reductionism, 153
Matrix of potential meanings, 89
Maturana, H.R., 52, 65, 76, 233
Meaning constructing, 211
Meaningful language, 59, 72
Meaning-making agents, 4
Meaning making processes, 212
Meaning response, 1–4, 96, 107, 127, 137, 139, 144, 146, 147, 195, 217, 230, 236, 237
Meaning-specific responses, 4
Meaning systems, 86, 236
Mechanical agent, 12
Mechanical dyadic model, 197
Mechanistic model, 196, 210, 211, 213
Medical discourse, 123, 217–219, 221, 233, 234, 237
Medical ethics, 3, 6, 17, 137, 138, 236
Medical interventions, 4, 88, 101, 152, 160, 170, 195
Medicalization, 18, 153–158, 185, 218
Medically unexplained symptoms (MUS), 141, 143, 169
Medical semiotics, 3
Medical utopia, 153
Medicinal interventions, 224
Mental constructs, 98
Merleau-ponty, M., 28, 34, 39, 45, 230, 237
Mesmer, F.A., 134
Metaphor, 13, 59, 70, 92, 103, 105, 106, 118–120, 122–130, 144, 145, 222, 225, 230, 234
Metaphoricity, 59
Metaphoric language, 119, 120
Mind-body coordination, 234
Mind-body dichotomy, 230
Mind-body dualism, 123
Mind-body interactions, 12
Mind-body interventions, 9
Mind-body system, 136
Mind-matter gap, 196
Miosphere, 235
Moerman, D.E., 2, 3, 7, 107, 137, 144, 195
Monod, J., 49

N
Nagel, T., 38
Narration, 85, 86, 88–91, 96, 98, 109, 222, 225, 237
Narrative, 34, 59, 85, 86, 89–92, 122, 123, 125, 127, 128, 130, 221, 222, 224, 226–230, 236
Narrative medicine, 85–93
Narrativity, 59
Nasavi, 228
Natural language, 46, 47, 62, 74
Negative expectation, 137, 138, 160, 171–174, 183–185
Negative suggestion, 138
Neuroimmune modulation, 96
Neuroimmune response, 100
Neuroimmunologic response, 12
Nietzsche, 152
Nocebo effect, 14–15, 137–139, 151–190, 227
Nocebo response, 139, 154, 160, 172–178, 181–183, 185, 187, 225, 234
Non-reductionistic model of consciousness, 15
Nonspecific effect, 2–4
Non-substance-bound healing, 86, 88
Non-verbal communication, 144
Nussbaum, M., 188

O
Objective reality, 63, 166
Obsessive-compulsive disorders, 189
Ockham, W.V., 208
Omnipotent doctor archetype, 167–171
Ontology, 30, 31, 33, 44, 48, 50, 58–60, 63, 68, 74, 75, 77
Open placebo, 211–213
Operant conditioning, 206
Organic unity theory, 12
Organismic sense, 228
Organon, 235
Ostenfeld-Rosenthal, A.M., 143

P
Parkin, D., 119
Parsimony principle, 12
Paternalistic medicine, 167
Paternalistic relationship, 140
Pathogenesis signs, 230
Pathogenic agent, 4
Patient-centered approach, 133
Patient's narrative, 89–91, 222, 224, 227
Patient's sign world, 231
Peircian (bio) semiotics, 16
Penrose, R., 62, 63
Perceived meaning, 197
Perceived reality, 166
Perceived self-efficacy, 101
Performative language, 119
Personal consciousness, 217
Personal context, 220, 221, 226, 230
Personality traits, 101, 102, 173–178, 180
Phaneroscopy, 28, 36
Pharmacological agent, 2, 4, 8, 199
Phatic communion, 58
Phenomenal context, 219, 224, 225
Phenomenological experience, 15, 17
Phenomenology, 24–26, 28, 29, 31–34, 36, 39, 40, 42, 44–47, 60, 63, 73, 78
Physical-behavioral world, 104
Physicochemical agents, 27
Physiological homeostasis, 87, 88
Physiological response, 87, 88, 106
Piaget, 42
Piercem, C.S., 233
Placebo effect, 1, 3–9, 23, 92, 98, 101, 102, 106–109, 136, 137, 139, 144, 146, 172, 175, 177, 195, 198–203, 205, 207, 210–213, 227
Placebo effectiveness, 92
Placebo response, 1–18, 101–103, 107–109, 128, 139, 143, 144, 146, 171, 173–177, 187, 196, 198, 211, 222
Popper, K.R., 33, 47, 48, 156
Principle of autonomy, 2
Prophylactic behaviors, 179, 180, 183
Pseudomachine, 195–213
Psychoactive interventions, 223
Psychobiological system, 87
Psychoimmune modulation, 187, 225
Psychological intervention, 8
Psychological symptoms, 89
Psychoneuroimmunological mechanisms, 8
Psychoneuroimmunologic function, 224
Psychoneuroimmunologic response, 106, 123, 127, 222
Psychoneuroimmunologic state, 4
Psychoneuroimmunologic systems, 217, 233
Psychoneuroimmunology, 18, 146
Psychophysical methods, 136
Psychophysical narratives, 230
Psychophysical pathways, 185, 230
Psychophysical response, 129, 222
Psychophysical signs, 95
Psychophysiological pathways, 8
Psychosocial agent, 92
Psychosocial context, 11, 17, 109, 135, 143
Psychosocial factors, 146, 185, 222
Psychosocial interactions, 138
Psychosocial stress, 172
Psycho-somatic-cultural interactions, 30
Psychosomatic dynamism, 5, 97, 219
Psychosomatic medicine, 3, 9, 12, 13, 85, 196
Psychosomatics, 26
Psychosomatic treatment, 225, 227, 229
Psychotherapeutic interventions, 197
Psychotherapeutic treatment, 121

Q
Qualia, 23, 33, 36, 38–40, 42, 43, 50, 52, 54, 56, 57, 60, 64, 65, 70
Qualitative organicism, 63

R
Randomized clinical trials (RCTs), 4, 5, 146, 203, 210
Rapport, 6, 14, 106, 141, 224
Rational emotive behavior therapy theory (REBT), 97
Reductionism, 12, 34, 123
Referentiality, 119
Regularity, 31, 119, 208
Relational space, 134
Relativity theories, 43, 51, 63
Resource-based approach, 187, 188
Response expectancy theory, 98
Rites of healing, 120–123
Ritual effect, 117–130
Rossano, M.J., 119

S
Salutogenesis, 185, 226, 230, 235
Salutogenic agent, 4
Sapir-Whorf hypothesis, 62
Saussurian semiology, 34, 35
Scotus, D., 69
Searle, J., 26, 37, 39, 54
Sebeok, T.A., 14, 56, 218

Secondness, 28, 31, 36, 37, 61, 64, 69–72, 74
Second-person experience, 72
Self-actualization, 187, 236
Self-coherence, 91
Self-consciousness, 25, 38, 70, 72, 189
Self-efficacy, 7, 95, 101, 108, 188
Self-healing processes, 8, 212
Self-image, 144, 185
Self-organisation, 52, 54, 55, 71, 77
Self-regulation, 176
Self-scrutiny, 138, 177
Semiosis, 13, 14, 16, 17, 35, 40, 41, 64, 67, 70–72, 75–77, 128, 129, 152, 185, 221, 232, 234
Semiosphere, 14, 18, 217–237
Semiotic agent, 130
Semiotic approach, 1–18, 41, 237
Semiotic interactions, 17, 24, 71
Semiotic model, 197, 212
Semiotic processes, 17, 33, 40, 66, 68, 76, 77, 197, 199–201, 208, 210–211, 217
Semiotic relationship, 196
Semiotic theory, 48, 67, 85
Sequentiality, 119
Sham control, 207
Shapiro, J., 145, 233
Shared language, 24
Shared meaning, 86
Sign games, 72, 74
Sign system, 62, 125, 217, 232, 235, 236
Social consciousness, 217
Social context, 64, 92, 97, 109, 141, 144, 226
Social iatrogenesis, 181–183
Social theory of identity, 85
Sociocultural belief system, 96
Sociocultural context, 104–106, 108, 109, 146, 222
Sociocultural iatrogenesis, 234
Sociogenic illness, 138
Soft data, 9, 17
Somatic experience, 154, 158
Somatic-social strategy, 156–157
Somatic symptoms, 89–90
Sonntag, 125, 126
Spatial-bodily reality, 120
Specific effect, 2, 3, 203, 207
Spinoza, 60
State of consciousness, 41, 71, 120, 134–136
Stereotypic language, 119
Stream of consciousness, 74
Stress response, 87, 88
Structuralistic viewpoint, 122
Stuart, S., 235, 236
Subjective experiences, 24, 26, 37, 63, 91, 143
Suggestibility, 102, 120, 139, 175
Suggestion, 3, 7, 8, 12, 29, 40, 41, 57, 62, 68, 96, 118, 120, 121, 123, 126–128, 134, 135, 137–139, 144, 145, 162, 209, 225
Surgical interventions, 3, 138
Symbolic agents, 12, 229, 233
Symbolic factors, 223, 225
Symbolic-intentional world, 104
Symbolic intervention, 230
Symbolic signs, 14, 95, 217, 222, 224, 229, 231
Symptomatology, 218
Symptom formation, 99, 222
Symptom signs, 218
Systemic approach, 10, 109, 187, 234
Systemic model, 15, 187
Systemic worldview, 187, 234
System's theory, 10, 53, 152

T

Temperamental traits, 176, 179
Temporality, 119
Theatre training, 145
Theory of autopoiesis, 52
Theory of somatic markers, 65
Therapeutic agent, 2
Therapeutic modality, 134
Therapeutocracy, 153
Therapist-patient relationship, 16
Therapist's narrative, 109
Thirdness, 28, 31, 61, 63, 64, 70–72, 74
Third-person experience, 41, 57
Top-down organization, 234
Transformative effect, 121
Transpersonal level, 235
Triadic semiotic model, 197
Turner, J.A., 100

U

Umwelt, 13, 66

V

Varela, F.J., 60, 65, 233
Verbal suggestions, 7, 118, 120
von Bertalanffy, L., 10
von Lucadou, W., 205, 206
von Uexküll, T., 3, 12, 13, 196

W

War metaphor, 125, 126, 234
Webs of belief, 95, 96, 99
Whitehead, A.N., 61, 67, 68
Wissenschaft, 30, 32, 36, 41, 46, 48, 63, 77
Wittgenstein, L., 34, 64
World models, 198, 212

MIX
Papier aus verantwortungsvollen Quellen
Paper from responsible sources
FSC® C105338

If you have any concerns about our products,
you can contact us on
ProductSafety@springernature.com

In case Publisher is established outside the EU,
the EU authorized representative is:
Springer Nature Customer Service Center GmbH
Europaplatz 3, 69115 Heidelberg, Germany

Printed by Libri Plureos GmbH
in Hamburg, Germany